W9-BNN-399

To Dr Alan Guttmacher
for his kind contribution
to the interview with
Croatian TV, July 10, 2002.

Slobodan Vukicevic

## Progress in Inflammation Research

**Series Editor**

Prof. Dr. Michael J. Parnham
PLIVA
Research Institute
Prilaz baruna Filipovica 25
10000 Zagreb
Croatia

**Forthcoming titles:**
*The Hereditary Basis of Allergic Diseases*, J. Holloway, S. Holgate (Editors), 2002
*Anti-Inflammatory or Anti-Rheumatic Drugs*, R.O. Day, D.E. Furst, P.L. van Riel (Editors), 2002
*Inflammatory Processes and Cancer*, D.W. Morgan, U. Forssmann, M. Nakada (Editors), 2003
*Inflammation and Cardiac Diseases*, G.Z. Feuerstein (Editor), 2003

(Already published titles see last page.)

# Bone Morphogenetic Proteins

## From Laboratory to Clinical Practice

Slobodan Vukicevic
Kuber T. Sampath

Editors

Birkhäuser Verlag
Basel · Boston · Berlin

Editors

Slobodan Vukicevic
Laboratory for Mineralized Tissues
School of Medicine
Salata 11
10000 Zagreb
Croatia

Kuber T. Sampath
Genzyme Corporation
One Mountain Road
Framingham, MA 01701-9322
USA

A CIP catalogue record for this book is available from the Library of Congress, Washington D.C., USA

Deutsche Bibliothek Cataloging-in-Publication Data
Bone morphogenetic proteins. From laboratory to clinical practice /
Slobodan Vukicevic ; Kuber T. Sampath, ed.. - Basel ; Boston ; Berlin : Birkhäuser, 2002
(Progress in inflammation research)
ISBN 3-7643-6509-9



ISBN 3-7643-6509-9 Birkhäuser Verlag, Basel – Boston – Berlin


Member of the BertelsmannSpringer Publishing Group
Printed on acid-free paper produced from chlorine-free pulp. TCF ∞
Cover design: Markus Etterich, Basel
Cover illustration: Image of the sagital cranial section of a 17.5-day rat embryo (see page 101)
Printed in Germany
ISBN 3-7643-6509-9

9 8 7 6 5 4 3 2 1

www.birkhauser.ch

# Contents

# List of contributors

Nikolina Basic, Zagreb University Clinical Center, University of Zagreb, Zagreb 10000, Croatia

Amy P. Baumann, Department of Cardiovascular and Metabolic Diseases, Pfizer Global Research and Development, Groton, CT 06340, USA;
e-mail: amy_baumann@groton.pfizer.com

Taco J. Blokhuis, Vrije Universiteit Medical Centre (VUmc), Department of Trauma Surgery, P.O. Box 7057, 1007 MB Amsterdam, The Netherlands;
e-mail: TJ.Blokhuis@vumc.nl

Fran Borovecki, Department of Anatomy, School of Medicine, Salata 11, 10000 Zagreb, Croatia; e-mail: FBOR@MEF.HR

Anton Brkic, Veterinary Faculty, University of Zagreb, 10000 Zagreb, Croatia

Tessa A. Castleberry, Department of Cardiovascular and Metabolic Diseases, Pfizer Global Research and Development, Groton, CT 06340, USA

Stephen D. Cook, Tulane University School of Medicine, Department of Orthopaedic Surgery, 1430 Tulane Avenue – SL32, New Orleans, LA 80112, USA;
e-mail: scook2@tulane.edu

Cosimo De Bari, UZ Gasthuisberg, Department of Rheumatology, Herestraat 49, 3000 Leuven, Belgium; e-mail: cosimo.debari@med.kuleuven.ac.be

Dirk De Valck, UZ Gasthuisberg, Department of Rheumatology, Herestraat 49, 3000 Leuven, Belgium; e-mail: dirk.devalck@med.kuleuven.ac.be

Francesco Dell'Accio, UZ Gasthuisberg, Department of Rheumatology, Herestraat 49, 3000 Leuven, Belgium; e-mail: francesco.dellaccio@med.kuleuven.ac.be

Peter ten Dijke, Division of Cellular Biochemistry, The Netherlands Cancer Institute, Plesmanlaan 121, 1066 CX Amsterdam, The Netherlands;
e-mail: p.t.dijke@nki.nl

Haimanti Dorai, Centocor, Inc., 200 Great Valley Parkway, Malvern, PA 19355, USA

Karen M. Drahushuk, Department of Pharmacology and Toxicology, 102 Farber Hall, State University of New York, 3455 Main Street, Buffalo, NY 14214, USA;
e-mail: drahushu@acsu.buffalo.edu

Matthias Dreyer, Physiologische Chemie II, Biozentrum der Universität Würzburg, Am Hubland, 97074 Würzburg, Germany;
e-mail: mad@biozentrum.uni-wuerzburg.de

Jonathan S. Erulkar, Department of Orthopedics & Rehabilitation, Yale University School of Medicine, P.O. Box 208071, New Haven, CT 06520-8071, USA;
e-mail: jonathan.erulkar@yale.edu

Gary E. Friedlaender, Department of Orthopedics & Rehabilitation, Yale University School of Medicine, P.O. Box 208071, New Haven, CT 06520-8071, USA;
e-mail: gary.friedlaender@yale.edu

Lex R. Giltaij, Stryker Biotech, Notengaard 13, 3941 LV Doorn, The Netherlands;
e-mail: lexgiltaij@CSI.COM

William A. Grasser, Department of Cardiovascular and Metabolic Diseases, Pfizer Global Research and Development, Groton, CT 06340, USA;
e-mail: William_A_Grasser@groton.pfizer.com

Jonathan N. Grauer, Department of Orthopedics & Rehabilitation, Yale University School of Medicine, P.O. Box 208071, New Haven, CT 06520-8071, USA;
e-mail: jonathan.grauer@yale.edu

Henk J. Th. M. Haarman, Vrije Universiteit Medical Centre (VUmc), Department of Trauma Surgery, P.O. Box 7057, 1007 MB Amsterdam, The Netherlands;
e-mail: HJTM.Haarman@vumc.nl

Miroslav Haspl, Departments of Orthopaedic Surgery, Clinics of Surgery, Orthopaedic Surgery and Ophthalmology, University of Zagreb, Salata 11, 10000 Zagreb, Croatia

Dennis Higgins, 102 Farber Hall, Department of Pharmacology and Toxicology, State University of New York, 3455 Main Street, Buffalo, NY 14214, USA; e-mail: higginsd@acsu.buffalo.edu

Mislav Jelic, Departments of Orthopaedic Surgery and Anatomy, School of Medicine, University of Zagreb, Salata 11, 10000 Zagreb, Croatia; e-mail: MJELIC@MEF.HR

Søren Jepsen, Department of Periodontics, University of Kiel, Arnold Heller Strasse 16, 24105 Kiel; e-mail: jepsen@konspar.uni-kiel.de

Olexander Korchynsky, Division of Cellular Biochemistry, The Netherlands Cancer Institute, Plesmanlaan 121, 1066 CX Amsterdam, The Netherlands; e-mail: o.korchynsky@nki.nl; olexkor@hotmail.com

Pamela Lein, Department of Environmental Health Sciences, School of Hygiene and Public Health, Johns Hopkins Unversity, Baltimore, MD 21205, USA; e-mail: plein@jhsph.edu

Rik Lories, UZ Gasthuisberg, Department of Rheumatology, Herestraat 49, 3000 Leuven, Belgium; e-mail: rik.lories@uz.kuleuven.ac.be

Frank P. Luyten, UZ Gasthuisberg, Department of Rheumatology, Herestraat 49, 3000 Leuven, Belgium; e-mail: frank.luyten@uz.Kuleuven.ac.be

Snjezana Martinovic, Department of Anatomy, School of Medicine, Salata 11, 10000 Zagreb, Croatia; e-mail: SMART@MEF.HR

Joachim Nickel, Physiologische Chemie II, Biozentrum der Universität Würzburg, Am Hubland, 97074 Würzburg, Germany; e-mail: nickel@biozentrum.uni-wuerzburg.de

Thomas A. Owen, Department of Cardiovascular and Metabolic Diseases, Pfizer Global Research and Development, Groton, CT 06340, USA

Vishwas Paralkar, Department of Cardiovascular and Metabolic Diseases, Pfizer Global Research and Development, Groton, CT 06340, USA; e-mail: vishwas_m_paralkar@groton.pfizer.com

Tushar Ch. Patel, Department of Orthopedics & Rehabilitation, Yale University School of Medicine, P.O. Box 208071, New Haven, CT 06520-8071, USA; e-mail: orthoped203@hotmail.com

Peter Patka, Vrije Universiteit Medical Centre (VUmc), Department of Trauma Surgery, P.O. Box 7057, 1007 MB Amsterdam, The Netherlands;
e-mail: P.Patka@vumc.nl

Marko Pecina, Department of Orthopaedic Surgery, School of Medicine, University of Zagreb, Salata 11, 10000 Zagreb, Croatia

Dunja Rogic, Zagreb University Clinical Center, University of Zagreb, Zagreb 10000, Croatia

David C. Rueger, Stryker Biotech, Research Department, 35 South Street, Hopkinton, MA 01748, USA; e-mail: rueger.david@strybio.com

Kuber T. Sampath, Genzyme Corporation, One Mountain Road, Framingham, MA 01701-9322, USA; e-mail: kuber.sampath@genzyme.com

Walter Sebald, Physiologische Chemie II, Biozentrum der Universität Würzburg, Am Hubland, 97074 Würzburg, Germany;
e-mail: sebald@biozentrum.uni-wuerzburg.de

Andrew Shimmin, Melbourne Orthopaedic Group, 33 The Avenue, Windsor Victoria 3181, Australia

Ana Stavljenic-Rukavina, Zagreb University Clinical Center, University of Zagreb, Zagreb 10000, Croatia

Hendrik Terheyden, Dept. of Oral and Maxillofacial Surgery, University of Kiel, Arnold Heller Strasse 16, 24105 Kiel; e-mail: terheyden@mkg.uni-kiel.de

Slobodan Vukicevic, Laboratory for Mineralized Tissues, School of Medicine, Salata 11, 10000 Zagreb, Croatia; e-mail:VUKICEV@MEF.HR

# Preface

Selection guides the evolution of bone in directions determined by pre-adaptation and adaptation to environments. As a pre-adaptive characteristic, bone could have evolved more than 500 million years ago in the Cambian period, with the parting of evolutional pathways between arthropods and chordates.

Recently, governmental agencies in the USA, Europe and Australia have approved the use of bone morphogenetic protein-7 (BMP-7/OP-1) and bone morphogenetic protein-2 (BMP-2) in humans for the treatment of long bone non-unions and spinal fusions. The BMP is the first recombinant protein to be used in orthopedic practice worldwide. Not since the discovery of vitamin D and PTH has biomedical research in the field of mineralized tissues led to knowledge as fundamental as that on the role of BMPs in nature.

Since the original description of the potential of demineralized bone matrix to induce bone by Marshall Urist in 1965, it has taken more than 30 years to bring BMP-gene products to clinical medicine. Those three decades have been filled with important discoveries from many researchers that contributed to several breakthrough findings and led to advanced understanding of bone repair mechanisms.

The clinical application of BMP, an ancient gene, nicely overlaps with the decade of bones and joints, as designated by the World Health Organization (WHO). As our civilization is aging and newly discovered medicines are continuously extending our lives, it is evident that living without a proper function of our locomotive system is impossible.

At the beginning of the 21st Biotech century, bone is the first human organ to be biologically regenerated by BMPs when normal physiological repair mechanisms fail. We dedicate this book to the late Marshall Urist, who made the initial discovery and gave the name BMPs in 1965 to the activity of demineralized bone matrix (DBM) to induce bone at an ectopic site in mammals.

The book covers the biochemistry, molecular and cell biology of BMPs, receptors and their nuclear effectors in bone formation. A detailed discussion on deciphering the binding code of BMP-receptor interaction is presented. We have included a detailed description of preclinical models of orthopedic, periodontal and maxillofacial defects treated by BMPs. Two chapters cover the use of BMPs in human bone defects, fractures and spinal fusion. The role of BMPs in the development of joints and their role in segmentation of articular cartilage is discussed in detail. We have also included a chapter on the recently discovered function of BMPs in kidney development and postnatal models af acute and chronic renal failure. The final chapter describes major advances in our understanding of effects of BMPs on neural tissues.

Our sincere appreciation is due the authors of the chapters for their profound dedication in making this project a reality. We acknowledge the help of Mr. Branko Šimat and Mrs. Morana Šimat for their technical support throughout the project. We also thank Dr. Hans Detlef Klueber and Ms Karin Neidhart of the publisher Birkhäuser Verlag for their patience in collecting the manuscripts and for the final editing of the book.

January 2002

Slobodan Vukicevic
Kuber T. Sampath

# Biochemistry of bone morphogenetic proteins

*David C. Rueger*

Stryker Biotech, Research Department, 35 South Street, Hopkinton, MA 01748, USA

## Introduction

Bone has a remarkable ability for regeneration and repair. The cellular events associated with this repair process mimic closely those events associated with embryonic bone development. In 1965, Dr. Marshall Urist showed that new bone formation could be induced using demineralized bone matrix [1]. By implanting demineralized bone particles intramuscularly in animals, he observed the formation of new bone. With these studies, Dr. Urist pioneered the concept that there is some substance naturally present in bone, which is responsible for the regeneration and repair activity. He called this substance bone morphogenetic protein (BMP) and initiated a search for these molecules.

In 1981, Sampath and Reddi made the observation that the bone formation induced by demineralized bone powder could be inactivated by extraction with denaturants and that this activity could be restored by reconstitution of the extract with the inactivated bone powder [2–4]. This observation supported the existence of BMP molecules and led to the development of an assay for the purification of these proteins. This assay, which is commonly referred to as the rat subcutaneous assay, measures bone formation in an ectopic site in the thorax region of the rat. Sampath and Reddi have shown that the cellular events which are produced sequentially in implants of extracts of demineralized bone in combination with the residual bone collagenous matrix in this assay are the same as those cellular events observed in embryonic bone development or in adult fracture repair.

## Discovery

In the late 1980s, using the rat assay of Sampath and Reddi along with advanced techniques of molecular biology and protein chemistry, the first genes believed to code for bone inductive proteins were identified [5–13]. They were named bone morphogenetic protein (BMP) or osteogenic protein (OP). In order to achieve this

Bone Morphogenetic Proteins, edited by Slobodan Vukicevic and Kuber T. Sampath

feat, bone inductive preparations were purified from bovine bone in sufficient quantity and purity to provide amino acid sequence data. Using these sequences, nucleic acid probes were generated and used for the identification and characterization of DNA sequences encoding these proteins. Eventually the complete human genes were identified. A list of the first BMP genes identified is presented in Table 1 along with alternative names shown in parentheses. A gene named BMP-1 was also described in these initial studies, but was unrelated to the other genes and eventually determined not to be a BMP. BMP-1 has since been identified as gene coding for a procollagen-C-proteinase; this protein is related to *Drosophila* tolloid and may be involved in the proteolysis of BMP binding proteins such as noggin and chordin [14].

Identification of BMPs was difficult due to the fact that such small amounts were present in bone and because of their limited solubility. As a result the development of laborious purification protocols in the presence of dissociating agents was necessary. In addition, the only assays known at that time were *in vivo* assays, principally the rat subcutaneous assay, and each step in the development process needed to be evaluated by these 2–3 week assays. However, given enough time and tenacity, some bone inductive preparations were purified from bovine bone in sufficient amounts to characterize.

As an example, osteoinductive preparations used for the discovery of OP-1 were extensively purified [9]. These preparations were highly active *in vivo* and composed of disulfide-linked dimers that migrated on sodium dodecyl sulfate gels as a diffuse band with an apparent molecular weight of 30–36 kDa. Upon reduction, the dimers yielded two subunits that migrated with molecular weights of about 18 kDa and 16 kDa, both of which were glycosylated. After chemical or enzymatic deglycosylation, the dimers migrated as a diffuse 27-kDa band that upon reduction yields two polypeptides that migrate at 16 kDa and 14 kDa, respectively. Analysis of the dimers revealed that they existed primarily, if not totally, as homodimers although the presence of a small amounts heterodimer could not be ruled out. The carbohydrate moiety did not appear to be essential for biological activity since the deglycosylated protein remained capable of inducing bone formation *in vivo*.

Protein sequence characterization was the primary goal in the early discovery research. Since only microgram amounts were available, success was achieved with much difficulty. Multiple proteases were used to cleave the osteoinductive preparations and micromethods were necessary for isolation of the peptides. Oligonucleotide probes based on peptide sequences from these preparations were constructed and used to screen human cDNA libraries [7]. Several genes were identified, including one named that was named OP-1 and another that had been named BMP-2. Using these data, as well as published data on other BMPs the 18-kDa subunit from the bovine osteoinductive preparations was identified as the bovine equivalent of mature human OP-1, whereas the 16-kDa subunit was the bovine equivalent of mature BMP-2.

*Table 1 - The initial bone morphogenetic proteins*

|  |
|---|
| BMP-2 (BMP-2A) |
| BMP-3 (osteogenin) |
| BMP-4 (BMP-2B) |
| BMP-5 |
| BMP-6 (Vgr-1) |
| BMP-7 (OP-1) |

Data from the natural bovine protein preparations did not prove that any of the initial BMP genes were indeed osteoinductive; impurities in these preparations could have been responsible for the activity. In fact during development of the purification procedure, other proteins were originally thought to be the active factors. However, the production of the subunits by recombinant DNA methods provided a means to clearly prove that these proteins were indeed BMPs. The use of separate recombinant proteins would also be necessary to determine if multiple BMPs were necessary for the observed activity. During the discovery research for OP-1, recombinant OP-1 and BMP-2 were individually produced in Chinese hamster ovary (CHO) cells [10, 12]. In order to achieve this, the full length cDNAs were inserted into mammalian expression vectors and transfected into CHO cells. After gene amplification, the selected clones were grown in flasks and the media collected. The recombinant proteins were found to be secreted into the culture media and thus were purified from those solutions and characterized. Purified OP-1 was produced as dimers of 34–38 kDa that, upon reduction, migrate as 23, 19 or 17 kDa monomers. This form of OP-1 which is referred to as the mature domain corresponded to the OP-1 sequence obtained from the bovine osteoinductive preparations. Digestion of the monomers with N-Glycanase reduced the 23, 19 and 17 kDa monomers to a single 14 kDa species indicating that the apparent molecular weight differences in recombinant OP-1 was due to glycosylation.

Using the rat subcutaneous assay, the purified OP-1 protein, by itself, was shown to be osteoinductive, capable of switching on the cascade of cellular events required for bone formation activity. This activity was dose dependent and similar to that observed for demineralized bone powder or purified preparations of bovine osteoinductive protein. Of the original bone-derived BMPs, BMP-2, -3, -4, -5, and -6 have also been expressed in CHO cells and the recombinant proteins purified [15]. All except BMP-3 have demonstrated osteoinductive activity in the rat subcutaneous assay. BMP-3 is now believed to be an inhibitor of osteoinductive BMP activity [16]. It is interesting in this respect that BMP-3 is the most abundant BMP in bone [17].

Continued discovery research has yielded additional related mammalian proteins, described under a variety of names, including BMPs, cartilage derived mor-

*Table 2 - BMP family*

| BMP number | Other names |
|---|---|
| BMP-2 | BMP-2A |
| BMP-3 | Osteogenin |
| BMP-3B | GDF-10 |
| BMP-4 | BMP-2B |
| BMP-5 | – |
| BMP-6 | Vgr-1 |
| BMP-7 | OP-1 |
| BMP-8 | OP-2 |
| BMP-8B | OP-3 |
| BMP-9 | – |
| BMP-10 | – |
| BMP-11 | GDF-11 |
| BMP-12 | GDF-7, CDMP-3 |
| BMP-13 | GDF-6, CDMP-2 |
| BMP-14 | GDF-5, CDMP-1, MP-52 |
| BMP-15 | – |
| BMP-16 | – |

phogenetic proteins (CDMPs), and growth and differentiation factors (GDFs) [18–21]. These are listed in Table 2. It should be noted that most of these proteins have multiple names, including some with three names. Fifteen mammalian BMP family members have been described.

## TGF-β superfamily

All of the BMPs are members of the TGF-β superfamily of genes [21–23]. This superfamily is quite large and currently includes approximately 45 genes. Members have been identified in most species including human and mouse, as well as *Drosophila*, *Xenopus*, zebrafish and *Caenorhabditis elegans*. The structure of the proteins of this superfamily is shown in Figure 1. Each of the proteins is produced as a N-terminal signal sequence, a prodomain and a mature domain at the carboxy-terminus. The structural hallmark of this superfamily is a highly conserved 7 cysteine motif in the mature domain. This domain also contains a relatively short amino terminal extension that exhibits considerably more evolutionary divergence. The BMP family is the largest subgroup in the TGF-β superfamily of molecules. The

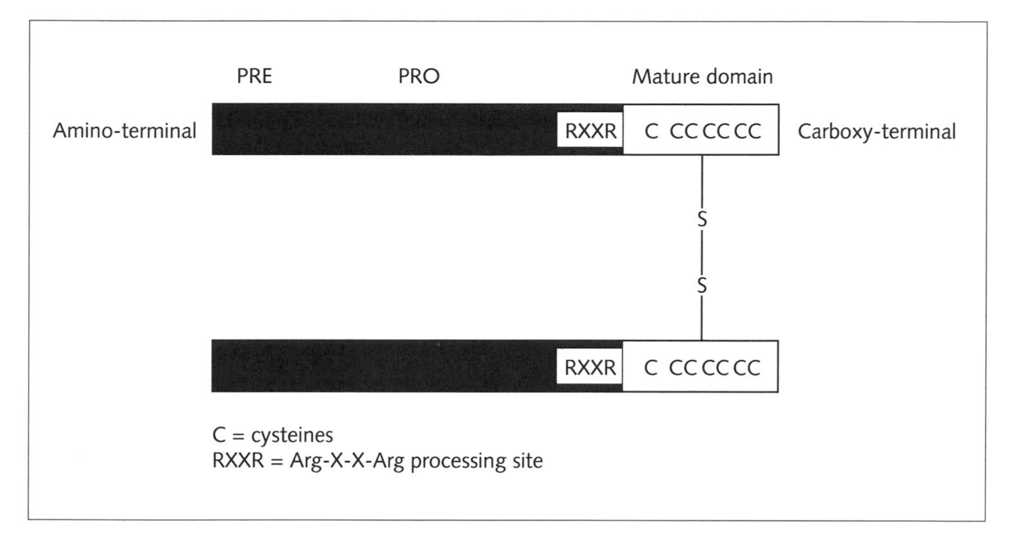


*Figure 1*
*Schematic of TGF-β superfamily protein structure.*

original characteristic of this family was the ability of its members to induce new bone formation. However, not all members in this family have demonstrated this activity. Moreover, this capability of forming new bone is shared by no non-BMP factors including the TGF-βs, themselves.

Alignment of protein sequences in the cysteine domains reveals striking sequence similarities and differences amongst the superfamily members. Table 3 shows the alignment calculated as percent of identical amino acid residues and compares OP-1/BMP-7 to the other members [24]. The comparison demonstrates the different family groupings, including, the BMP (generally those having 50% or more homology with OP-1), the activin (inhibin) and the TGF-β families. When compared to other BMPs, OP-1/BMP-7 is most closely related by sequence to BMP-5 and BMP-6 with 88% and 87% amino acid sequence identity, respectively, in the cysteine rich C-terminus. OP-1 is more distantly related to BMP-2 and BMP-4, having 60 and 58% identity, respectively. GDF-5, another BMP that has been extensively evaluated, has even less similarity, showing 51% identity. As discussed above, BMP-3 is not osteoinductive and, in fact, is more distant having 42% identity. The TGF-βs, themselves, are quite distant from most BMPs and do not exhibit bone inductive activity. Many of the newer proteins in the list have not yet been expressed as recombinant proteins and thus it is not known if they possess osteoinductive activity. Species comparisons show a strong evolutionary conservation. For example, there is a 98% identity in amino acid sequences in OP-1 between the human and mouse genes in the conserved mature domain [25].

*Table 3 - Comparison of representative TGF-β superfamily members: percent identity in 7-cysteine region*

| Family member | Percent | Family member | Percent |
|---|---|---|---|
| OP-1/BMP-7 | 100 | GDF-1 | 45 |
| BMP-5 | 88 | SCREW | 47 |
| BMP-6 | 87 | BMP-3B | 42 |
| BMP-8 | 74 | BMP-3 | 42 |
| BMP-8B | 67 | NODAL | 41 |
| 60A | 69 | InhibinβA | 43 |
| UNIVIN | 63 | InhibinβB | 38 |
| BMP-2 | 60 | InhibinβC | 39 |
| BMP-4 | 58 | TGF-β4 | 38 |
| dpp | 58 | TGF-β5 | 37 |
| Vg-1 | 57 | TGF-β1 | 35 |
| GDF-6 | 53 | TGF-β3 | 37 |
| GDF-7 | 53 | TGF-β2 | 36 |
| GDF-5 | 51 | BMP-11 | 36 |
| BMP-9 | 51 | GDF-9 | 30 |
| DORSALIN | 49 | MIS | 26 |
| BMP-10 | 47 | GDNF | 22 |
| GDF-3 | 49 | | |

The members of the TGF-β superfamily are signaling molecules that are responsible for specific morphogenetic events involved in tissue and organ development. A number of the members of this superfamily have been identified based on tissue-specific functional assays and molecular cloning approaches in various developmental systems. The decapentaplegic gene, DPP, and the 60-A gene are responsible for proper development of *Drosphila melanogaster* embryos. In *Xenopus laevis*, vegetal pole-derived transcripts, Vgr-1, and activins have been demonstrated to play a critical role in mesoderm induction. In addition, other members of the TGF-β superfamily include: Mullerian inhibiting substance (MIS), which causes regression of the Mullerian duct in the development of the male reproductive tract; inhibins and activins, which act together to regulate the release of follicle-stimulating hormone in the pituitary gland; and growth and differentiation factors (GDFs), which are thought to be involved in many aspects of tissue morphogenesis.

## Expression systems

Expression of recombinant forms of the early BMPs was accomplished using mammalian cell lines, most particularly CHO cells [10, 12]. With this expression system, the BMPs are produced and properly refolded inside the cells and then secreted into the media in an active form. The BMPs are then purified from the media and characterized. Preparations of both OP-1 and BMP-2 used in clinical testing and currently in various stages of the regulatory approval process were produced by this methodology. However, more recently, several BMPs have been produced in bacterial cells. Recombinant preparations of GDF-5, 6 and 7 have been produced in active form from *Escherichia coli* [19, 26]. With this expression system, the BMPs are produced in a randomly folded state inside the cells and after lysing the cells, the BMPs are purified and refolded into the proper conformation. BMP-2 has also been reported to be successfully produced in *E. coli* [28]. On the other hand OP-1 has been extensively evaluated in *E. coli* expression systems but appears to be a more formidable refolding challenge and only small amounts of active protein have been produced. Other expression systems such as yeast, plants and transgenic animals have not been reported for expression of BMPs.

## Protein structure

As first described for TGF-β, members of the BMP family of proteins are synthesized as large precursors that are approximately three times larger than the mature protein and are eventually proteolytically processed to yield mature disulfide-linked dimers [27]. The expression and processing has been extensively examined for OP-1/BMP-7. The OP-1 gene predicts a polypeptide of 431 amino acids with a 29 amino acid signal sequence. Residues 293 through 431 comprise the mature domain and residues 29 through 292 comprise the prodomain. OP-1 is initially synthesized in the cell as a monomeric 50 kDa pro-protein that is dimerized, glycosylated, and then proteolytically cleaved at the Arg-Xaa-Xaa-Arg maturation site in an acidic cellular compartment before secretion into the medium. Of the four potential N-linked glycosylation sites two are used, one in the mature domain and one in the pro-domain. Secreted OP-1 demonstrates an apparent molecular weight of 110–120 k, indicating that after proteolytic processing the two pro-domains remain non-covalently associated with the disulfide linked mature. During the purification procedure for mature OP-1/BMP-7, the prodomain is normally separated from the complex by the use of dissociating conditions. However, the intact complex can be purified in the absence of these agents and has been characterized. This purified complex is termed soluble or proOP-1 and is significantly more soluble in physiological buffers than the mature OP-1. The function or functions of the prodomain has not been elucidated. However, in addition to its presumed role in protein folding and transport, the prodomain may participate in tissue targeting or receptor specificity.

The crystal structure has now been reported for the mature dimers of OP-1/BMP-7 and BMP-2. The structure to 2.8 Å resolution has shown that OP-1/BMP-7 like TGF-β, itself, is in a "hand" structure consisting of two fingers of antiparallel beta strands and an alpha helical region at the heel of the palm [24]. The central core of the hand or palm is the site of a threaded ring structure created by the internal disulfide bonds also known as the cysteine knot. An intermolecular disulfide bond in this "palm" region forms the dimer. Envisioning a handshake provides a conceptual picture of this dimer interaction. The three-dimensional structure to 2.7 Å has recently been published for the BMP-2 molecule and shown to be very similar to that of OP-1 [28]. Most recently, the crystal structure of a BMP-2 : BMPR-1A receptor ectodomain complex was solved [29]. This data revealed a hydrophobic area of the type 1 receptor that fit into a hydrophobic pocket composed of residues of both BMP-2 monomers.

## Signaling pathways

Bone morphogenetic proteins (BMPs) exert their effects through complex formation with a heteromeric receptor complex [30–34]. The complex consists of two type I and two type II polypeptides which are transmembrane serine/threonine kinases. At present, 12 type I receptors have been identified for the TFG-β superfamily. Three of these type I receptors (ActR-1, BMPR-IA and BMPR-IB) have been shown to bind to one or more members of the BMP family. Seven type II receptors have been identified for the TGF-β superfamily. Of the Type II receptors, ActR-II and –IIB, and BMPR-II can bind different members of the BMP subfamily. Table 4 lists the type I and type II receptors that are known to be present in mammalian tissues. The asterisks in the table refer to the receptors that bind BMPs.

The current consensus is that BMPs can bind to type I or II receptors alone, but with a weak affinity. Both types of receptors are required to be present for high affinity binding and signaling. Different BMPs recognize different type I and type II receptors with different affinities. The BMP ligand also appears to enforce specificity of receptor pairing and thus determining in part the nature of the resulting signaling. It is also known that molecules such as noggin, chordin and DAN bind to BMPs with high affinity and prevent their interaction with the receptors. Furthermore, the binding of BMPs to extracellular matrix components such as collagen and heparin sulfate probably influence their ability to interact with the receptors.

The intracellular signaling pathways that are induced by the serine/threonine kinase receptors involve a family of signaling molecules called Smad proteins. Currently, eight different Smad proteins have been identified in mammals. They can be divided into three subclasses: R-Smads (receptor-activated Smads), Co-Smads

*Table 4 - Mammalian members of the TFG-β superfamily receptors*

| Type I receptors | Type II receptors |
|---|---|
| ALK-1 | BMPR-II* |
| ActR-I (ALK-2)* | TBR-II |
| BMPR-IA (ALK-3)* | ActR-IIB* |
| ActR-IB (ALK-4) | ActR-II* |
| TBR-I (ALK-5) | |
| BMPR-IB (ALK-6)* | |

(common partner Smads), and anti-Smads (inhibitory Smads). Smads 1, 2, 3, 5 and 8 are R-Smads. Smad 4 is a Co-Smad. Smads 6 and 7 are anti-Smads. The R-Smads are phosphorylated by specific type I receptors. ALK-1 and the BMP type I receptors interact with Smads 1, 5 and 8 and the TGF-β and activin type I receptors bind to Smads 2 and 3. Following phosphorylation, R-Smads dissociate from the receptor, bind to Smad 4 and enter the nucleus. Inside the nucleus heteromeric complexes of Smads regulate transcription of the BMP genes by utilizing DNA-binding proteins to target specific genes. Smads bind DNA alone but with low affinity and specificity.

## Chromosomal localization

Several members of the BMP gene family have been mapped to their human chromosome locations [18]. These results demonstrate that these genes are widely dispersed in the human genome similar to the other members of the TGF-β superfamily. Chromosomal dispersion may have facilitated the development of tissue specific functions for the various family members. The following genes have been localized to specific chromosomes: BMP-2 (chromosome 20), BMP-3 (chromosome 4), BMP-3B (chromosome 10), BMP-4 (chromosome 14), BMP-5 (chromosome 6), BMP-6 (chromosome 6) and OP-1 (chromosome 20), BMP-8 (chromosome 1), GDF-5 (chromosome 22), GDF-6 (chromosome 8), GDF-7 (chromosome 2) and BMP-15 (chromosome X). It is not known whether the genes occupying the same chromosome are clustered or are nonlinked.

## BMP localization

Although BMPs were originally isolated and identified from bone, it was soon discovered by a variety of studies that BMPs are expressed in most other tissues of the human body. The expression has been found in many adult tissues, but also, surprisingly, throughout embryonic development [21,35]. During embryogenesis BMPs serve as important inductive signals for tissue development and have been shown to have a pivotal role in development of the musculoskeletal system, the nervous system, the heart, kidney, skin, eyes, and teeth. After birth, the BMPs play roles in tissue repair and regeneration. As an example, numerous analytical procedures have been utilized to localize OP-1. The first indication that OP-1 had a more widespread localization than bone occurred when OP-1 cDNA was found in a cDNA library generated from hippocampus [7]. Subsequently, the mRNA was extracted from a variety of adult mouse tissues and evaluated for the presence of OP-1 mRNA [8, 25]. Large amounts of OP-1 mRNA were found in the kidney and significant amounts were found in the bladder, adrenal tissue, brain and calvaria. No detectable OP-1 mRNA was found in the heart and liver. Mouse embryos were also evaluated and found to contain OP-1 mRNA in multiple organs at levels that varied depending upon the time after conception. In gene knockout studies, mice lacking the OP-1 gene displayed severe defects in the developing kidney and eye and appeared polydactyl [36]. These studies clearly demonstrated that the absence of OP-1 disrupts the cellular interactions required for the growth and development of these organs. Finally immunolocalization studies have demonstrated that the OP-1 protein is also present in multiple tissues in both adult animals and during embryonic development. Detailed histological analyses have been done with bone and cartilage, brain and kidney tissues.

## Biological activities

The biological activities of BMPs have been evaluated *in vivo* using a variety of animal models and *in vitro* using a variety of cell lines [37–43]. Because of their discovery in bone, most of these studies have been done using bone cells and bone defect animal models. To a lesser extent, related tissues such as cartilage and other hard tissues such as dentin have also been examined. More recently the biological activities are being evaluated in soft tissues, particularly brain, kidney and muscle.

The rat subcutaneous bone formation assay has been the standard method used to evaluate the osteoinductive potential of BMPs. Implantation of purified recombinant BMP with bone collagen matrix in subcutaneous sites in rats induces a sequence of cellular events that leads to the formation of new bone complete with bone marrow elements [44]. Only osteoinductive BMPs have this activity. During this process the first step is the recruitment by the BMP of nearby mesenchymal stem cells into the collagen matrix. The BMP stimulates the stem cells to proliferate and then triggers their differentiation into chondrocytes in 5 to 7 days. Cartilage is formed and on capillary invasion, the chondrocytes hypertrophy, become calcified, and osteoblasts appear in the implant site. Newly formed bone is present in 9 to 12 days. Subsequently, the bone is remodeled extensively and becomes occupied by ossicles filled with bone marrow elements in 14 to 21 days. This cellular process is referred to as "endochondral bone formation". Osteoinductive BMPs are also observed to form bone by "intramembraneous bone formation" whereby the BMP triggers the mesenchymal cells to differentiate directly to osteoblasts and thus bypass the cartilage stage. Little is known about how one route is chosen over the other.

The critical activity of implanted BMPs during the bone formation process occurs at the beginning of the biological cascade. These activities involve the interactions with the mesenchymal cells including chymotaxis, proliferation and differentiation into bone and/or cartilage cells. The subsequent steps appear to rely on the local induction of a range of factors, including other BMPs. For instance, OP-1 has recently been shown to induce numerous growth factors and multiple BMPs, including itself, during the bone induction process [45]. Additional support for the action of OP-1 throughout the bone formation process is also provided by *in vitro* studies [46–52]. These data have demonstrated that OP-1 has multiple chondrogenic effects; the protein can (1) induce the chondrogenic phenotype in chondrocyte precursor cells, (2) induce chondrogenesis in non-cartilage stem cells, (3) promote re-expression of chondrocyte phenotype by dedifferentiated articular chondrocytes and (4) enhance mature chondrocyte characteristics in normal articular chondrocytes. Similarly, OP-1 also interacts with bone cells; the protein can (1) induce osteoblast phenotype expression by osteoprogenitor cells, (2) induce osteogenesis in non-bone stem cells and 3) enhance the osteoblastic characteristics of normal osteoblast cells.

To date many studies have been published evaluating the efficacy of BMPs in conjunction with matrix materials for local repair of bone defects [53]. These include most long bones, various cranialfacial bones and the spinal column. For the most part these studies have utilized either OP-1 or BMP-2 but recent studies have also examined GDF-5 [26]. In general the BMPs have been shown to be highly efficacious in repairing bones in many animal species, including rat, rabbit, dog, sheep, goat, monkey and baboon. More recently OP-1 and BMP-2 have also been shown to be efficacious in initial testing in humans [54–57].

Cartilage is observed as an intermediate step during the BMP-induced bone induction process. Furthermore *in vitro* studies have demonstrated that BMPs can promote chondrogenic differentiation, maturation and maintenance of chondrocyte phenotype and BMPs have been localized to cartilage [58]. These observations suggested that BMPs might be useful for healing cartilage. Studies have been reported evaluating BMPs in *in vivo* models of both osteochondral and chondral defects. Both OP-1 and BMP-2 formulated with collagen have been evaluated in osteochondral defects and shown to be efficacious [38, 59–60]. These studies have demonstrated that the BMPs can improve both the bone and cartilage healing in the defects, but the repair appears to be variable amongst species and the specific type and stability of the cartilage has not fully been evaluated. In one sheep study using a chondral defect, OP-1 was shown to induce substantial healing [61]. In this model which used mini-osmotic pumps to slowly deliver the OP-1 into the articular joint, no healing was observed in the control defects. Although in an early stage of development, the data suggest that BMPs have an exciting potential to heal cartilage, a tissue that unlike bone is not known to repair itself.

Brain tissue has been one of the first non-bone tissues investigated for the biological activities of BMPs [62]. *In vitro* studies using OP-1 have demonstrated that this BMP increased expression of the adrenergic phenotype in neural crest cells and OP-1 regulated expression of L1 and neural cell adhesion molecules in a neural cell line. In further studies it was discovered that OP-1 selectively induces dendritic growth in cultured rat neurons and the dendrites correctly segregate, modify cytoskeletal and membrane proteins, and form synaptic contacts of appropriate polarity [63]. Based on these observations OP-1 was evaluated *in vivo* for the repair of nerve tissue in stroke models. In rat models of cerebral hypoxial ischemia OP-1 was injected intracisternally into the brain and shown to protect against damage, as well as to facilitate the recovery from damage caused by experimental stroke [64].

The biological activities of BMPs have also been evaluated using the kidney [65]. This organ has been identified as the major site for synthesis of OP-1 during embryonic development as well as in adulthood. In addition, numerous *in vitro* studies have suggested that OP-1 is required for metanephric mesenchyme differentiation and can effect kidney cells in culture. The results from both chronic and acute disease models in rats have demonstrated that systemic (IV) administration of OP-1 can protect against damage as well as facilitate recovery from this damage [66].

## Delivery materials

The study of BMPs has required a large amount of support research into the means to deliver these proteins [29, 37 ,67]. However, BMP delivery research has never been given the priority that has been given to the BMPs, themselves. Hopefully, the increased availability and the ever expanding therapeutic potential in bone as well as other tissues will give impetus to this important area. Most studies have focused on biomaterials to deliver BMPs for use in the original therapeutic indication, local implantation of an osteoinductive device for repair of bony defects. This use has required a solid-phase matrix that must function as an appropriate cellular scaffold during the bone formation cascade. More recently, studies have been reported examining BMP delivery in soluble formulations without these matrixes. The goal of these studies has been to locally inject the proteins into bone or cartilage defects. Finally, in studies being done in soft tissues, delivery is being evaluated in much more complex systems, such as systemic delivery and intracistermal delivery into the brain.

The initial delivery discovery work utilized particles of guanidine-extracted, demineralized bone powder as the carrier for BMPs. This material is primarily Type I collagen and has served as the "gold standard" by which all other materials have been compared. The sequential cellular response at the interface of the BMP matrix implants includes a multistep cascade: binding of fibrin to implanted matrix, chemotaxis of cells, proliferation of progenitor cells, differentiation into chondroblasts, cartilage calcification, vascular invasion, bone formation, remodeling and bone marrow differentiation. Ideally the carrier needs to perform several important functions: provide a substrate for the recruitment and attachment of progenitor cells, bind the BMP, accommodate each step of the cellular response during bone formation, and protect the BMP from non-specific proteolysis. In addition, selected materials must be biocompatible and preferably biodegradable; the carrier should act as a temporary scaffold until replaced completely by new bone. In some cases, slow degrading materials may be useful where solid, load-bearing characteristics are required.

A variety of biocompatible biomaterials have been evaluated for local delivery of BMPs for new bone formation. These include various extracellular matrix components, alone and in combination (different collagens, fibrin, fibronectin, hyaluronic acids, glycosaminoglycans), ceramics (hydroxyapatites, tricalcium phosphates, cements), synthetic polymers (particularly polylactic and polyglycollic acid polymers) and bone graft materials (both autograft and allograft). Most of these materials, have been shown to support bone formation. However, in general, none have produced comparable results to that achieved with Type I collagen. For instance, calcium phosphates are slow to resorb and synthetic polymer degradation products can be inhibitory. It is also clear that different defect sites in the body have different environments and will need specially designed materials for many of these sites. However, for the present, type I collagen is the material of choice for clinical development of BMPs. The initial BMP product (OP-1 Implant) to receive regulatory approval for sale uses highly purified bone-derived type I collagen as the delivery matrix. In addition, the only other BMP that has been extensively evaluated and is in the late-stage regulatory approval process also used type I collagen for delivery; BMP-2 utilizes a skin-derived collagen in a sponge formation.

In recent years several studies have been reported using formulations without solid-phase matrices for local delivery of BMPs into bone defects [68,69]. These studies have demonstrated that injectable BMPs can be used to speed the rate of fracture repair in various animal studies. Both OP-1 and BMP-2 have been injected into defects in buffer solutions and remain in the defect area long enough to stimulate the bone formation process. Possibly the BMPs are able to use the natural fracture callous as a scaffold and their limited solubility under physiological conditions may involve a precipitation event at the site. Nevertheless the data appear to be quite promising and a wide variety of materials to facilitate this type of delivery needs to be developed.

## Conclusion

It has been over a decade since the first BMP genes were reported. Over this time recombinant BMPs have been produced from these genes and extensively characterized biochemically and biologically. A large variety of animal efficacy models has been utilized to evaluate the therapeutic potential, particularly using two of the early BMPs, BMP-2 and OP-1 (BMP-7). More recently the efficacy of these BMPs to repair bone has been demonstrated in humans. Finally, in 2001, the first BMP, OP-1 (BMP-7) received regulatory approval for marketing and sales. However, this is only the beginning. Many more BMPs or BMP-like molecules have been discovered and are being produced in recombinant form for evaluation. Although most knowledge has been gained in the bone field, these proteins are also important in most if not all tissues and little is known about most of them. BMPs have proven to be an important new area of developmental biology and have clearly become an important new tool in the field of tissue engineering.

## References

1 Urist MR (1965) Bone formation by autoinduction. *Science* 150: 893–899

2 Sampath TK, Reddi AH (1981) Dissociative extraction and reconstitution of extracellular matrix components involved in local bone differentiation. *Proc Natl Acad Sci USA* 78 (12): 7599–7602

3 Sampath TK, Reddi AH (1983) Homology of bone-inductive proteins from human, monkey, bovine, and rat extracellular matrix. *Proc Natl Acad Sci USA* 80: 6591–6595

4 Sampath TK, Muthukumaran N, Reddi AH (1987) Isolation of osteogenin, an extracellular matrix-as-sociated bone inductive protein by heparin affinity chromatography. *Proc Natl Acad Sci USA* 84: 7109–7113

5 Celeste AJ, Lannazzi JA, Taylor RC, Hewick RM, Rosen V, Wang EA, Wozney JM (1990) Identification of transforming growth factor-β superfamily members present in bone-inductive protein purified from bovine bone. *Proc Natl Acad Sci USA* 87 (24): 9843–9847

6 Hammonds RG, Schwall R, Dudley A (1991) Bone inducing activity of mature BMP-2b produced from a hybrid BMP-2a/2b precursor. *Mol Endocrinol* 5: 149–155

7 Özkaynak E, Rueger DC, Drier EA, Corbett C, Ridge RJ, Sampath TK, Oppermann H (1990) OP-1 cDNA encodes an osteogenic protein in the TGF-β family. *EMBO J* 9: 2085–2093

8 Özkaynack E, Schnegelsberg PN, Jin DF, Clifford GM, Warren FD, Drier EA, Oppermann H (1992) Osteogenic protein-2 a new member of the transforming growth factor-beta superfamily expressed early in embryogenesis. *J Biol Chem* 267 (35): 220–227

9 Sampath TK, Coughlin JE, Whetstone RM, Banach D, Corbett C, Ridge RJ, Özkaynak,E, Oppermann H, Rueger DC (1990) Bovine osteogenic protein is composed of

dimers of OP-1 and BMP-2A, two members of the transforming growth factor-β superfamily. *J Biol Chem* 265 (22): 13198–13205

10 Sampath TK, Maliakal JC, Hauschka PV, Hones WK, Sasak H, Tucker RF, White KH, Coughlin JE, Tucker MM, Pang RH (1992) Recombinant human osteogenic protein-1 (hOP-1) induces new bone formation *in vivo* with a specific activity comparable with natural bovine osteogenic protein and stimulates osteoblast proliferation and differentiation *in vitro*. *J Biol Chem* 267 (28): 20352–20362

11 Wang EA, Rosen V, Cordes P, Hewich RM, Kriz MF, Luxenberg DP, Sibley BS, Wozney, JM (1988) Purification and Characterization of other distinct bone-inducing factors. *Proc Natl Acad Sci USA* 85: 9484–9488

12 Wang EA, Rosen V, D'Alessandro JS, Bauduy M, Cordes P, Harada T, Israel DI, Hewich RM, Kerns KM. LaPan P et al (1990) Recombinant human bone morphogenetic protein induces bone formation. *Proc Natl Acad Sci* 87: 2220–2224

13 Wozney JM, Rosen V, Celeste AJ, Mitsock LM, Whitters MJ, Kriz RW, Hewick RM, Wang EA (1988) Novel regulators of bone formation: Molecular clones and activities. *Science* 242: 1528–1534

14 Hofbauer LC, Heufelder AE (1996) Updating the metalloprotease nomenclature: bone morphogenetic protein 1 identified as procollagen C proteinase. *Eur J Endocrin* 135: 35–36

15 Wozney JM, Rosen V (1998) Bone morphogenetic protein and bone morphogenetic protein gene family in bone formation and repair. *Clin Ortho Related Res* 346: 26–37

16 Bahamonde ME, Lyons KM (2001) BMP3: to be or not to be a BMP. *JBJS* 83-A (Supp 1, Part 1): 56–62

17 Reddi AH (2001) Bone morphogenetic Proteins: From basic science to clinical applications. *JBJS* 830-A (Supp 1, Part 1): 1–6

18 Demers C, Hamdy R (1999) Bone morphogenetic proteins. *Science Medicine* 6 (6): 8–17

19 Erlacher L, McCartney J, Piek E, ten Dijke P, Yanagishita M, Oppermann H, Luyten FP (1998) Cartilage-derived morphogenetic proteins and osteogenic protein-1 differentially regulate osteogenesis. *J Bone Min Res* 13 (3): 383–392

20 McPherron AC, Lawler AM, Lee S-J (1997) Regulation of skeletal muscle mass in mice by a new TGF-beta superfamily member. *Nature* 387: 83–90

21 Wozney JM (1998) The bone morphogenetic protein family: Multifunctional cellular regulators in the embryo and adult. *Eur J Oral Sci* 106 (1): 160–166

22 Reddi AH(1998) Role of morphogenetic proteins in skeletal tissue engineering and regeneration. *Nature Biotech* 16: 247–252

23 Sakou T (1998) Bone morphogenetic proteins: From basic studies to clinical approaches. *Bone* 22: 591–603

24 Griffith DL, Keck PC, Sampath TK, Rueger DC, Carlson WD (1996) Three-dimensional structure of recombinant human osteogenic protein 1: Structural paradigm for the transforming growth factor ß superfamily. *Proc Natl Acad Sci* 93: 878–883

25 Özkaynack E, Schnegelsberg PN, Oppermann H (1991) Murine osteogenic protein (OP-1): High levels of mRNA in kidney. *Biochem Biophy Res Comm* 179 (1): 116–123

26 Spiro RC, Liu LS, Heidaran MA, Thompson AY, Ng CK, Pohl J, Poser J (2000) Inductive activity of recombinant human growth and differentiation factor-5. *Biochem Soc Trans* 28: 362–368
27 Jones WK, Richmond EA, White K, Sasak H, Kusmik W, Smart,J, Oppermann H, Rueger DC, Tucker RF (1994) Osteogenic Protein-1 (OP-1) expression and processing in Chinese hamster ovary cells: Isolation of a soluble complex containing the mature and pro-domains of OP-1. *Growth Fac* 11: 215–225
28 Scheufler C, Sebald W, Hulsmeyer M (1999) Crystal structure of human bone morphogenetic protein-2 at 2.7 Å resolution. *J Mol Biol* 287: 103–115
29 Kirsch T, Sebald W, Dreyer MK (2000) Crystal structure of the BMP-2-BRIA ectodomain complex. *Nature Structural Bio* 7 (6): 492–496
30 Massague J (1998) TGF-β Signal transduction. *Annu Rev Biochem* 67: 753–791
31 Miyazono K (2000) TGF-β signaling by Smad proteins. *Cytokine Growth Fac Rev* 11: 15–22
32 Piek E, Heldin CH, ten Dijke P (1999) Specificity, diversity, and regulation in TFG-ß superfamily signaling. *FASEB J* 13: 2105–2124
33 Ray RP, Wharton KA (2001) Minireview: Twisted perspective: New insights into extracellular modulation of BMP signaling during development. *Cell* 104: 801–804
34 Wrana JL (2000) Regulation of Smad activity. *Cell* 100: 190–192
35 Hogan, BLM (1996) Bone morphogenetic proteins: multifunctional regulators of vertebrate development. *Genes Dev* 10: 1580–1594
36 Karsenty G, Luo G, Hofmann C, Bradley A (1996) BMP-7 is required for nephrogenesis, eye development and skeletal patterning. *Annals NY Acad Sci* 785: 98–107
37 Azari K, Doll BA, Sfeir C, Mu Y, Hollinger OJ (2001) Therapeutic potential of bone morphogenetic proteins. *Expert Opin Investig Drugs* 10 (9): 1677–1686
38 Cook S.D, Rueger DC (1996) Osteogenic Protein-1: Biology and applications. *Clin Ortho* 324: 29–38
39 Kirker-Head CA (2000) Potential applications and delivery strategies for bone morphogenetic proteins. *Advanced Drug Delivery Rev* 43: 65–92
40 Reddi AH (1998) Fracture Repair Process: Initiation of fracture repair by bone morphogenetic proteins. *Clin Orthop Rel Res* 355S: S66–S72
41 Ripamonti U, Duneas N (1998) Tissue morphogenesis and regeneration by bone morphogenetic proteins. *Plast Recon Surg* 101 (1): 227–239
42 Ripamonti U, Ramoshebi LN, Matsaba T, Tasker J, Crooks J, Teare J (2001) Bone induction by BMPs/OPs and related family members in primates. *JBJS* 83-A(Supp 1, Part 2): 116–127
43 Schmitt JM, Hwang K, Winn SR, Hollinger JO (1999) Bone morphogenetic proteins: and update on basic biology and clinical relevance. *J Orthop Res* 17 (2): 269–278
44 Sampath TK, Rueger DC (1994) Structure, function, and orthopedic applications of osteogenic protein-1 (OP-1). *Complications in Orthopedics* 9 (Winter): 101–107
45 Patel TC, Weinstein MA, White AP, Grauer J, Horowitz HC, Friedlaender GE (2001)

Autologous growth factor gene expression in a rabbit model: an evaluation of recombinant human osteogenic protein-1. Eurospine Meeting, Sept., Gothenberg, Sweden.
46 Asahina I, Sampath TK, Nishimura I, Hauschka PV (1993) Human osteogenic protein-1 induces both chondroblastic and osteoblastic differentiation of osteoprogenitor cells derived from newborn rat calvaria. *J Cell Biol* 123 (4): 921–933
47 Asahina I, Sampath TK, Hauschka P (1996) Human osteogenic protein-1 induces chondroblastic, osteoblastic, and/or adipocytic differentiation of clonal murine target cells. *Exp Cell Res* 222: 38–47
48 Cheifetz S, Li I, McCulloch C, Sampath K, Sodek J (1996) Influence of osteogenic protein-1 (OP-1; BMP-&) and transforming growth factor β1 on bone formation *in vitro*. *Connective Tissue Res* 35: 71–78
49 Kitten AM, Lee JC, Olson M (1995) Osteogenic protein-1 enhances phenotypic expression in ROS 17/2.8 cells. *Am J Physiol* 269: E917–E926
50 Kitten AM, Kreisberg JK, Olson MS (1999) Expression of osteogenic protein-1 mRNA in cultured kidney cells. *J Cellul Physiol* 181: 410–415
51 Knutsen R, Wergedal JE, Sampath TK, Baylink DJ, Mohan S (1993) Osteogenic protein-1 stimulates proliferation and differentiation of human bone cells *in vitro*. *Biochem Biophys Res Comm* 194: 1352–1358
52 Maliakal, JC, Asahina I, Hauschka PV, Sampath TK (1994) Osteogenic protein-1 (BMP-7) inhibits cell proliferation and stimulates the expression of markers characteristic of osteoblast phenotype in rat osteosarcoma. *Growth Factors* 11: 227–234
53 Pecina M, Giltaij LR, Vukicevic S (2001) Orthopaedic applications of osteogenic protein-1 (BMP-7). *Internation Orthopaedics* 25: 203–208
54 Fiedlaender GE, Perry CR, Cole JD, Cook SD, Cierny G, Muschler GF, Zych GA, Calhound JH, LaForte AJ, Yin S (2001) Osteogenic protein-1 (bone morphogenetic protein-7) in the treatment of tibial nonunions. *JBJS* 83-A (Supp 1, Part 2): 151–164
55 Geesink RGT, Moefnagels NHM, Bulstra SK (1999) Osteogenic activity of OP-1 bone morphogenetic protein (BMP-7) in a human fibular defect. *JBJS* 81-B (4): 710–718
56 Groenveld EHJ, Burger EH (2000) Bone morphogenetic proteins in human bone regeneration. *Eur J Endocrin* 142: 9–21
57 Riedel GE, Valentin-Opran (1999) Clinical evaluation of rhBMP-2/ACS in orthopedic trauma: A progress report. *Orthopedics* 22 (7): 663–665
58 Chubinskaya S, Merrihew C, Cs-Szabo G, Mollenhauer J, McCartney J, Rueger D, Kuettner K (2000) Human articular chondrocytes express osteogenic protein-1. *J Histoch Cytoch* 48(2): 239–250
59 Sellers RS, Peluso D, Morris EA (1997) The effect of recombinant human bone morphogenetic protein-2 (rhBMP-2) on the healing of full-thickness defects of articular cartilage. *J Bone Joint Surg* 79-A (10): 1452–1463
60 Sellers RS, Zhang R, Glasson SS, Kim HD, Peluso D, D'Augusta DA, Beckswith K, Morris EA (2000) Repair of articular cartilage defects one year after treatment with recombinant human bone morphogenetic protein-2 (rhBMP-2). *J Bone Joint Surgery* 82-A (2): 151–160

61 Jelic M, Pecina M, Haspl M, Kos J, Taylor K, Maticic D, McCartney J, Yin S, Rueger D, Vukicevic S (2001) Regeneration of articular cartilage chondral defects by osteogenic protein-1 (bone morphogenetic protein-7) in sheep. *Growth Factors* 19: 101–113
62 Helm GA, Alden TD, Sheehan JP, Kallmes D (2000) Bone morphogenetic proteins and bone morphogenetic protein gene therapy in neurological surgery: A review. *Neurosurgery* 46: 1213–1222
63 Wither GS, Higgins D, Charette M, Banker G (2000) Bone morphogenetic protein-7 enchances dendritic growth and receptivity to innervation in cultured hippocampal neurons. *Eur J Neurol* 12: 106–116
64 Liu Y, Belayev L, Zhao W, Busto R, Saul I, Alonso O, Ginsberg MD (2001) The effect of bone morphogenetic protein-7 (BMP-7) on functional recovery, local cerebral glucose utilization and blood flow after transient focal cerebral ischemia in rats. *Brain Res* 905: 81–90
65 Reddi A (2000) Bone morphogenetic proteins and skeletal development: the kidney-bone connection. *Pediatr Nephrol* 14: 598–601
66 Vukicevic S, Basic V, Rogic D, Basic N, Shih M-S, Shepard A, Jin D, Dattatreyamurty B, Jones W, Dorai H et al (1998) Osteogenic protein-1 (bone morphogenetic protein-7) reduces severity of injury after ischemic acute renal failure in rat. *J Clin Invest* 102 (1): 202–214
67 Uludag H, Gao T, Porter TJ, Friess W, Wozney JM (2000) Delivery systems for BMPs: factors contributing to protein retention at an application site. *JBJS* 83-A (Suppl 1, Part 2): 128–135
68 Blokhuis T, den Boer F, Bramer J, Jenner J, Bakker F, Patka P, Haarman H (2001) Biomechanical and histological aspects of fracture healing, stimulated with osteogenic protein-1. *Biomaterials* 22: 725–730
69 Welch R, Jones A, Buchloz R, Reinert C, Tjia J, Pierce W, Wozney J, Li X (1998) Effect of recombinant human bone morphogenentic protein-2 on fracture healing in a goat tibial fracture model. *J Bone Min Res* 13: 1483–1490

# Prostate-derived factor and growth and differentiation factor-8: Newly discovered members of the TGF-β superfamily

*Vishwas M. Paralkar[1], William A. Grasser[1], Amy P. Baumann[1], Tessa A. Castleberry[1], Thomas A. Owen[1] and Slobodan Vukicevic[2]*

[1]Department of Cardiovascular and Metabolic Diseases Pfizer Global Research and Development, Groton, CT 06340, USA; [2]University of Zagreb, School of Medicine, Zagreb, Croatia

## Introduction

The transforming growth factor-β (TGF-β) superfamily is a large group of structurally related proteins that play various important roles during embryonic development, as well as in adult life. This superfamily in addition to TGF-βs also contains the inhibins, activins, Mullerian inhibiting substance, and bone morphogenetic proteins (BMPs), as well as the various growth and differentiation factors (GDFs). Members of the TGF-β superfamily are highly conserved, secreted molecules whose biologically active C-terminal domains play a variety of roles in embryonic pattern formation, body plan establishment and organogenesis in numerous species from *Drosophila* and *C. elegans* through humans [1–3]. Animals and humans lacking or having mutations in various TGF-β family members exhibit a wide variety of phenotypes, ranging from early embryonic death due to lack of mesodermal development to viable, but severely compromised animals with a variety of skeletal defects, to human diseases such as fibrodysplasia ossificans progressiva and dentinogenesis imperfecta. Among the TGF-β family members, the BMPs form a large subgroup of proteins, which were originally named on the basis of their ability as components of demineralized bone matrix to induce ectopic bone formation. Subsequently, classical protein chemistry in conjunction with molecular biology resulted in the cloning and expression of a number of BMPs. Their extensive homology to each other, in addition to highly conserved structural features, places them in the TGF-β superfamily [4]. Conversely, BMPs have been shown to be involved in bone and cartilage repair in animals and humans and have been demonstrated to lessen the severity of damage in animal models of kidney failure and stroke [5]. Following the identification and cloning of the various BMPs and TGF-βs, Lee and co-workers used degenerate oligonucleotides made against sequences that were conserved among various members of the TGF-β superfamily to identify new members of the family. These newly identified members have been named growth and differentiation factors (GDFs) [6]. At present, extensive work is underway to identify additional BMP family members, to further characterize their secondary signaling pathways and to

Bone Morphogenetic Proteins, edited by Slobodan Vukicevic and Kuber T. Sampath

explore additional clinical applications for these proteins. Such an approach in our laboratory resulted in the cloning and characterization of another member of the superfamily, which we designated as prostate-derived factor (PDF) [7]. This name was based on the high expression of this protein in the prostate. Extensive work has been done on TGF-β and its family members such as the BMPs and they have been the subjects of numerous reviews. However, very little is known about some of the newly identified members of the TGF-β superfamily. This chapter will focus on two interesting but relatively newly identified members of the TGF-β superfamily, namely, PDF and GDF-8/myostatin [8]. As stated earlier, members of the TGF-β superfamily have been implicated in organogenesis (based on localization studies and gene deletion experiments). Although a role for PDF in other organs cannot be ruled out, this review will focus on its role and expression in the prostate and its regulation by androgens. Current literature shows that myostatin is involved in regulation of skeletal muscle growth, and this review will summarize the current data on myostatin and its role in the skeletal muscle.

## Prostate-derived factor (PDF)

Members of the TGF-β superfamily have been shown to play important roles in embryonic development and epithelial-mesenchymal interactions during embryonic tissue differentiation. We were interested in identifying novel members of the TGF-β superfamily, which *via* their pattern of expression might give us clues regarding their role in tissue differentiation and or embryonic development. This work resulted in the cloning and characterization of PDF [7]. The name PDF was based on its high expression in the prostate and a large body of literature indicating the importance of members of the TGF-β/BMP superfamily in prostate cancer. Others have also identified and cloned this molecule simultaneously and named it placental bone morphogenetic protein (PLAB), macrophage inhibitory cytokine-1 (MIC-1) and growth and differentiation factor-14 (GDF-14) based on its structural similarity to the TGF-β superfamily [9–11]. Given the high expression of PDF in the prostate, the known role of TGF-β in regulating growth of normal prostate cells, and the lack of growth inhibition of cancerous prostate cells when treated with TGF-β we were interested in characterizing the role of PDF in the normal prostate and in prostate cancer. Northern blot analysis of PDF expression revealed that the two organs with the highest levels of expression were the placenta and the prostate. In the prostate, PDF expression was localized to the epithelium of the main prostatic glands by immunohistochemistry using affinity-purified polyclonal antisera. The expression was similar in the hypertrophic prostate and again was restricted to the epithelial cells with a lack of expression in the fibromuscular stroma. Within the male urogenital tract, the specificity of PDF expression in the prostate was determined by immunolocalization in other accessory male genital

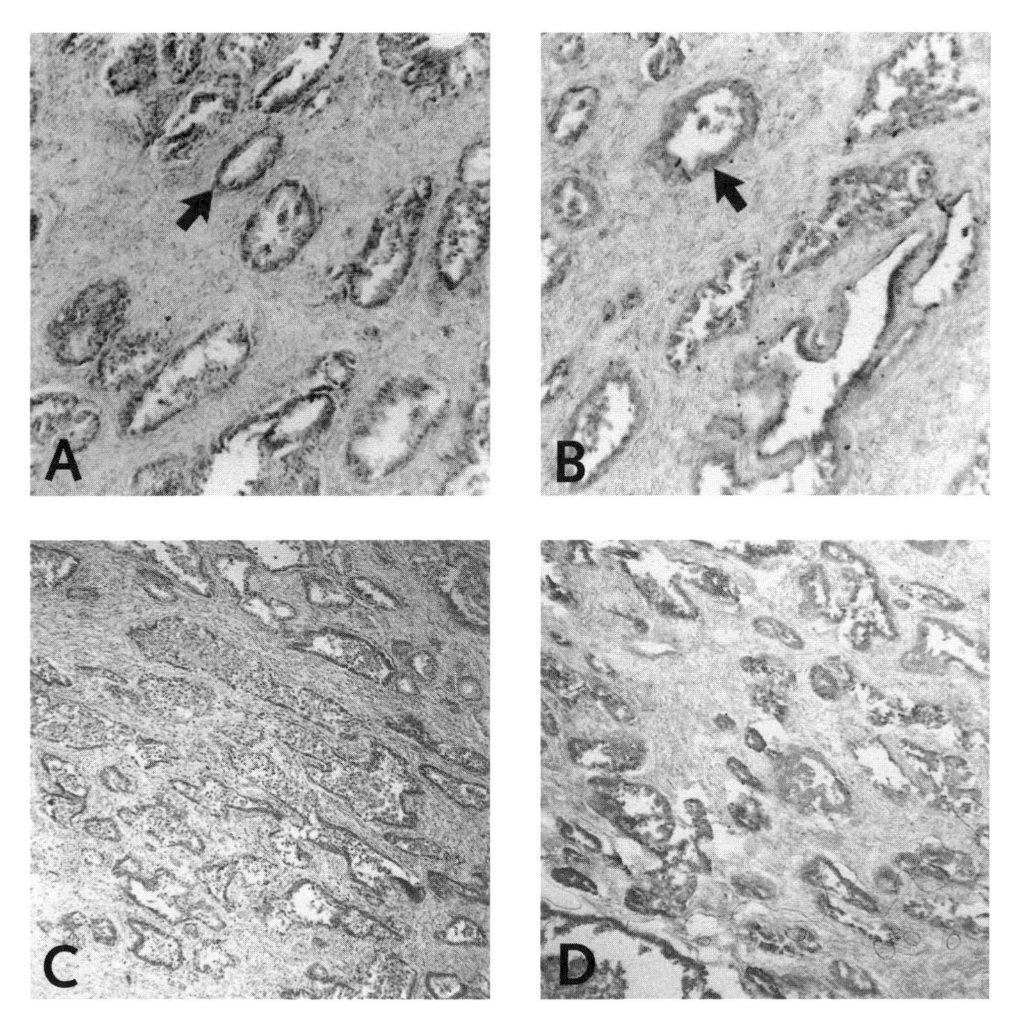


*Figure 1*
*Immunolocalization of PDF in prostate. (A) Control section of the normal prostate from the area of main glands incubated with the anti-PDF primary antibody alone. (B) Serial section from the main prostatic gland showed intense PDF staining. (C) Section from a prostate cancer sample showing a lack of PDF staining. (D) Staining for PDF in the hypertrophic prostate removed from a 57-year-old patient.*

glands. These data showed that PDF was expressed only in the epithelial cells of the main prostatic glands but not in the seminal vesicles or the bulbourethral glands. However, much to our surprise when we tried to localize PDF in samples from primary prostate tumors, we could not detect any protein expression by

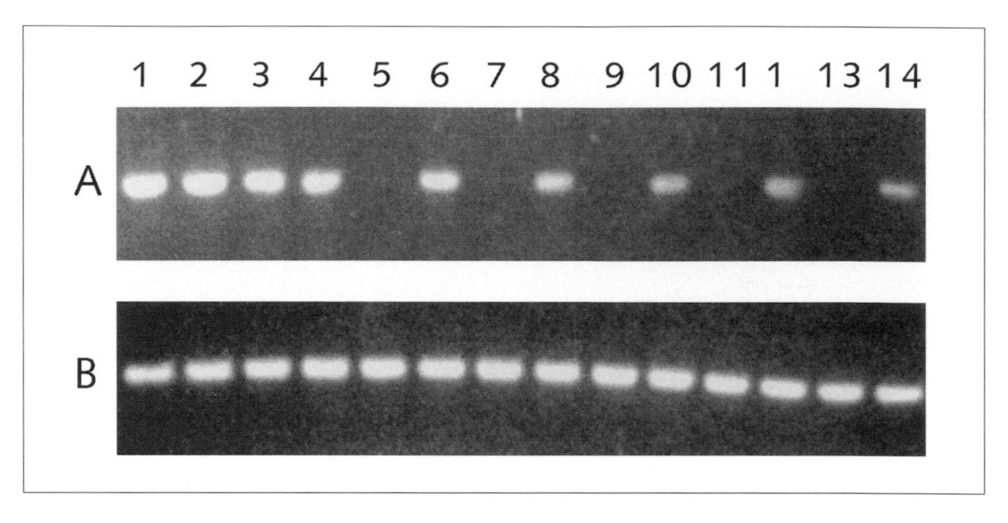


*Figure 2*
*Expression of PDF in prostate cancer. RT-PCR analysis was used to confirm that PDF expression was not detectable in prostate cancer. Lanes 1–4 show data from the normal prostate the single band obtained by PCR corresponds to PDF. Lanes 5, 7, 9, and 11 contained RNA from prostate cancer. Lanes 6, 8, 12 contained RNA from hypertrophic prostate, whereas lane 10 contained RNA from normal prostate and lane 13 was empty. Corresponding PCR done for 18S ribosomal RNA showed no difference (data not shown).*

immunohistochemistry (Fig. 1). Previous data had shown that TGF-β expression increases in prostate cancer (for review see [12]). We had expected similar results with PDF, but after repeated attempts, we could not detect any PDF protein expression in samples of prostate cancer unresponsive to androgens. To confirm these data, we obtained additional prostate tumor samples and used RT-PCR to detect PDF mRNA in these samples. As it can be seen from Figure 2, even by RT-PCR analysis we could not detect the presence of PDF in prostate tumors, confirming our earlier data which showed a lack of expression of the protein. The data on 10 prostate cancer samples so far shows a complete lack of expression of PDF protein or mRNA in prostate tumors unresponsive to androgens. Further evidence for lack of expression of PDF was obtained by examining its expression in prostate cancer-derived cell lines, where it was observed that with the exception of LNCaP cells, no other cell line examined expressed PDF (Fig. 3A). This result was also of interest since we had shown earlier that *in vivo*, PDF is regulated by androgens [7]. In LNCaP cells, dihydrotestosterone (DHT) increased PDF expression about two-fold over a 72-h treatment period, a magnitude of increase similar to that seen *in vivo* (Fig. 3B). This suggests that the androgen regulation of PDF in the prostate is not

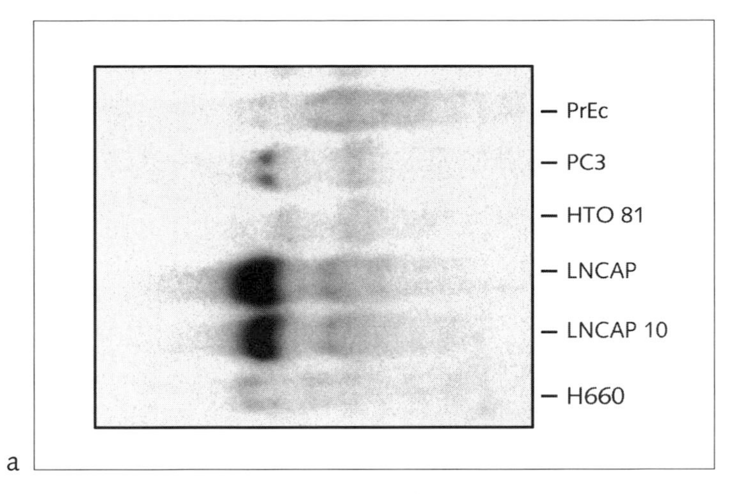


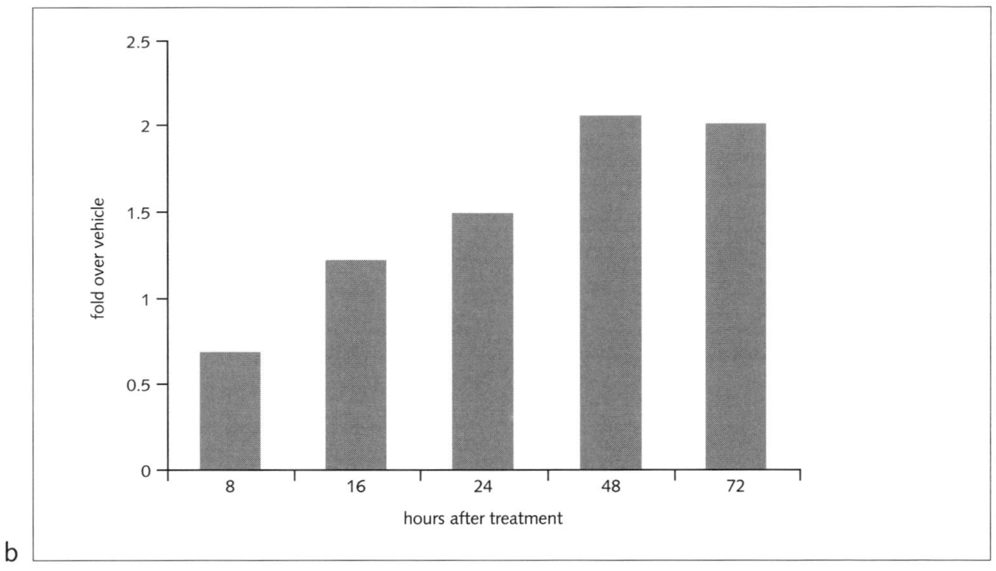


*Figure 3*
*Expression of PDF* in vitro. *(A) RNA was prepared from various prostate cancer cell lines. Northern blot analysis showed that only cells expressing androgen receptor, namely LNCaP and H660 express PDF, whereas all other cell lines switched off PDF expression where it functions to regulate gene expression. (B) Confluent LNCaP were treated for the indicated times with vehicle (0.1% ethanol) or DHT at a final concentration of $10^{-7}$ M in serum-free medium. After treatment, cells were scraped, frozen in liquid nitrogen and RNA prepared. Northern blot analysis was performed for PDF and 18S ribosomal RNA and the data expressed as fold increase in PDF expression as compared to vehicle following normalization to the 18S signal.*

a direct transcriptional effect. These data make us believe that PDF has an important role in prostate development and perhaps in the development of prostate cancer. Since TGF-β has been shown to inhibit the growth of normal prostate cells, data-showing overexpression of TGF-β in prostate cancer was a paradox. However, this was explained by the observations that prostate cancer cells were less sensitive to TGF-β than their normal counterparts. Our data, showing a lack of expression of PDF in prostate cancer unresponsive to androgens suggest that in addition to a reduced sensitivity of prostate cancer to TGF-β itself, in at least some cases prostate cancers may also down-regulate expression of other members of the TGF-β superfamily, thereby enhancing the potential for proliferation and subsequent metastases of prostate cancer cells.

## Growth and differentiation factor-8 (GDF-8)

Various GDFs have been cloned by using degenerate oligonucleotides made against conserved sequences among known members of the TGF-β family. GDF-8 is one of the many GDFs cloned and identified using this approach, and based on its amino acid sequence it belongs to the TGF-β superfamily [8]. However, like PDF, GDF-8 does not fall into any of the known sub-families such as the BMPs or the TGF-β subfamily. Northern blot analysis and *in situ* hybridization in developing embryos showed that GDF-8 expression was localized to developing somites in early stages, while in later stages of embryogenesis and in adults it is found in most muscles. GDF-8 was originally thought to be expressed specifically in skeletal muscle, but it has since been detected in mammary tissue [20] and cardiac muscle using RT-PCR [21]. GDF-8 null mice are larger than wild type mice and individual muscles are nearly 200% heavier than those of wild type littermates [8]. This identifies GDF-8 as a negative regulator of skeletal muscle growth during development. Support for the murine data came from historical studies showing that natural mutations of the GDF-8 gene in certain breeds of cattle result in the double muscling phenotype similar to that of the GDF-8 null mouse [13]. In the Belgian Blue, an 11 base pair deletion in the active C-terminus occurs; in Piedmontese there is a G→A substitution in the active region [21]. Other inactivating mutations have been identified demonstrating allelic heterogeneity at this locus. Thus, the role of GDF-8 in skeletal muscle growth during embryogenesis is very well documented and it most likely plays a role during normal growth and differentiation.

What is not well understood is the mechanism by which GDF-8 regulates muscle growth. The increase in muscle weight in the GDF-8 null mouse is a result of both skeletal muscle hyperplasia (increase in the number of cells) and skeletal muscle hypertrophy (increase in the size of skeletal muscle cells). Does GDF-8 inhibit growth of muscle cells at all stages of differentiation during development? Does it reduce the number of muscle cells by inducing apoptosis? Another possible mechanism would be that GDF-8 plays a role in regulating skeletal muscle size. In this case, GDF-8 would only be a negative regulator when skeletal muscle has reached a certain size, which is proportional to the rest of the body during development. Other questions that are equally interesting are: What is the role of GDF-8 in the adult? Is it expressed in all muscle types and fibers or is it preferentially expressed in certain muscle subtypes? What happens to GDF-8 expression and what is its role in diseases, which lead to skeletal muscle atrophy? Researchers around the world are beginning to answer some of these questions. The role of GDF-8 in the adult and in diseases that lead to skeletal muscle atrophy is important from a clinical viewpoint because only these data will determine clinical applications of GDF-8, if any. To study the role of GDF-8, various researchers have examined its expression in murine or rat model systems of injury or atrophy. Typical atrophy models include hindlimb unloading, denervation by sciatic neurectomy, microgravity and immobilization. The most common method of inducing injury/regeneration is bupivicaine injection. In hindlimb unloading, the animals' rear limbs are suspended and thus rendered unusable, whereas in sciatic neurectomy the lower limb muscles are unable to contract. Using hindlimb unloading in mice, Carlson et al. [14] found a 67% increase in GDF-8 mRNA expression at day 1 of unloading in the gastrocnemius/plantaris complex (fast twitch muscle). By day 7, GDF-8 levels were still elevated, but to a lesser extent (33%). No GDF-8 expression was detected in the soleus (slow muscle) [14]. Using the same atrophy model and a 10-day time point in Wistar rats, Wehling et al. [15] not only demonstrated an increase in plantaris GDF-8 mRNA; a concomitant 37% increase in protein was also found. With bupivicaine injection in Wistar rats, Sakuma et al. [16] and Yamanouchi et al. [17] have reported conflicting results in fast muscle. Using unilateral sciatic neurectomy Sakuma's data is in agreement with the emerging picture for fast muscle, that is, an increase in GDF-8 mRNA expression in response to atrophy.

Conflicting results have been reported for slow fiber types such as the soleus. Since it seems to be a consensus that expression in the soleus is low at best, perhaps using more sensitive techniques such as kinetic PCR additional data will yield a clearer pattern of GDF-8 expression. It must be pointed out that different studies have encompassed different time points (1 day–28 days), different fast muscles, and different models. If GDF-8 is acting as both a negative and positive regulator as proposed by Lee and Mc Pherron [30], it would be expected that levels would fluctuate over time. In addition, since atrophy increases over time in disuse models, the degree of regulation observed must be correlated with the amount of atrophy. Additional data is clearly needed to sort out this interesting picture. To truly understand the mechanism by which this molecule works, it will be necessary to demonstrate whether changes in GDF-8 levels precede or follow the observed changes in individual muscle weights. Although it is very difficult to conclude from these data any definitive role of GDF-8, it seems to be preferentially expressed in fast or mixed fiber types rather than in slow muscle fibers.

The expression of GDF-8 in diseases that lead to muscle wasting in humans such as HIV has been recently studied [18]. These researchers tested the hypothesis that if GDF-8 plays a negative role in skeletal muscle mass, then under conditions that lead to muscle wasting, its expression would be upregulated. This study was done by obtaining muscle biopsies and serum from healthy and HIV-infected men, and examining GDF-8 expression in the biopsies by Western blot analysis and in the serum by a RIA. Their findings agree with their hypothesis that GDF-8 levels are higher in HIV patients with muscle wasting than in healthy individuals. However, it should be noted that the antibody developed by these researchers recognizes a protein band of 26 kDa even under stringent reducing conditions. Based on GDF-8 amino acid sequence, the size of the active C-terminal GDF-8 monomer should be about half of what these researchers found. Although it is possible that glycosylation could account for some of the size difference, it seems unlikely that it alone would account for the size of the band seen in this paper. It is also possible that in spite of their best efforts, GDF-8 obtained from skeletal muscle samples might be extremely difficult to reduce and the researchers are actually detecting the GDF-8 dimer. To explain the size discrepancies the paper calls the protein myostatin–immunoreactive protein instead of myostastin. Although there are questions about these data, it is an extremely encouraging study that has attempted to truly characterize the role of GDF-8 in adult humans under conditions of muscle degeneration. Once the immunoreactive band seen by these researchers is identified or as more tools are generated such studies characterizing the role of GDF-8 in other patients with muscle atrophy will shed more light on the role of GDF-8 in the adult, and more specifically in muscle degeneration/atrophy.

Recent observations suggest that myostatin functions partially by inhibiting myoblast proliferation *via* p21 upregulation and G1 arrest [22–24]. To delineate other potential actions of myostatin, such as regulation of apoptosis, more data are needed. The growing body of literature on this topic is encouraging, but suggests a complex picture.

## Cell surface receptors and intracellular signaling by GDF-8 and PDF

Members of the TGF-β superfamily initiate intracellular effects by binding to and activating specific cell surface transmembrane receptors. These receptors, designated as type I and type II, possess intrinsic serine-threonine kinase activity. The receptors transmit signals to a family of transducers known as Smads [25]. Nuclear localization of Smads and subsequent activation of activation of target genes can be attributed to different response elements in the TGF-β and BMP subfamilies. The Smad binding element CAGA activates TGF-β signaling [26, 27], and GCCG appears to be specific to BMP signaling [28]. Based on our data showing the ability of PDF to activate the p3TP-Lux promoter reporter construct, PDF seems to uti-

*Table 1 - Putative cell surface receptor for TGF-β superfamily members*

| Ligands | Type II receptor | Type I receptor |
|---|---|---|
| BMP-7 | ActRII, BMPR-II | ActRI, BMPR-1A, BMPR-1B |
| GDF-8 | ActR-IIb | TGF-βR-I |
| PDF | Unknown | Unknown |

lize a signaling pathway similar to other members of the TGF-β superfamily [7]. Similarly, GDF-8 also activates p3TP-Lux and other TGF-β stimulated promoter reporter constructs (unpublished observations). The data of Celeste et al. point to the usage of ActR-IIb (Tab. 1) as a receptor combination for GDF-8 (Celeste et al., personal communication). To date, the exact receptors utilized by PDF remain unknown, as does the potential use of additional receptor subtype combinations by GDF-8. This would not be unexpected as BMP-7 can utilize combinations of three of the six different type I and three of the four different type II receptors for the TGF-β superfamily. GDF-5, a member of the BMP subfamily, has been shown to use combinatorial signaling to mediate digit formation. The outcome of signaling through BMPR1b is initiation of chondrogenesis, and apoptosis is effected through an alternative receptor [29]. The BMPs, unlike other superfamily members such as TGF-β, have the ability to bind to either the type I or the type II receptors. However, their binding to the type I receptor is of low affinity and high affinity binding is only observed in the presence of both types I and types II receptors. Whether such binding or combinatorial signaling is also utilized by PDF or GDF-8 remains to be determined.

## Potential clinical applications

One of the most advanced clinical applications of TGF-β superfamily members is the use of BMPs in osteoinduction during fracture repair or during bone reconstructive surgery. The expression of GDF-8 has been shown to negatively regulate skeletal muscle mass during development, it is thus interesting to speculate clinical application for a GDF-8 inhibitor in diseases which lead to muscle wasting such as cancer or AIDS, or for the treatment of loss of muscle mass due to aging. Similarly, a potential application for PDF in prostate cancer also seems plausible since the expression of PDF is androgen-regulated and PDF appears to be absent in prostate cancer. It must be pointed out that the data on PDF and prostate cancer is very preliminary at this point. It is also possible that future work will show a potential for either PDF or GDF-8 as a therapeutic agent in tissues other than muscle or prostate.

## Conclusion

Members of the highly conserved TGF-β superfamily play many roles both during development and in the adult animal. These molecules are synthesized and secreted by cells within a wide variety of tissues and affect gene expression by signaling through combinations of type I and type II transmembrane receptors and intracellular effector proteins known as Smads. The current elucidation of the mechanism of action of PDF and GDF-8 may lead to the development of therapeutic uses for these molecules. Prostate cancer is the most common cancer to strike men. Prostatic disease in general accompanies aging as do a variety of other elements that lead to muscular atrophy. The role of TGF-β superfamily members in the aged are very poorly understood and have not yet been the focus of many laboratories. Although the role of GDF-8 in embryonic muscle development is apparent, very little is known about GDF-8 expression during aging or during cancer-induced cachexia. The data so far in various models of skeletal muscle injury and GDF-8 expression are unclear and it is difficult to identify an *in vivo* role for GDF-8 in stages beyond embryogenesis. From a clinical application it is going to be extremely critical to characterize the function of these proteins in the aged. This will help in the possible clinical development of PDF, GDF-8, or other members of the TGF-β family.

## References


1 Kingsley DM (1994) The TGF-B superfamily: new members, new receptors, and new genetic tests of function in different organisms. *Genes Dev* 8: 133–146

2 Graff JM (1997) Embryonic patterning: to BMP or not to BMP, that is the question. *Cell* 89: 171–174

3 Reddi AH (1997) Bone Morphogenetic Proteins: an unconventional approach to isolation of first mammalian morphogens. *Cytokine Growth Factor Rev* 8: 11–20

4 Wozney J, Rosen V, Celeste AJ, Mitsock LM, Kriz RW, Hewick RM, Wang EA (1988) Novel regulators of bone formation: molecular clones and activities. *Science* 242: 1528–1534

5 Ebendal T, Bengtsson H, Söderströmet S (1998) Bone morphogenetic proteins and their receptors: potential functions in the brain. *J Neuro Res* 51: 139–146

6 Lee SJ (1990) Identification of a novel member (GDF-1) of the transforming growth factor-beta superfamily. *Mol Endocrinol* 4: 1034–1040

7 Paralkar VM, Vail AL, Grasser WA, Brown, TA, Xu, H, Vukicevic S, Ke HZ, Qi H, Owen TA, Thompson DD (1998) Cloning and characterization of a novel member of the transforming growth factorβ/bone morphogenetic protein family. *J Biol Chem* 273: 13760–13767

8 McPherron AC, Lawler AM, Lee SJ (1997) Regulation of skeletal muscle mass in mice by a new TGF-beta superfamily member. *Nature* 387: 83–90

9 Yokoyama-Kobayahi M, Saeki M, Sekine S, Kato S (1997) Human cDNA encoding a novel TGF-beta superfamily protein highly expressed in placenta. *J Biochem* 122: 622–626

10 Hromas R, Broxmeyer HE, Kim C, Christopherson K 2nd, Hou YH (1997) PLAB, a novel placental bone morphogenetic protein. *Biochem Biophys Acta* 1354: 40–44

11 Bootcov MR, Bauskin AR, Valenzuela SM, Moore AG, Bansal M, He XY, Zhang HP, Donnellan M, Mahler S, Pryor K et al (1997) MIC-1, a novel macrophage inhibitory cytokine, is a divergent member of the TGF-beta superfamily. *Proc Natl Acad Sci USA* 94: 11514–11519

12 Barrack E (1997) TGF beta in prostate cancer: a growth inhibitor that can enhance tumorigenicity. *Prostate* 31 (1): 61–70

13 McPherron AC, Lee SJ (1997) Double muscling in cattle due to mutations in the myostatin gene. *Proc Natl Acad Sci USA* 94: 12457–12461

14 Carlson CJ, Booth FW, Gordon SE (1999) Skeletal muscle myostatin mRNA expression is fiber-type specific and increases during hindlimb unloading. *Am J Physiol* 277: R601–R606

15 Wehling M, Cai B, Tidball JG (2000) Modulation of myostatin expression during modified muscle use. *FASEB J* 14:103–110

16 Sakuma K, Watanabe K, Sano M, Uramoto I, Totsuka T (2000) Differential adaptation of growth and differentiation factor 8/myostatin, fibroblast growth factor 6 and leukemia inhibitory factor in overloaded, regenerating and denervated rat muscles. *Biochem Biophys Acta* 1497:77–88

17 Yamanouchi K, Soeta C, Naito K, Tojo H (2000) Expression of myostatin gene in regenerating skeletal muscle of the rat and its localization. *Biochem Biophys Res Comm* 270: 510–516

18 Gonzales-Cadavid NF, Taylor WE, Yarasheski K, Sinha-Hikim I, Ma K, Ezzat S, Shen R, Lalani R, Asa S, Mamita M et al (1998) Organization of the human myostatin gene and expression in healthy men and HIV-infected men with muscle wasting. *Proc Natl Acad Sci USA* 95: 14398–14943

19 Sharma M, Kambadur R, Matthews KG, Somers WG, Devlin GP, Conaglen JV, Fowke PJ, Basset JJ (1999) Myostatin, a transforming growth factor-beta superfamily member, is expressed in heart muscle and is upregulated in cardiomyocytes after infarct. *J Cellular Physiol* 180: 1–9

20 Ji S, Losinski RL, Cornelius SG, Frank GR, Willis GM, Gerrard DE, Depreux FF, Spurlock ME (1998) Myostatin expression in porcine tissues: tissue specificity and developmental and postnatal regulation. *Am J Physiol* 275 (2): R1265–R1273

21 Kambadur R, Sharma M, Smith TP, Bass JJ (1997) Mutations in myostatin (GDF8) in double-muscled Belgian Blue and Piedmontese cattle. *Genome Res* 7: 910–915

22 Thomas M, Langley B, Berry C, Sharma M, Kirk S, Bass J, Kambadur R (2000) Myostatin, a negative regulator of muscle growth, functions by inhibiting myoblast proliferation. *J Biol Chem* 275 (51): 40235–40243

23 Taylor WE, Bhasin S, Artaza J, Byhower F, Azam M, Willard DH, Kull FC, Gonzalez-

Cadavid N (2001) Myostatin inhibits cell proliferation and protein synthesis in C2C12 muscle cells. *Am J Physiol Endocrinol Metab* 280: E221–E228
24 Rios R, Carneiro I, Arce VM, Devesa J (2001) Myostatin regulates cell survival during C2C12 myogenesis. *Biochem Biophys Res Comm* 280: 561–566
25 Miyazono Kl, ten Dijke P, Heldin CH (2000) TGF-beta signaling by Smad proteins. *Cytokine Growth Factor Rev* 11 (1–2): 15–22
26 Zawel L, Dai JL, Buckhaults P, Zhou S, Kinzler KW, Vogelstein B, Kern SE (1998) Human Smad3 and Smad4 are sequence-specific transcription activators. *Molecular Cell* 1: 611–617
27 Dennler S, Itoh S, Vivien D, ten Dijke P, Huet S, Gauthier JM (1998) Direct binding of Smad3 and Smad4 to critical TGF beta-inducible elements in the promoter of human plasminogen activator inhibitor-type 1 gene. *EMBO J* 17 (11): 3091–3100
28 Kusanagi K, Inoue H, Ishidou Y, Mishima HK, Kawabata M, Miyazono K (2000) Characterization of a bone morphogenetic protein-responsive Smad-binding element. *Mol Biol Cell* 11: 555–565
29 Baur ST, Mai JJ, Dymecki SM (2000) Combinatorial signaling through BMP receptor IB and GDF5: shaping of the distal mouse limb and the genetics of distal limb diversity. *Development* 127:605–619
30 Lee SJ and McPherron AC (1999) Myostatin and the control of skeletal muscle mass. *Curr Op Genet Devel* 9: 604–607

# Bone morphogenetic protein receptors and their nuclear effectors in bone formation

*Olexander Korchynsky and Peter ten Dijke*

Division of Cellular Biochemistry, The Netherland Cancer Institute, Plesmanlaan 121, 1066 CX Amsterdam, The Netherlands

## Introduction

Pioneering studies on the ability of extracts from decalcified bone matrix to promote ectopic bone and cartilage formation [1] led to searches for the identity of these morphogens which define skeletal patterning. With the advent of powerful methods for protein purification, capability to determine amino acid sequences on small amounts of protein and DNA cloning, bone morphogenetic proteins (BMPs) were discovered [2–4]. The amino acid sequences predicted from their cDNA sequences revealed that BMP-2, BMP-3 and BMP-4 (BMP-1 is a member of the astacin family of metalloproteases) are members of the TGF-β superfamily, which also includes the TGF-βs and activins [5]. Mainly through their sequence homology with other BMPs approximately 20 members in the BMP subgroup have now been identified and can be divided in multiple groups of structurally related proteins, e.g. BMP2 and BMP-4 are highly related, BMP-6, BMP-7 and BMP-8 form another subgroup, and growth and differentiation factor (GDF)-5 (also termed cartilage-derived morphogenetic protein (CDMP)-1, GDF-7 (also termed CDGF-2) and GDF-6 are similar to each other. *In vitro* BMPs were found to have potent effects on various cells implicated in cartilage and bone formation, e.g. induce proteoglycan synthesis in chondroblasts and stimulate alkaline phosphatase activity and type I collagen synthesis in osteoblasts [4]. When injected into muscle of rats, BMPs can induce a biological cascade of cellular events leading to ectopic bone formation [3, 4]. GDF-5, GDF-6 and GDF-7 induce more efficiently tendon and cartilage-like structures [6, 7]. Preclinical studies of certain BMPs in primates and other mammals have demonstrated their effectiveness in restoring large segmental bone defects [8, 9].

Like other members of the TGF-β family, BMPs are multifunctional proteins with effects on cell types not related to bone formation, e.g. epithelial cells, monocytes and neuronal cells [10, 11]. In addition, BMPs were found to be expressed not only in skeletal tissues, but also in many soft tissues. Consistent with these results, phenotypes of mice with mutated BMP genes revealed that they are multifunctional proteins that possess distinct roles in bone formation and many other mor-

Bone Morphogenetic Proteins, edited by Slobodan Vukicevic and Kuber T. Sampath

phogenic processes (Tab. 1) [12]. Interestingly, several different mouse and human skeletal disorders have been linked to genetic alterations in BMP genes. The mouse skeletal disorders short ear and brachipodism are caused by a null mutations in BMP-5 [13] and GDF-5 [14], respectively. Double muscle cattle were found to have mutations in GDF-8 (also called myostatin) [15]. Hunter-Thompson type chondrodysplasia has been linked to mutations in human cartilage-derived morphogenetic protein [16].

Here we review the BMP signal transduction pathways leading to bone formation. In particular, we will discuss the latest advances towards our understanding of the function of BMP receptors and their nuclear effector proteins, termed Smads, in controlling target gene expression.

## Identification and structure of BMP receptors

TGF-β family members, which include BMPs, elicit their cellular effects by inducing specific heteromeric complexes of two related serine/threonine kinase receptors, i.e. type I receptor and type II receptors [17, 18]. Among the TGF-β family of receptors, the cDNAs encoding mouse activin and human TGF-β type II receptor were isolated first by an expression cloning strategy [19, 20]. Subsequently, other mammalian type II and type I receptors, including those for BMPs, were isolated based upon their sequence similarity with other serine/threonine kinase receptors [21–30, 30–32]. Both receptor types contain glycosylated cysteine-rich extracellular ligand-βinding domains, short transmembrane domains and intracellular serine-threonine kinase domains (Fig. 1) [17, 18]. A shared feature for type I receptors is that they have a glycine/serine residue-rich stretch in the juxtamembrane region, which is essential for type I receptor activation [17, 18, 32]. Three mammalian BMPR-Is have been described to date [24, 29], i.e. activin receptor-like kinase (ALK)2, BMPR-IA (also termed ALK3) and BMPR-IB (also termed ALK6). Initially, ALK2 has been referred to as a type I receptor for TGF-β [33] or activin (ActR-I) [22], but recent studies suggest that ALK2 is most important in BMP signaling [24, 29, 34, 35]. Different BMPs bind with different affinity to the type I receptors. For example, BMP-4 binds preferentially to BMPR-A and -IB [24], BMP-7 binds with higher affinity to ALK2 and BMPR-IB than to BMPR-IA [24], and GDF-5 binds preferentially to BMPR-IB, when compared with other type I receptors [36]. Functional importance of BMPR-Is in bone formation was shown by the induction of chondroblast and osteoblast differentiation upon ectopic expression of mutant constitutively active BMPR-Is in mesenchymal precursor cells, and by observations that overexpression of dominant negative BMPR-Is interfered with BMP-induced osteoblast differentiation [37–40]. Surprisingly, BMP-3, which is one of the most abundant BMPs in adult bone, functions as an antagonist of BMP signaling, and is claimed to signal *via* the activin type IB receptor (ActR-IB)/ALK4 [41].

*Table 1 - Phenotypes of organisms with disruption of genes for BMPs, their receptors or their downstream Smads**

| Mutated gene | Phenotype | Reference |
|---|---|---|
| *BMP* | | |
| BMP-2 | Embryonic death (E7.5-E10.5). Defects in amnion/chorion formation and cardiac development. | [169] |
| BMP-3 | Viable. Increased bone mass. | [41] |
| BMP-4 | Embryonic death (E7.5-E9.5). Block of mesoderm formation. | [58] |
| BMP-5 | Viable. Skeletal abnormalities, short ear, brachypodism. | [13] |
| BMP-6 | Viable. Delay in developing sternum ossification. | [170] |
| BMP-7/OP-1 | Perinatal lethality. Severe defects in kidney and eyes. | [171] |
| | Abnormalities of rib cage, skull and hindlimbs. | [172] |
| BMP5/7 | Embryonic death (E10.5). Retarded heart development. | [173] |
| BMP-8B | Viable. Defects in spermatogenesis. | [174] |
| BMP-15 (GDF9B) | Viable. Increased ovulation rate leading to twins and triple births in heterozygotes and infertility in homozygotes. | [175] |
| GDF-5 | Viable. Skeletal abnormalities, short ear, brachypodism. | [14] |
| GDF-8 | Viable. Increased skeletal muscle mass and body size. | [15] |
| *Receptors* | | |
| ActR-IA/ALK-2 | Embryonic death (E9.5). Block of mesoderm formation. | [42] |
| BMPR-IA/ALK3 | Embryonic death (E7.5-E9.5). Block of mesoderm formation. | [57] |
| BMPR-IB/ALK6 | Viable. Defects in limb development. | [46] |
| BMPR-II | Embryonic death (E9.5). Block of mesoderm formation. | [49] |
| ActR-IIA & ActR-IIB | Embryonic death (E9.5). Arrest at the egg cylinder stage and block of mesoderm formation. | [60] |
| *Smads* | | |
| Smad1 | Embryonic death (E9.5). Defects in allantois formation. | Lechleider et al., pers. comm. |
| Smad4 | Embryonic death (E6.5-E8.5). Block of mesoderm formation. | [176] |
| Smad5 | Embryonic death (E9.5-E10.5). Defects in angiogenesis. | [177] |

**All gene mutations are in mice except for sheep BMP-15 and bovine GDF-8.*

Three distinct type II receptors have been implicated in BMP signaling: BMPR-II, activin type II receptor (ActR-II) and ActR-IIB [26, 27, 30]. However, binding affinities of ActR-II and ActR-IIB for BMPs are lower than those for activins [30].

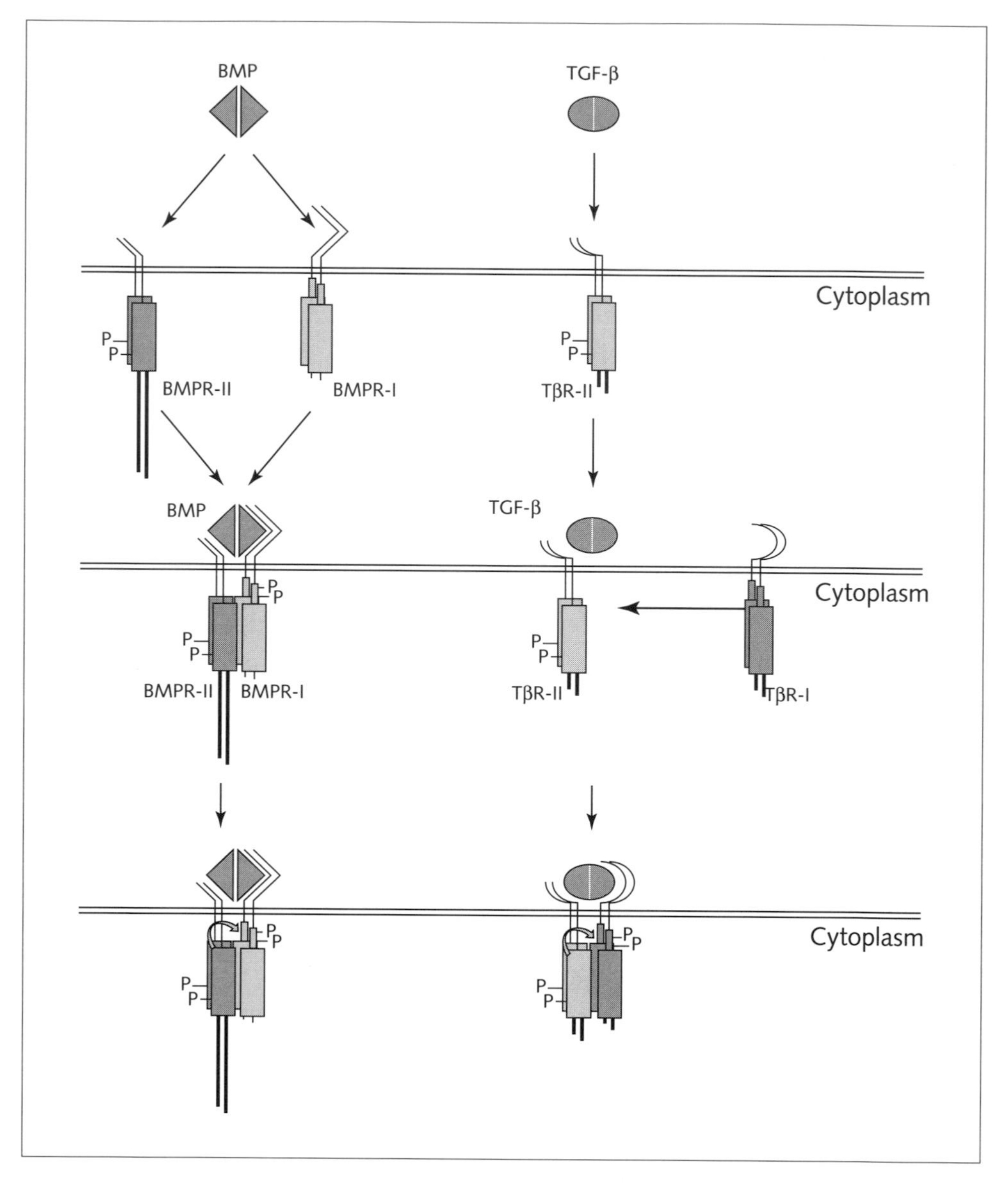


*Figure 1*
*Activation of BMP and TGF-β receptors. BMPs bind with weak affinity to type I or type II receptors alone, but with high affinity to type I/type II heteromeric complex. Upon BMP-induced heteromeric complex formation, the constitutively active type II serine/threonine kinase of type II receptor phosphorylates type I receptor in its GS-domain. TGF-β binds first to TGF-β type II receptor, and subsequently recruits TGF-β type I receptor and initiates signaling in a similar fashion as described for BMP receptor activation.*

Type II receptors, but not type I receptors, have extensions rich in serine and threonine residues distal from the kinase domains. In particular, BMPR-II has a very long extension of which the function is unknown.

## Expression of BMP type I and type II receptors

During mouse embryogenesis ALK2 is expressed primarily in the extraembryonic visceral endoderm before gastrulation and it is widely expressed in midgestation embryos [42, 43]. BMPR-IA was also found broadly expressed, but it is absent in the liver during embryogenesis [44]. Among the three BMPR-Is, BMPR-IB expression is the one that is most tissue or developmental stage restricted in its expression pattern [44, 45]; BMPR-IB is predominantly expressed in mesenchymal cells representing the primordia of long bones and later in development it is widely expressed in skeleton components [46]. During chicken limb development BMPR-IB is strongly expressed in undifferentiated mesechymal cells condensations prefiguring the future cartilage primordium. Expression of chicken BMPR-IA, however, is restricted to the prehypertrophic chondrocytes [45].

All three type II receptors (BMPR-II, ActR-II and ActR-IIB) are differentially expressed during mouse embryogenesis [47–49]. BMPR-II mRNA is detected in one-cell, two-cell and blastocyst stage embryos [50] and it is present in both embryonic and extraembryonic regions [49]. ActR-II and ActR-IIB, however, are mainly expressed in extraembryonic ectoderm [47]. All three BMP type II receptors are expressed in hypertrophic cartilage and ossified tissue [51, 52]. Interesting, BMP receptor expression is enhanced at sites of fracture repair [53]. Furthermore, during pathological ossification in the spinal ligaments, hypertrophic chondrocytes were found to express high levels of BMP receptors, and these sites colocalized with high levels of BMP expression [51, 54, 55]. Aberrant expression of BMPs and their receptors, possibly induced by mechanical stress, may be involved in the pathogenesis of orthotopic ossification [56].

## Determination of *in vivo* function of BMP type I and type II receptors through gene targeting approaches

BMP type I and type II receptors were found to be critically important for embryogenesis (Tab. 1) [12]. Mice lacking ALK3 and BMPR-II are lethal due to absence of mesodermal development [49, 57] and have a phenotype similar to BMP-4 knockout mice [58]. ALK2-deficient embryos are much smaller than their normal littermates, and lack a morphologically discernible primitive streak and die prior to or during early gastrulation [42]. ALK6-deficient mice are viable and exhibit mainly appendicular skeleton defects [46]. Mice lacking ActR-II or ActR-IIB are viable and were found to have a milder phenotype compared to a deficiency of one of their ligands. Some of ActR-II-deficient animals had mandibular hypoplasia and other skeletal and facial abnormalities [48]. ActR-IIB knockout mice showed cardiac defects, abnormal anteroposterior and left-right body axis patterning [59]. However, ActR-II and ActR-IIB double-knockout homozygous showed strong lethal embryonic abnormalities; these mice were growth arrested at the egg cylinder stage and did not form mesoderm [60]. The stronger phenotype in the double knock-out *versus* the single knock-outs suggests a functional redundancy for ActR-II and ActR-IIB in the mouse.

## Mechanism of BMP receptor activation

Like other TGF-β family members, both type I and type II receptors are required for BMP signaling [17, 18]. BMPs bind with weak affinity to type II or type I receptors alone and with high affinity to a heteromeric complex of the two receptor types [24, 26–30] (Fig. 1). The affinity of BMPR-I for ligand binding is higher that of BMPR-II and it is thus plausible that BMPR-I binds ligand initially and recruits then BMPR-II into the ligand-receptor complex [61]. This is in contrast to TGF-β and activin, which first bind to type II receptors and subsequently recruit type I receptors [21–23, 62] (Fig. 1). The mechanism of receptor activation has been best characterized for TGF-β [32], but it is likely to occur in an analogous fashion for BMPs [17, 18]. Upon BMP-induced heteromeric complex formation, the constitutively active type II receptor kinase phosphorylates type I receptor predominantly in its GS domain. The type I receptor acts thus downstream of type II, and consistent with this notion has been shown to confer signaling specificity to the type I/type II heteromeric complex [63] (Fig. 1). The activated type I receptor initiates intracellular signaling by phosphorylating downstream components, including the nuclear effector proteins known as Smads. The L45 loop regions in the kinase domain of type I receptors were found to be important determinants for signaling specificity [64–66].

BMPR-II is distinct from the other type II receptors in that it has a long carboxy-terminal (C-) tail extension [26, 27]. Functional importance of this tail is not known; BMPR-II lacking this C-tail is fully functional in transactivating BMPR-I [28]. However, patients with familial primary pulmonary hypertension syndrome have been genetically linked to mutations in BMPR-II, and certain of these mutations result in a partial truncation of the C-tail [67–70].

## Identification and structure of Smad proteins

Our understanding of BMP intracellular signaling has dramatically increased through genetic studies in *Drosophila* and *Caenorhabditis* (*C.*) *elegans*, in which

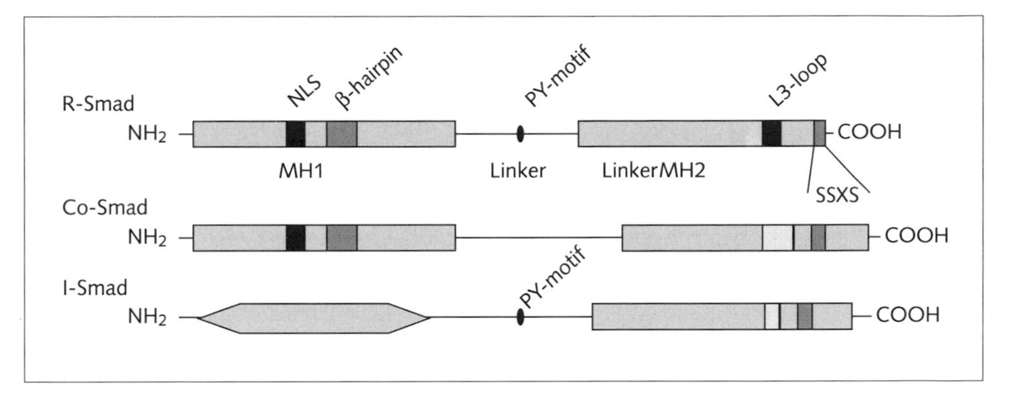


*Figure 2*
*Structure of Smad proteins. Receptor-regulated Smads (R-Smads) and common partner Smads (Co-Smads) consist of two highly conserved MH1 and MH2 domains that are separated by a proline-rich linker region. The amino-terminal region of inhibitory Smads (I-Smads) has only weak similarity to MH1 domains of R- and Co-Smads. The L3-loop in R-Smads interacts with activated type I receptors. Activated BMPR-I phosphorylates R-Smads in their C-terminal SXS motif, which is not present in Co-Smads and I-Smads. Nuclear localization signal (NLS) and DNA binding motif (β-hairpin) are conserved in the MH1 domains of R-Smads and Co-Smad. The PY motif is important for interaction with WW-containing HECT E3 ligases.*

*Mothers against DPP* (MAD) [71] and small body size (SMA) genes [72], respectively, were identified. MAD and SMA proteins were found to possess a critical role downstream of BMP-like proteins in these organisms. Thus far nine mammalian MAD and SMA related (Smad) proteins have been identified, which perform a pivotal function in TGF-β family intracellular signaling [17, 18]. Based upon their functional properties, Smads can be divided into three distinct subclasses: signal transducing receptor-regulated Smads (R-Smads) and common-mediator Smads (Co-Smads, i.e. Smad4) and inhibitory Smads (I-Smads, i.e. Smad6 and Smad7) which inhibit the activation of R- and Co-Smads [73–75] (Fig. 2). R- and Co-Smads have conserved amino and carboxy regions, known as MAD homology (MH1) domain and MH2 domains, respectively. Both domains are separated by a variable proline-rich linker region. Whereas the I-Smads have an MH2 domain, their amino-terminal regions show only weak sequence similarity to the MH1 domains (Fig. 2) [17, 18].

## Activation and function of Smad proteins

R-Smads interact transiently with and become phosphorylated by the activated type I receptor (Fig. 3); whereas Smad1, Smad5 and Smad8 act in the BMP pathway and

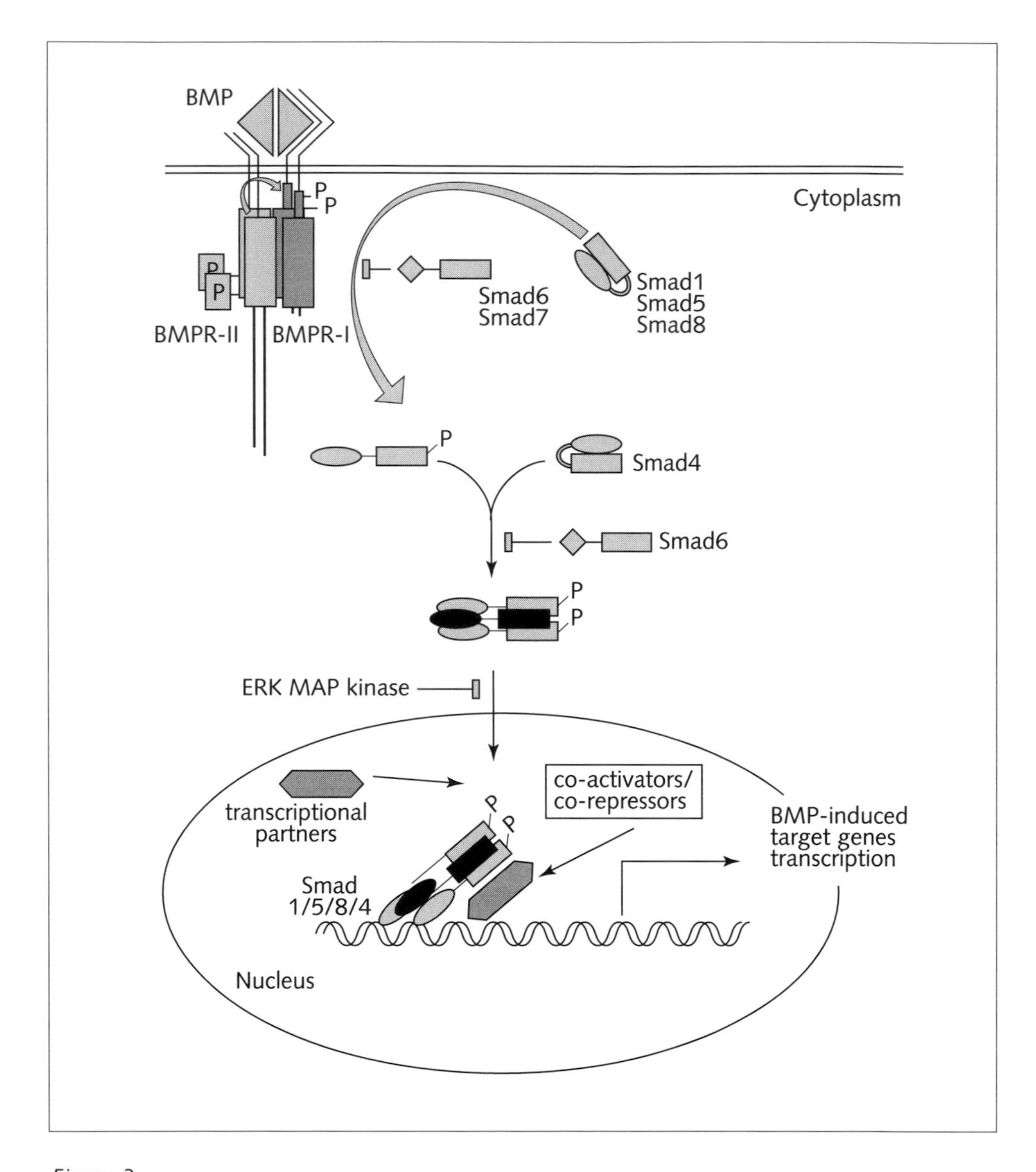


*Figure 3*
*Signaling from activated BMP receptors to nucleus by Smad proteins. Upon BMP receptor activation R-Smads are phosphorylated by the activated BMP type I receptor. Activated R-Smads can form a heteromeric complex with Co-Smad and translocate into the nucleus where they can directly or through their transcriptional partners bind to specific sequences in the promoters of BMP target genes and activate transcription of those genes. I-Smads block BMP signaling. Phosphorylation of Smads by Erk/MAPK into linker region inhibits nuclear translocation of Smads.*

Smad2 and Smad3 are activated by TGF-β and activin type I receptors [17, 18]. The L3 loop of R-Smad was shown to interact with the L45 loop in TGF-β and BMP type I receptors, a region which determines signaling specificity among different type I receptors [76]. Smad2 and Smad3 have been shown to be presented to TGF-β receptor complex through phospholipid binding FYVE-domain containing proteins, termed Smad anchor for receptor activation (SARA) [77] and Hrs [78]. However, SARA/Hrs-like proteins that facilitate BMP type I receptor-mediated activation of R-Smads remain to be identified. R-Smad phosphorylation by the activated type I receptor occurs at the two most carboxy-terminal serine residues in a SSXS motif [79–82]. In osteoblasts, BMP was found to induce the C-terminal phosphorylation of Smad5, and to a lesser extent Smad1 [83–85].

Upon BMP receptor activation BMP R-Smads form heteromeric complexes with Co-Smad4, i.e. Smad4 [86]. Preferentially trimeric Smad complexes are formed [87–89] (of which the exact stoichiometry needs further investigation) that efficiently translocate to the nucleus (Fig. 3) [17, 18]. Nuclear accumulation of BMP R-Smads and Smad4 was observed in osteoblasts after stimulation with BMP [83, 84]. The osteoblast-induced differentiation of mesenchymal precursor cell lines by ectopic expression of Smad1 or Smad5 became more pronounced when co-expressed with Smad4 and greatly enhanced by addition of BMP, which strongly promotes R-Smad/Co-Smad nuclear accumulation [40].

A nuclear localization signal (NLS)-like sequence in the MH1 domain of Smad3 that is conserved among all R- and Co-Smads was shown to be required for TGF-β-induced nuclear import [90, 91]. In Smad4 a functional leucine-rich nuclear export sequence (NES) was identified that ensures cytoplasmic location of Smad4 in unstimulated cells. TGF-β-induced complex formation of Smad4 with R-Smads was found to inactivate the NES [92, 93]. Nuclear entry of the Smad4/R-Smad complex may be stimulated upon unmasking of the NLS on the R-Smad and/or Co-Smad upon heteromeric complex formation. Within the nucleus, R-Smad/Co-Smad complexes act directly and/or in cooperation with other transcription factors, to regulate the transcription of target genes (see below) [94, 95].

Gene disruption of Smad genes in mice has begun to reveal specific and developmental functions of Smads that are implicated in BMP signaling. Whereas mice lacking Smad1, Smad4 or Smad5 are developmentally arrested, Smad6 mice make it to term (Tab. 1) [12]. To study the role of Smads in cartilage and bone formation, conditional knock-outs in, for example, mesenchymal precursor cells and osteoblasts are eagerly awaited.

## Expression and stability of Smad proteins

In a recent study the expression of Smad1 to Smad6 was examined in the 15th day of gestation of the mouse embryo. All tissues were found to express Smad4 and at least one of the R-Smads. Among the Smads, Smad6 expression was found most restricted [96]. At sites of endochondral ossification expression patterns of BMPs and their receptors were found to overlap with Smad1, Smad5 and Smad4 expression in proliferating chondrocytes and in the maturing chondrocytes [52, 96]. Highest expression of inhibitory Smads was shown in zones of mature chondrocytes. These findings suggest that Smad expression is an important determinant in regulating BMP signaling during the different phases of the bone forming process [52, 96].

The stability of Smad proteins appears also to be carefully regulated. Smad ubiquitination regulatory factor 1 (Smurf-1) was identified as a HECT domain containing E3 ubiquitin ligase for BMP R-Smads [97]. The WW motifs in Smurf1 interact with the PY motif (PPXY) in the linker regions of Smad1 and Smad5. Increased expression of Smurf1 leads to a selective decrease in BMP R-Smads thereby decreasing the cellular competence to BMP-mediated responses [97]. The proteasome-mediated degradation of Smad1/5 by Smurf1 is independent from their activation by ligands. Whether Smads can also be modified in order to make them more stable is an interesting area for future research; e.g. the conjugation of ubiquitin to lysine residues in Smads may be blocked by acetylation of those same residues, and ligation of small ubiquitin-related and modifier (SUMO) to Smads may inhibit their ubiquitin-mediated degradation.

## Smads are transcription factors

R-Smads (except for Smad2) and Smad4 were found to recognize specific sequences *via* their MH1 domains in the promoters of Smad target genes [98-100]. The affinity of Smad3 and Smad4 to DNA is much higher than BMP R-Smads. An *in vitro* screen of random DNA oligonucleotides that specifically bound to MH1-linker domain subdomains of Smad3 and Smad4 revealed that these Smads bind with highest affinity to sequences containing GTCT sequence (called also Smad-binding element, SBE) [101]. Multimers of SBE when placed in front of a minimal promoter reporter construct provide a strong enhancer function for TGF-β family members [98, 99, 101]. SBE-like sequences have been shown to be critically important for TGF-β-inducibility of multiple TGF-β responsive genes [94, 95]. TGF-β induced activation of several TGF-β-induced genes, including Smad7 [102-106], plasminogen activator inhibitor-1 [98, 107], α2(I) collagen [108] and type VII collagen [109] is critically dependent on SBE sequences, which have been found in multiple copies in promoters of these genes. The Smad1 MH1 domain was shown to bind SBE [110] and a reporter construct containing a multimerized SBE present in JunB promoter is activated by BMP [110]. BMP R-Smads (and also Smad3 and Smad4) also have been shown to bind to GCAT motifs [111] or to GC-rich sequences present in promoters of different BMP target genes [112, 113]. Mutation of these sequences significantly decreased BMP-induced response [112, 114]. BMP-inducibility of reporter constructs containing multimerized GC-rich sequences is very low [112, 114, 115] and requires high levels of Smad overexpression [112, 114]. The true physiological significance of the low affinity interaction of BMP R-Smads with GC-rich sequences or GCAT motifs remains to be shown.

The DNA affinity of Smads, and in particular BMP R-Smads, is weak. Smads thus need to cooperate with other DNA binding factors in order to bind efficiently to the promoters of target genes [94, 95, 116]. The 30-zinc finger nuclear protein OAZ was the first identified DNA-binding factor that associates with BMP R-Smads in response to BMP [116, 117]. OAZ interacts with the MH2 domains of Smad1 and also Smad4. Expression of OAZ is tissue and cell type-specific and OAZ cannot be detected in different cells, including mesenchymal precursors [117]. Interestingly, a member of core binding factor (CBF) family of transcriptional factors Cbfa1 (also called osteoblast-specific factor (Osf) 2, *Runt*-related gene 2 (RUNX2) acute myeloid leukemia (AML) protein 3 (AML3) or polyomavirus enhancer core-binding protein-2αA (PEBPα2A) and its homologues Cbfa2 and Cbfa3 were shown to interact directly with Smad1/5 (as well as Smad2 and Smad3) [118, 119]. Cbfa1 precedes the appearance of osteoblasts and mice deficient in Cbfa1 lack osteoblasts and the bone ossification is completely blocked [120]. Cbfa1 is also critically important for already differentiated osteoblasts and acts as a maintenance factor for mature osteoblasts by regulating the rate of bone matrix deposition [121]. The Cbfa1 genetic locus has been linked to one of most frequent human skeletal disorders termed *cleidocranial dysplasia* (CCD) syndrome [122]. CCD patients express truncated mutant Cbfa1 proteins that retain the ability to bind DNA by their *runt* domains, but fail to interact with Smads. These data suggest that Cbfa1 and Smad cooperate in BMP-induced osteoblast differentiation [123].

Initially the MH2 domains of R- and Co-Smads were found to have transactivation properties when fused to a GAL4-DNA binding domain [124, 125]. Subsequent studies have provided a mechanistic explanation for this; Smad1 as well as Smad2 and Smad4 were found to interact with transcriptional co-activators CBP/p300 which possess intrinsic acetyltransferase activity [126]. P300 and CBP facilitate transcription by decreasing the chromosome condensation through histone acetylation and by increasing the accessibility of Smad with components of the basal transcriptional machinery. CBP/p300 interact with many different transcription factors. The synergy between BMP and leukemia inhibitory factor (LIF) in the induction of differentiation of neuronal progenitors into astrocytes was shown to be mediated by cooperative binding of Smad1 and STAT3 to CBP/p300 [127].

## Negative regulation of BMP/Smad pathway

Negative regulation occurs at nearly every step in the BMP/Smad pathway. Several extracellular proteins, e.g. noggin and chordin, can bind to BMPs and inhibit their interaction with BMP receptors [128]. Bambi (for BMP and activin membrane-bound inhibitor was found to act as pseudo type I receptor and inhibits signaling possibly by preventing type I receptor homomeric complex formation [129, 130]. At the intracellular level activation of extracellular signal-regulated kinase (ERK) can lead to inhibition of BMP signaling; ERK MAPkinase mediated phosphorylation of Smad1 in its linker region was found to inhibit BMP-induced Smad1 nuclear accumulation [131]. I-Smads, i.e Smad6 and Smad7, potently interfere with TGF-β family intracellular signaling [73-75], whereas Smad7 functions as a general inhibitor of TGF-β, activin and BMP pathways, Smad6 specifically inhibits the BMP signaling [132]. Overexpression of I-Smads in mesenchymal precursor cells potently interfered with BMP-induced osteoblast differentiation [40]. I-Smads interact efficiently with activated type I receptors, and the initial mechanism described for I-Smad antagonism was by competing with R-Smads for type I receptor interaction [73-75]. However, other mechanisms by which I-Smads antagonize TGF-β family/Smad pathways have now been described. Smad7 has been found to constitutively interact with HECT-domain ubiquitin ligase, Smurf2 [133] and more recently with Smurf1 as well [134]. Binding of Smad7 to Smurf induces the export of Smad7/Smurf complex from the nucleus. Upon recruitment of the complex to the activated TGF-β receptor, Smurf1 or Smurf2 induces TGF-β receptor degradation through proteosomal and lysosomal pathways. Smad7 may thus function as an adapter protein to mediate degradation of TGF-β receptor complex [133, 134]. Smurf2 has also been reported to bind Smad6 and target the BMP receptor for degradation [133]. Other mechanisms for Smad6-inhibition of BMP signaling have been proposed: (i) by competing with Smad4 for heteromeric complex formation with activated R-Smads [135], (ii) by acting as a direct transcriptional corepressor [136], and (iii) by inhibiting the action of TAK1, a MAPKKK implicated downstream of BMP receptor signaling to apoptosis [137]. Further studies are needed to determine the physiological importance of these inhibitory mechanisms for I-Smads.

Tob, a member of an emerging family of antiproliferative proteins, was shown to bind R-Smads and to negatively regulate osteoblast proliferation and differentiation by suppressing the BMP R-Smads' transcriptional activity [115]. Mice deficient in Tob showed increased bone mass due to increased numbers of osteoblasts. Another negative regulator is the transcriptional corepressor Ski, which can interact with Smad4 [138] and Smad1 or Smad5 through their MH2 domains [139]. Ectopic expression of Ski was found to inhibit BMP-2-induced osteoblast differentiation of murine W-20-17 cells [139].

## BMP receptor-initiated signaling distinct from Smad activation

Ectopic expression of BMP R-Smads can recapitulate osteoblast differentiation, but not chondrogenic differentiation [40]. Thus, BMP-induced osteoblast differentiation

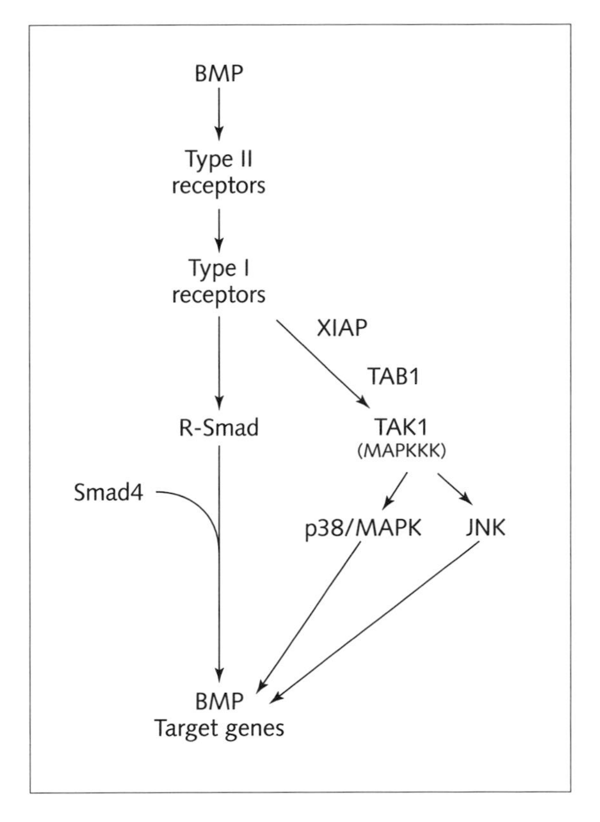


*Figure 4*
*BMP signaling through Smad-dependent and Smad-independent pathways. BMP activates the Smad pathway as well as other signaling pathways. Abbreviations: JNK, c-Jun N-terminal kinase; TAB, TAK-1 binding protein; TAK, TGF-β activated kinase; XIAP, X-linked inhibitor of apoptosis protein.*

appears to occur mainly *via* the Smad pathway, whereas BMP-induced chondrogenic differentiation is mediated *via* Smad-dependent and Smad-independent pathways [40]. Other pathways distinct from Smad pathway that are initiated downstream of ligand-induced activation of BMP receptor complex have been identified (Fig. 4). TGF-β-activated kinase 1 (TAK1), a MAP kinase kinase kinase (MKKK), can be activated by TAK1 binding protein (TAB1) in response to BMP and activate both SAPK and p38 pathways [140, 141]. X-chromosome-linked inhibitor of apoptosis (XIAP) may provide the direct link between TAB1 and type I receptor as it was shown to interact with both proteins [141]. p38 MAP kinase activation induces the phosphorylation of transcriptional factor ATF-2, and both ATF-2 and Smads were shown to act synergistically in transcriptional regulation [142]. BMP-induced apoptosis was shown to be mediated by TAK1-p38 kinase pathway [137]. In ATDC5 cells activation of p38 kinase by GDF-5 contributes to chondrogenesis [143]. Further studies are needed to demonstrate the physiological and general importance of Smad independent pathways in BMP signal transduction.

## BMP target genes

A number of extracellular matrix proteins, including osteocalcin, collagen type α, bone sialoprotein and decorin are potently induced by BMP [144–146]. Some of them may be direct targets for BMPs (such as collagen), whereas others (such as osteocalcin) are indirect and are only induced after prolonged exposure to BMPs (Tab. 2). The BMP-induced expression of alkaline phosphatase, a late BMP target gene, is often used as a read-out for BMP-induced osteoblast differentiation [40, 144–147]. BMPs were shown to activate osteopontin gene expression by preventing the binding of transcriptional repressor Hoxc-8 to the osteopontin promoter. Activated Smads can bind to Hoxc-8 and dislodge the inhibitory Hoxc-8 from the DNA [148, 149]. In addition, a Smad binding region was identified in osteopontin promoter, and shown to be involved in BMP-mediated activation of this promoter [150]. BMP-induces expression of osteoprotegrin (OPG), an osteoblast-secreted decoy receptor, which specifically binds to the osteoclast differentiation factor and inhibits osteoclast maturation [151]. Interestingly, characterization of the OPG promoter revealed two homeodomain transcriptional factor Hoxc-8 binding sites that are essential for OPG promoter activation by BMP [151].

Connective tissue growth factor (CTGF), an important regulator of extracellular matrix formation, is also induced by BMP [152]. In the rat long bone growth plate the CTGF expression in chondroblasts is restricted to hypertrophic region [152], which overlaps with the expression of BMP signaling components [52]. Recombinant CTGF promotes the proliferation and differentiation of chondrocytes and induces the expression of osteoblast-specific genes and bone mineralization [153].

In many cell types (including osteoblasts) the expression of inhibitory Smads (Smad6 and Smad7) are potently induced by BMPs [75, 154, 155]. In the Smad6 promoter a BMP responsive GC-rich elements has been identified [112]. BMP-responsive elements in Smad7 promoter remain to be elucidated. The BMP-induced I-Smads may serve a role in a negative feedback loop in Smad signaling to control the intensity and duration of BMP signaling response [75, 154, 155].

BMPs have also been found to induce many transcription factors (Tab. 2). JunB was shown in osteoblast precursor cells as a direct early BMP-2 target gene involved in the inhibition of myogenic differentiation [156]. Investigation of JunB promoter revealed the importance of multiple Smad binding elements through which this gene can be activated by ectopic BMP R-Smad expression [99]. BMPs induce the expres-

*Table 2 - Genes induced by BMPs in osteoblasts or their precursors*

| BMP target gene | Defects resulting from gene inactivation |
|---|---|
| *Components of ECM* | |
| Ostecalcin [145] | Viable. Osteopetrosis [178]. |
| Ostepontin [144] | Viable. Altered collagen fibrillogenesis and wound healing [179].<br>Resistant to to ovariectomy-induced osteoporosis [180]. |
| Collagen I$\alpha$1 & $\alpha$2 [146] | Viable. Osteogenesis imperfecta [181]. |
| Bone sialoprotein [182] | Not determined. |
| *Decoy receptor* | |
| Osteoprotegrin [148,150] | Not determined. |
| *Enzymes* | |
| Alkaline phosphatase [145] | Metabolic and skeletal defects. Infantile hypophosphatasia [183]. |
| *Growth factors* | |
| CTGF* [152] | Not determined. |
| *Inhibitors of BMP function* | |
| Smad6* [75, 112, 155] | Cardiovascular abnormalities. Defects in endocardial cushion transformation [184]. |
| Smad7* [74, 155] | Not determined. |
| *Transcriptional regulators* | |
| Msx-2* [157, 164] | Viable. Defects in craniafacial bone ossification and endochondral bone formation. Tooth, mammary gland, cerebellum defects. Mutated in cranysynostosis patients. Haploinsufficiency causes parietal foramina [165]. |
| Dlx-5* [162, 165] | Viable. Delayed membraneous ossification [163]. |
| Id1*, Id2*, Id3* [157, 158] | Id1$^{-/-}$ Id3$^{-/-}$ and Id2$^{-/-}$ Id3$^{-/-}$ are not viable [185]. Haematopoietic and neural abnormalities (Storm, Huynh et al., 1994, 74 /id). |
| JunB* [99, 156] | Embryonic death (E8.5-E10). Multiple defects in placental neovascularisation [186]. |
| Cbfa1 [39, 147, 187] | Death after birth. No ossification. Skeleton made from chondrocytes only [120]. Mutated in CCD patiens [122]. |

*For this gene it has been demonstrated that it is a direct BMP target.*

sion of helix-loop-helix proteins inhibitors of differentiation (Id) in osteoblasts and their precursors in part *via* transcriptional and post-transcriptional events [157, 158]. The induction of Id proteins by BMPs may indirectly support osteoblast differentiation of mesenchymal precursor cells by blocking their adipocyte [159] and myoblast differentiation [160, 161].

Mammalian homologs of the *Drosophila distalles* (dll) *Dlx5* and *Dlx6* are direct gene targets for BMP [162]. Overexpression of Dlx5 in cells induces their osteoblast differentiation while disruption of Dlx5 exhibits defects in the ossification of the membranous bones [163]. BMPs can directly induce the Msx-1 and Msx-2 homeobox genes. Mice deficient in Msx-2 [164] or Msx-1 have defects in the skull bones and show an overall decrease in bone mass [165]. Albeit not a direct BMP target, Cbfa1 induction by BMP is critical for BMP-induced osteogenesis [147]. Cbfa1 can induce extracellular matrix proteins, but Cbfa1 is not sufficient to induce the whole onset of osteoblastic differentiation without cooperation with Smad5 [147]. Many new (in)direct target genes for BMP are likely to be reported as a result of cDNA micro array studies that are currently ongoing in many laboratories

## Perspectives

Recent studies have demonstrated the pivotal role of BMP type I and type II receptors and their downstream Smad effectors in BMP-induced osteoblast differentiation. However, the molecular mechanisms that govern BMP-induced osteogenic differentiation need further study. In particular, physiological interactions between BMP family members with their receptors and Smads, and downstream gene targets in osteoblasts remain to be validated by comparing the phenotypes of mice deficient in a particular BMP ligand, receptor, Smad or target gene. In many cases a null mutation of a particular BMP signaling component leads to an embryonal lethal phenotype. Conditional knock-out approaches will therefore be required to study the role of these molecules in bone formation. The repertoire of BMP Smad interacting proteins in different osteoblast (precursor) cell types or at different states of their differentiation need to be elucidated. In addition, the genetic programs that are initiated in mesenchymal precursor cells, chondroblasts and osteoblasts upon BMP stimulation *via* the various BMP intracellular pathways need to be determined. To analyze this efficiently functional genomics technologies will be useful. This approach should provide an answer to the question of why different BMP type I receptors, although activating the same set of Smad proteins, can induce distinct biological responses. Stimulating mesenchymal precursor cells with a constitutively active (ca) BMPR-IA induces adipocyte differentiation, whereas (ca) BMPR-IB induces osteoblast differentiation and apoptosis [39]. During limb bud morphogenesis in the chick, BMPR-IA was found to mediate osteogenesis whereas BMPR-IB induced preferentially chondrocyte differentiation [45].

BMPs in animal models have shown to be very effective in bone repair [8, 9, 166]. Adenoviral BMP7 gene transfer [167] and BMP4 plasmid implantation into bone [168] have been succesfully used in mouse models of osteogenic induction. However, clinical use of BMPs as regenerative agents in humans has thus far been limited; there is a need of using high doses of BMPs to get specific effects, if any. With the elucidation of the BMP/Smad pathway numerous inhibitors of BMP signaling have been identified. An interesting possibility, which remains to be explored, is that by inhibiting the action of antagonists, like extracellular noggin and the intracellular I-Smad, BMP signaling can be potentiated, thereby making BMPs more effective therapeutic agents.

*Acknowledgements*
Our studies on BMP receptor and Smad signal transduction are supported by the Netherlands Organization for Scientific Research (ALW 809.67.024) and the Dutch Cancer Society (NKI 2000-2217).

## References

1 Urist MR (1965) Bone: formation by autoinduction. *Science* 150: 893–899
2 Wozney JM, Rosen V, Celeste AJ, Mitsock LM, Whitters MJ, Kriz RW, Hewick RM, Wang EA (1988) Novel regulators of bone formation: molecular clones and activities. *Science* 242: 1528–1534
3 Sampath TK, Maliakal JC, Hauschka PV, Jones WK, Sasak H, Tucker RF, White KH, Coughlin JE, Tucker MM, Pang RH et al (1992) Recombinant human osteogenic protein-1 (hOP-1) induces new bone formation *in vivo* with a specific activity comparable with natural bovine osteogenic protein and stimulates osteoblast proliferation and differentiation *in vitro*. *J Biol Chem* 267: 20352–20362
4 Vukicevic S, Luyten FP, Reddi AH (1989) Stimulation of the expression of osteogenic and chondrogenic phenotypes *in vitro* by osteogenin. *Proc Natl Acad Sci USA* 86: 8793–8797
5 Massagué J (1990) The transforming growth factor-β family. *Annu Rev Cell Biol* 6: 597–641
6 Hotten GC, Matsumoto T, Kimura M, Bechtold RF, Kron R, Ohara T, Tanaka H, Satoh Y, Okazaki M, Shirai T et al (1996) Recombinant human growth/differentiation factor 5 stimulates mesenchyme aggregation and chondrogenesis responsible for the skeletal development of limbs. *Growth Factors* 13: 65–74
7 Wolfman NM, Hattersley G, Cox K, Celeste AJ, Nelson R, Yamaji N, Dube JL, DiBlasio-Smith E, Nove J, Song JJ et al (1997) Ectopic induction of tendon and ligament in rats by growth and differentiation factors 5, 6, and 7, members of the TGF-β gene family. *J Clin Invest* 100: 321–330

8 Reddi AH (1994) Symbiosis of biotechnology and biomaterials: applications in tissue engineering of bone and cartilage. *J Cell Biochem* 56: 192–195

9 Reddi AH (1998) Role of morphogenetic proteins in skeletal tissue engineering and regeneration. *Nat Biotechnol* 16: 247–52

10 Cunningham NS, Paralkar V, Reddi AH (1992) Osteogenin and recombinant bone morphogenetic protein 2B are chemotactic for human monocytes and stimulate transforming growth factor β1 mRNA expression. *Proc Natl Acad Sci USA* 89: 11740–11744

11 Hogan BL (1996) Bone morphogenetic proteins in development. *Curr Opin Genet Dev* 6: 432–438

12 Goumans M-J, Mummery C (2000) Functional analysis of the TGFβ receptor/Smad pathway through gene ablation in mice *Int J Dev Biol* 44: 253–265

13 Kingsley DM, Bland AE, Grubber JM, Marker PC, Russell LB, Copeland NG, Jenkins NA (1992) The mouse short ear skeletal morphogenesis locus is associated with defects in a bone morphogenetic member of the TGFβ superfamily. *Cell* 71: 399–410

14 Storm EE, Huynh TV, Copeland NG, Jenkins NA, Kingsley DM, Lee SJ (1994) Limb alterations in brachypodism mice due to mutations in a new member of the TGFβ-superfamily. *Nature* 368: 639–643

15 McPherron AC, Lee SJ (1997) Double muscling in cattle due to mutations in the myostatin gene. *Proc Natl Acad Sci USA* 94: 12457–12461

16 Thomas JT, Lin K, Nandedkar M, Camargo M, Cervenka J, Luyten FP (1996) A human chondrodysplasia due to a mutation in a TGF-β superfamily member. *Nat Genet* 12: 315–317

17 Massagué J (1998) TGF-β signal transduction. *Annu Rev Biochem* 67: 753–791

18 Heldin C-H, Miyazono K, ten Dijke P (1997) TGF-β signalling from cell membrane to nucleus *via* Smad proteins. *Nature* 390: 465–471

19 Mathews LS, Vale WW (1991) Expression cloning of an activin receptor, a predicted transmembrane serine kinase. *Cell* 65: 973–982

20 Lin HY, Wang X-F, Ng-Eaton E, Weinberg RA, Lodish HF (1992) Expression cloning of the TGF-β type II receptor, a functional transmembrane serine/threonine kinase. *Cell* 68: 775–785

21 Ebner R, Chen RH, Shum L, Lawler S, Zioncheck TF, Lee A, Lopez AR, Derynck R (1993) Cloning of a type I TGF-β receptor and its effect on TGF-β binding to the type II receptor. *Science* 260: 1344–1348

22 Attisano L, Cárcamo J, Ventura F, Weis FM, Massagué J, Wrana JL (1993) Identification of human activin and TGFβ type I receptors that form heteromeric kinase complexes with type II receptors. *Cell* 75: 671–680

23 Franzén P, ten Dijke P, Ichijo H, Yamashita H, Schulz P, Heldin C-H, Miyazono K (1993) Cloning of a TGFβ type I receptor that forms a heteromeric complex with the TGFβ type II receptor. *Cell* 75: 681–692

24 ten Dijke P, Yamashita H, Sampath TK, Reddi AH, Estevez M, Riddle DL, Ichijo H, Heldin C-H, Miyazono K (1994) Identification of type I receptors for osteogenic protein-1 and bone morphogenetic protein-4. *J Biol Chem* 269: 16985–16988

25 ten Dijke P, Yamashita H, Ichijo H, Franzén P, Laiho M, Miyazono K, Heldin C-H (1994) Characterization of type I receptors for transforming growth factor-β and activin. *Science* 264: 101–104
26 Nohno T, Ishikawa T, Saito T, Hosokawa K, Noji S, Wolsing DH, Rosenbaum JS (1995) Identification of a human type II receptor for bone morphogenetic protein-4 that forms differential heteromeric complexes with bone morphogenetic protein type I receptors. *J Biol Chem* 270: 22522–22526
27 Rosenzweig BL, Imamura T, Okadome T, Cox GN, Yamashita H, ten Dijke P, Heldin C-H, Miyazono K (1995) Cloning and characterization of a human type II receptor for bone morphogenetic proteins. *Proc Natl Acad Sci USA* 92: 7632–7636
28 Liu F, Ventura F, Doody J, Massagué J (1995) Human type II receptor for bone morphogenic proteins (BMPs): extension of the two-kinase receptor model to the BMPs. *Mol Cell Biol* 15: 3479–3486
29 Koenig BB, Cook JS, Wolsing DH, Ting J, Tiesman JP, Correa PE, Olson CA, Pecquet AL, Ventura F, Grant RA et al (1994) Characterization and cloning of a receptor for BMP-2 and BMP-4 from NIH 3T3 cells. *Mol Cell Biol* 14: 5961–5974
30 Yamashita H, ten Dijke P, Huylebroeck D, Sampath TK, Andries M, Smith JC, Heldin C-H, Miyazono K (1995) Osteogenic protein-1 binds to activin type II receptors and induces certain activin-like effects. *J Cell Biol* 130: 217–226
31 ten Dijke P, Ichijo H, Franzen P, Schulz P, Saras J, Toyoshima H, Heldin C-H, Miyazono K (1993) Activin receptor-like kinases: a novel subclass of cell-surface receptors with predicted serine/threonine kinase activity. *Oncogene* 8: 2879–2887
32 Wrana JL, Attisano L, Wieser R, Ventura F, Massagué J (1994) Mechanism of activation of the TGF-β receptor. *Nature* 370: 341–347
33 Miettinen PJ, Ebner R, Lopez AR, Derynck R (1994) TGF-β induced transdifferentiation of mammary epithelial cells to mesenchymal cells: involvement of type I receptors. *J Cell Biol* 127: 2021–2036
34 Macías-Silva M, Hoodless PA, Tang SJ, Buchwald M, Wrana JL (1998) Specific activation of Smad1 signaling pathways by the BMP7 type I receptor, ALK2. *J Biol Chem* 273: 25628–25636
35 Armes NA, Smith JC (1997) The ALK-2 and ALK-4 activin receptors transduce distinct mesoderm-inducing signals during early Xenopus development but do not co-operate to establish thresholds. *Development* 124: 3797–3804
36 Nishitoh H, Ichijo H, Kimura M, Matsumoto T, Makishima F, Yamaguchi A, Yamashita H, Enomoto S, Miyazono K (1996) Identification of type I and type II serine/threonine kinase receptors for growth/differentiation factor-5. *J Biol Chem* 271: 21345–21352
37 Akiyama S, Katagiri T, Namiki M, Yamaji N, Yamamoto N, Miyama K, Shibuya H, Ueno N, Wozney JM, Suda T (1997) Constitutively active BMP type I receptors transduce BMP-2 signals without the ligand in C2C12 myoblasts. *Exp Cell Res* 235: 362–369
38 Namiki M, Akiyama S, Katagiri T, Suzuki A, Ueno N, Yamaji N, Rosen V, Wozney JM, Suda T (1997) A kinase domain-truncated type I receptor blocks bone morphogenetic

protein-2-induced signal transduction in C2C12 myoblasts. *J Biol Chem* 272: 22046–22042

39 Chen D, Ji X, Harris MA, Feng JQ, Karsenty G, Celeste AJ, Rosen V, Mundy GR, Harris SE (1998) Differential roles for bone morphogenetic protein (BMP) receptor type IB and IA in differentiation and specification of mesenchymal precursor cells to osteoblast and adipocyte lineages. *J Cell Biol* 142: 295–305

40 Fujii M, Takeda K, Imamura T, Aoki H, Sampath TK, Enomoto S, Kawabata M, Kato M, Ichijo H, Miyazono K (1999) Roles of bone morphogenetic protein type I receptors and Smad proteins in osteoblast and chondroblast differentiation. *Mol Biol Cell* 10: 3801–3813

41 Daluiski A, Engstrand T, Bahamonde ME, Gamer LW, Agius E, Steven SL. (2001) Bone morphogenetic protein-3 is a negative regulator of bone density. *Nature Genetics* 27: 84–88

42 Gu Z, Reynolds EM, Song J, Lei H, Feijen A, Yu L, He W, MacLaughlin DT, van den Eijnden-van Raaij J, Donahoe PK et al (1999) The type I serine/threonine kinase receptor ActRIA (ALK2) is required for gastrulation of the mouse embryo. *Development* 126: 2551–2561

43 Verschueren K, Dewulf N, Goumans MJ, Lonnoy O, Feijen A, Grimsby S, Vande Spiegle K, ten Dijke P, Morén A, Vanscheeuwijck P et al (1995) Expression of type I and type IB receptors for activin in midgestation mouse embryos suggests distinct functions in organogenesis. *Mech Dev* 52: 109–123

44 Dewulf N, Verschueren K, Lonnoy O, Morén A, Grimsby S, Vande Spiegle K, Miyazono K, Huylebroeck D, ten Dijke P (1995) Distinct spatial and temporal expression patterns of two type I receptors for bone morphogenetic proteins during mouse embryogenesis. *Endocrinology* 136: 2652–2663

45 Zou H, Wieser R, Massagué J, Niswander L (1997) Distinct roles of type I bone morphogenetic protein receptors in the formation and differentiation of cartilage. Genes Dev 11: 2191–2203

46 Yi SE, Daluiski A, Pederson R, Rosen V, Lyons KM (2000) The type I BMP receptor BMPRIB is required for chondrogenesis in the mouse limb. *Development* 127: 621–630

47 Manova K, De L, Angeles M, Kalantry S, Giarre M, Attisano L, Wrana J, Bachvarova RF (1995) mRNAs for activin receptors II and IIB are expressed in mouse oocytes and in the epiblast of pregastrula and gastrula stage mouse embryos. *Mech Dev* 49: 3–11.

48 Matzuk MM, Kumar TR, Bradley A (1995) Different phenotypes for mice deficient in either activins or activin receptor type II. *Nature* 374 (6520): 356–360

49 Beppu H, Kawabata M, Hamamoto T, Chytil A, Minowa O, Noda T, Miyazono K (2000) BMP type II receptor is required for gastrulation and early development of mouse embryos. *Dev Biol* 221: 249–258

50 Roelen BA, Goumans M-J, van Rooijen MA, Mummery CL (1997) Differential expression of BMP receptors in early mouse development. *Int J Dev Biol* 41: 541–549

51 Yonemori K, Imamura T, Ishidou Y, Okano T, Matsunaga S, Yoshida H, Kato M, Sampath TK, Miyazono K, ten Dijke P et al (1997) Bone morphogenetic protein receptors

and activin receptors are highly expressed in ossified ligament tissues of patients with ossification of the posterior longitudinal ligament. *Am J Pathol* 150: 1335–1347
52 Sakou T, Onishi T, Yamamoto T, Nagamine T, Sampath Tk, ten Dijke P (1999) Localization of Smads, the TGF-β family intracellular signaling components during endochondral ossification. *J Bone Miner Res* 14: 1145–1152
53 Ishidou Y, Kitajima I, Obama H, Maruyama I, Murata F, Imamura T, Yamada N, ten Dijke P, Miyazono K, Sakou T (1995) Enhanced expression of type I receptors for bone morphogenetic proteins during bone formation. *J Bone Miner Res* 10: 1651–1659
54 Hayashi K, Ishidou Y, Yonemori K, Nagamine T, Origuchi N, Maeda S, Imamura T, Kato M, Yoshida H, Sampath TK et al (1997) Expression and localization of bone morphogenetic proteins (BMPs) and BMP receptors in ossification of the ligamentum flavum. *Bone* 21: 23–30
55 Okano T, Ishidou Y, Kato M, Imamura T, Yonemori K, Origuchi N, Matsunaga S, Yoshida H, ten Dijke P, Sakou T (1997) Orthotopic ossification of the spinal ligaments of Zucker fatty rats: a possible animal model for ossification of the human posterior longitudinal ligament. *J Orthop Res* 15: 820–829
56 Sakou T (1998) Bone morphogenetic proteins: from basic studies to clinical approaches. Bone 22: 591–603
57 Mishina Y, Suzuki A, Ueno N, Behringer RR (1995) Bmpr encodes a type I bone morphogenetic protein receptor that is essential for gastrulation during mouse embryogenesis. *Genes Dev* 9: 3027–3037
58 Winnier G, Blessing M, Labosky PA, Hogan BL (1995) Bone morphogenetic protein-4 is required for mesoderm formation and patterning in the mouse. *Genes Dev* 9: 2105–2116
59 Oh SP, Li E (1997) The signaling pathway mediated by the type IIB activin receptor controls axial patterning and lateral asymmetry in the mouse. *Genes Dev* 11: 1812–1826
60 Song J, Oh SP, Schrewe H, Nomura M, Lei H, Okano M, Gridley T, Li E (1999) The type II activin receptors are essential for egg cylinder growth, gastrulation, and rostral head development in mice. *Dev Biol* 213: 157–169
61 Gilboa L, Nohe A, Geissendorfer T, Sebald W, Henis YI, Knaus P (2000) Bone morphogenetic protein receptor complexes on the surface of live cells: a new oligomerization mode for serine/threonine kinase receptors. *Mol Biol Cell* 11: 1023–1035
62 Wrana JL, Attisano L, Cárcamo J, Zentella A, Doody J, Laiho M, Wang X-F, Massagué J (1992) TGF β signals through a heteromeric protein kinase receptor complex. Cell 71: 1003–1014
63 Cárcamo J, Weis FM, Ventura F, Wieser R, Wrana JL, Attisano L, Massagué J (1994) Type I receptors specify growth-inhibitory and transcriptional responses to transforming growth factor β and activin. *Mol Cell Biol.* 14(6): 3810–3821.
64 Chen YG, Hata A, Lo RS, Wotton D, Shi Y, Pavletich N, Massagué J (1998) Determinants of specificity in TGF-β signal transduction. *Genes Dev* 12: 2144–2152
65 Feng X-H, Derynck R (1997) A kinase subdomain of transforming growth factor-β

(TGF-β) type I receptor determines the TGF-β intracellular signaling specificity. *EMBO J* 16: 3912–3923

66 Persson U, Izumi H, Souchelnytskyi S, Itoh S, Grimsby S, Engström U, Heldin C-H, Funa K, ten Dijke P (1998) The L45 loop in type I receptors for TGF-β family members is a critical determinant in specifying Smad isoform activation. *FEBS Lett* 434: 83–87

67 Lane KB, Machado RD, Pauciulo MW, Thomson JR, Phillips JA, Loyd JE, Nichols WC Trembath RC, Micheala A, Brannon CA (2000) Heterozygous germline mutations in BMPR2, encoding a TGF-β receptor, cause familial primary pulmonary hypertension *Nat Genet* 26: 81–84

68 Machado RD, Pauciulo MW, Thomson JR, Lane KB, Morgan NV, Wheeler L, Phillips III, Newman J, Williams D, Galie N et al (2001) BMPR2 haploinsufficiency as the inherited molecular mechanism for primary pulmonary hypertension. *Am J Hum Genet* 68: 92–102

69 Thomson JR, Machado RD, Pauciulo MW, Morgan NV, Humbert M, Elliott GC, Ward K, Yacoub M, Mikhail G, Rogers P (2000) Sporadic primary pulmonary hypertension is associated with germline mutations of the gene encoding BMPR-II, a receptor member of the TGF-β family. *J Med Genet* 37: 741–745

70 Wilkins MR, Gibbs JS, Shovlin CL. (2000) A gene for primary pulmonary hypertension *Lancet* 356: 1207–1208

71 Savage C, Das P, Finelli AL, Townsend SR, Sun CY, Baird SE, Padgett RW (1996) *Caenorhabditis elegans* genes sma-2, sma-3, and sma-4 define a conserved family of transforming growth factor β pathway components. *Proc Natl Acad Sci USA* 93: 790–794

72 Sekelsky JJ, Newfeld SJ, Raftery LA, Chartoff EH, Gelbart WM (1995) Genetic characterization and cloning of mothers against dpp, a gene required for decapentaplegic function in *Drosophila melanogaster*. *Genetics* 139: 1347–1358

73 Hayashi H, Abdollah S, Qiu Y, Cai J, Xu YY, Grinnell BW, Richardson MA, Topper JN, Gimbrone MA, Wrana JL (1997) The MAD-related protein Smad7 associates with the TGFβ receptor and functions as an antagonist of TGFβ signaling. *Cell* 89: 1165–1173

74 Imamura T, Takase M, Nishihara A, Oeda E, Hanai J-I, Kawabata M, Miyazono K (1997) Smad6 inhibits signalling by the TGF-β superfamily. *Nature* 389: 622–626

75 Nakao A, Afrakhte M, Morén A, Nakayama T, Christian JL, Heuchel R, Itoh S, Kawabata M, Heldin N-E, Heldin C-H et al (1997) Identification of Smad7, a TGFβ-inducible antagonist of TGF-β signalling. *Nature* 389: 631–635

76 Lo RS, Chen YG, Shi Y, Pavletich NP, Massagué J (1998) The L3 loop: a structural motif determining specific interactions between SMAD proteins and TGF-β receptors. *EMBO J* 17: 996–1005

77 Tsukazaki T, Chiang TA, Davison AF, Attisano L, Wrana JL (1998) SARA, a FYVE domain protein that recruits Smad2 to the TGFβ receptor. *Cell* 95: 779–791

78 Miura S, Takeshita T, Asao H, Kimura Y, Murata K, Sasaki Y, Hanai JI, Beppu H, Tsukazaki T, Wrana JL et al (2000) Hgs (Hrs), a FYVE domain protein, is involved in Smad signaling through cooperation with SARA. *Mol Cell Biol* 20: 9346–9355

79 Abdollah S, Macías-Silva M, Tsukazaki T, Hayashi H, Attisano L, Wrana JL (1997) TβRI phosphorylation of Smad2 on Ser465 and Ser467 is required for Smad2-Smad4 complex formation and signaling. *J Biol Chem* 272: 27678–27685

80 Kretzschmar M, Liu F, Hata A, Doody J, Massagué J (1997) The TGF-β family mediator Smad1 is phosphorylated directly and activated functionally by the BMP receptor kinase. *Genes Dev* 11: 984–995

81 Macías-Silva M, Abdollah S, Hoodless PA, Pirone R, Attisano L, Wrana JL (1996) MADR2 is a substrate of the TGFβ receptor and its phosphorylation is required for nuclear accumulation and signaling. *Cell* 87: 1215–1224

82 Souchelnytskyi S, Tamaki K, Engström U, Wernstedt C, ten Dijke P, Heldin C-H (1997) Phosphorylation of Ser465 and Ser467 in the C terminus of Smad2 mediates interaction with Smad4 and is required for transforming growth factor-β signaling. *J Biol Chem* 272: 28107–28115

83 Ebisawa T, Tada K, Kitajima I, Tojo K, Sampath TK, Kawabata M, Miyazono K, Imamura T (1999) Characterization of bone morphogenetic protein-6 signaling pathways in osteoblast differentiation. *J Cell Sci* 112: 3519–3527

84 Nishimura R, Kato Y, Chen D, Harris SE, Mundy GR, Yoneda T (1998) Smad5 and DPC4 are key molecules in mediating BMP-2-induced osteoblastic differentiation of the pluripotent mesenchymal precursor cell line C2C12. *J Biol Chem* 273: 1872–1879

85 Tamaki K, Souchelnytskyi S, Itoh S, Nakao A, Sampath K, Heldin C-H, ten Dijke P (1998) Intracellular signaling of osteogenic protein-1 through Smad5 activation. *J Cell Physiol* 177: 355–363

86 Lagna G, Hata A, Hemmati-Brivanlou A, Massagué J (1996) Partnership between DPC4 and SMAD proteins in TGF-β signalling pathways. *Nature* 383: 832–836

87 Correia JJ, Chacko BM, Lam SS, Lin K (2001) Sedimentation studies reveal a direct role of phosphorylation in Smad3:Smad4 homo- and hetero-trimerization. *Biochemistry* 40: 1473–1482

88 Kawabata M, Inoue H, Hanyu A, Imamura T, Miyazono K (1998) Smad proteins exist as monomers *in vivo* and undergo homo- and hetero-oligomerization upon activation by serine/threonine kinase receptors. *EMBO J* 17: 4056–4065

89 Shi Y, Hata A, Lo RS, Massagué J, Pavletich NP (1997) A structural basis for mutational inactivation of the tumour suppressor Smad4. *Nature* 388: 87–93

90 Xiao Z, Liu X, Lodish HF (2000) Importin β mediates nuclear translocation of Smad3. *J Biol Chem* 275: 23425–23428

91 Xiao Z, Liu X, Henis YI, Lodish HF (2000) A distinct nuclear localization signal in the N terminus of Smad3 determines its ligand-induced nuclear translocation. *Proc Natl Acad Sci USA* 97: 7853–7858

92 Pierreux CE, Nicolas FJ, Hill CS (2000) Transforming growth factor β independent shuttling of Smad4 between the cytoplasm and nucleus. *Mol Cell Biol* 20: 9041–9054

93 Watanabe M, Masuyama N, Fukuda M, Nishida E (2000) Regulation of intracellular dynamics of Smad4 by its leucine-rich nuclear export signal. *EMBO Reports* 1: 176–182

94 Massagué J, Wotton D (2000) Transcriptional control by the TGF-β/Smad signaling system. *EMBO J* 19: 1745–1754
95 ten Dijke P, Miyazono K, Heldin C-H. (2000) Signaling inputs converge on nuclear effectors in TGF-β signaling. *Trends Biochem Sci* 25: 64–70
96 Flanders KC, Kim ES, Roberts AB (2001) Immunohistochemical expression of smads 1-6 in the 15-day gestation mouse embryo: signaling by BMPs and TGF-βs. *Dev Dyn* 220: 141–154
97 Zhu H, Kavsak P, Abdollah S, Wrana JL, Thomsen GH (1999) A SMAD ubiquitin ligase targets the BMP pathway and affects embryonic pattern formation. *Nature* 400: 687–693
98 Dennler S, Itoh S, Vivien D, ten Dijke P, Huet S, Gauthier JM (1998) Direct binding of Smad3 and Smad4 to critical TGF β-inducible elements in the promoter of human plasminogen activator inhibitor-type 1 gene. *EMBO J* 17: 3091–3100
99 Jonk LJ, Itoh S, Heldin C-H, ten Dijke P, Kruijer W (1998) Identification and functional characterization of a Smad binding element (SBE) in the JunB promoter that acts as a transforming growth factor-β, activin, and bone morphogenetic protein-inducible enhancer. *J Biol Chem* 273: 21145–21152
100 Yingling JM, Datto MB, Wong C, Frederick JP, Liberati NT, Wang X-F (1997) Tumor suppressor Smad4 is a transforming growth factor β-inducible DNA binding protein. *Mol Cell Biol* 17: 7019–7028
101 Zawel L, Dai JL, Buckhaults P, Zhou S, Kinzler KW, Vogelstein B, Kern SE (1998) Human Smad3 and Smad4 are sequence-specific transcription activators. *Mol Cell* 1: 611–617
102 Brodin G, Åhgren A, ten Dijke P, Heldin C-H, Heuchel R (2000) Efficient TGF-β induction of the Smad7 gene requires cooperation between AP-1, Sp1, and Smad proteins on the mouse Smad7 promoter. *J Biol Chem* 275: 29023–29030
103 Denissova NG, Pouponnot C, Long J, He D, Liu F (2000) Transforming growth factor β-inducible independent binding of SMAD to the Smad7 promoter. *Proc Natl Acad Sci USA* 97: 6397–6402
104 Nagarajan RP, Zhang J, Li W, Chen Y (1999) Regulation of Smad7 promoter by direct association with Smad3 and Smad4. *J Biol Chem* 274: 33412–33418
105 Stopa M, Anhuf D, Terstegen L, Gatsios P, Gressner AM, Dooley S (2000) Participation of Smad2, Smad3, and Smad4 in transforming growth factor β (TGF-β)-induced activation of Smad7. The TGF-β response element of the promoter requires functional Smad binding element and E-box sequences for transcriptional regulation *J Biol Chem* 275: 29308–29317
106 von Gersdorff G, Susztak K, Rezvani F, Bitzer M, Liang D, Bottinger EP (2000) Smad3 and Smad4 mediate transcriptional activation of the human Smad7 promoter by transforming growth factor β. *J Biol Chem* 275: 11320–11326
107 Stroschein SL, Wang W, Luo K (1999) Cooperative binding of Smad proteins to two adjacent DNA elements in the plasminogen activator inhibitor-1 promoter mediates

transforming growth factor β-induced Smad-dependent transcriptional activation. *J Biol Chem* 274: 9431–9441
108 Chen SJ, Yuan W, Lo S, Trojanowska M, Varga J (2000) Interaction of Smad3 with a proximal smad-binding element of the human α2(I) procollagen gene promoter required for transcriptional activation by TGF-β. *J Cell Physiol* 183: 381–392
109 Vindevoghel L, Lechleider RJ, Kon A, de Caestecker MP, Uitto J, Roberts AB, von Gersdorff G, Susztak K, Rezvani F, Bitzer M et al (2000) Smad3 and Smad4 mediate transcriptional activation of the human Smad7 promoter by transforming growth factor β. *J Biol Chem* 275: 11320–11326
110 Shi Y, Wang YF, Jayaraman L, Yang H, Massagué J, Pavletich NP (1998) Crystal structure of a Smad MH1 domain bound to DNA: insights on DNA binding in TGF-β signaling. *Cell* 94: 585–594
111 Henningfeld KA, Rastegar S, Adler G, Knochel W (2000) Smad1 and Smad4 are components of the bone morphogenetic protein-4 (BMP-4)-induced transcription complex of the Xvent-2B promoter. *J Biol Chem* 275: 21827–21835
112 Ishida W, Hamamoto T, Kusanagi K, Yagi K, Kawabata M, Takehara K, Sampath TK, Kato M, Miyazono K (2000) Smad6 is a Smad1/5-induced Smad inhibitor. Characterization of bone morphogenetic protein-responsive element in the mouse Smad6 promoter. *J Biol Chem* 275: 6075–6079
113 Kim J, Johnson K, Chen HJ, Carroll S, Laughon A (1997) *Drosophila* Mad binds to DNA and directly mediates activation of vestigial by Decapentaplegic. *Nature* 388: 304–308
114 Kusanagi K, Inoue H, Ishidou Y, Mishima HK, Kawabata M, Miyazono K (2000) Characterization of a bone morphogenetic protein-responsive Smad-binding element. *Mol Biol Cell* 11: 555–565.
115 Yoshida Y, Tanaka S, Umemori H, Minowa O, Usui M, Ikematsu N, Hosoda E, Imamura T, Kuno J, Yamashita T (2000) Negative regulation of BMP/Smad signaling by Tob in osteoblasts. *Cell* 103: 1085–1097
116 Derynck R, Zhang Y, Feng X-H (1998) Smads: transcriptional activators of TGF-β responses. *Cell* 95: 737–740
117 Hata A, Seoane J, Lagna G, Montalvo E, Hemmati-Brivanlou A, Massagué J (2000) OAZ uses distinct DNA- and protein-binding zinc fingers in separate BMP-Smad and Olf signaling pathways. *Cell* 100: 229–240
118 Hanai J-i, Chen LF, Kanno T, Ohtani-Fujita N, Kim WY, Guo WH, Imamura T, Ishidou Y, Fukuchi M, Shi MJ et al (1999) Interaction and functional cooperation of PEBP2/CBF with Smads. Synergistic induction of the immunoglobulin germline Cα promoter. *J Biol Chem* 274: 31577–31582
119 Pardali E, Xie XQ, Tsapogas P, Itoh S, Arvanitidis K, Heldin C-H, ten Dijke P, Grundstrom T, Sideras P (2000) Smad and AML proteins synergistically confer transforming growth factor β1 responsiveness to human germ-line IgA genes. *J Biol Chem* 275: 3552–3560
120 Komori T, Yagi H, Nomura S, Yamaguchi A, Sasaki K, Deguchi K, Shimizu Y, Bronson

RT, Gao YH, Inada M (1997) Targeted disruption of Cbfa1 results in a complete lack of bone formation owing to maturational arrest of osteoblasts. *Cell* 89: 755–764

121 Ducy P, Starbuck M, Priemel M, Shen J, Pinero G, Geoffroy V, Amling M, Karsenty G (1999) A Cbfa1-dependent genetic pathway controls bone formation beyond embryonic development. *Genes Dev* 13: 1025–1036

122 Mundlos S, Mulliken JB, Abramson DL, Warman ML, Knoll JH, Olsen BR (1995) Genetic mapping of cleidocranial dysplasia and evidence of a microdeletion in one family. *Hum Mol Genet* 4: 71–75

123 Zhang YW, Yasui N, Ito K, Huang G, Fujii M, Hanai J, Nogami H, Ochi T, Miyazono K, Ito Y (2000) A RUNX2/PEBP2αA/CBFA1 mutation displaying impaired transactivation and Smad interaction in cleidocranial dysplasia. *Proc Natl Acad Sci USA* 97: 10549–10554

124 Liu F, Hata A, Baker JC, Doody J, Carcámo J, Harland RM, Massagué J (1996) A human Mad protein acting as a BMP-regulated transcriptional activator. *Nature* 381: 620–623

125 Meersseman G, Verschueren K, Nelles L, Blumenstock C, Kraft H, Wuytens G, Remacle J, Kozak CA, Tylzanowski P, Niehrs C et al (1997) The C-terminal domain of Mad-like signal transducers is sufficient for biological activity in the *Xenopus* embryo and transcriptional activation. *Mech Dev* 61: 127–140

126 Pouponnot C, Jayaraman L, Massagué J (1998) Physical and functional interaction of SMADs and p300/CBP. *J Biol Chem* 273: 22865–22868

127 Nakashima K, Yanagisawa M, Arakawa H, Kimura N, Hisatsune T, Kawabata M, Miyazono K, Taga T (1999) Synergistic signaling in fetal brain by STAT3-Smad1 complex bridged by p300. *Science* 284: 479–482

128 Massagué J, Chen YG (2000) Controlling TGF-beta signaling. *Genes Dev* 14: 627–644

129 Onichtchouk D, Chen YG, Dosch R, Gawantka V, Delius H, Massagué J, Niehrs C (1999) Silencing of TGF-β signalling by the pseudoreceptor BAMBI. *Nature* 401: 480–485

130 Tsang M, Kim R, de Caestecker MP , Kudoh T, Roberts AB, Dawid IB (2000) Zebrafish nma is involved in TGFβ family signaling. *Genesis* 28: 47–57

131 Kretzschmar M, Doody J, Massagué J (1997) Opposing BMP and EGF signalling pathways converge on the TGF-β family mediator Smad1. *Nature* 389: 618–622

132 Ishisaki A, Yamato K, Hashimoto S, Nakao A, Tamaki K, Nonaka K, ten Dijke P, Sugino H, Nishihara T (1999) Differential inhibition of Smad6 and Smad7 on bone morphogenetic protein- and activin-mediated growth arrest and apoptosis in B cells. *J Biol Chem* 274: 13637–13642

133 Kavsak P, Rasmussen RK, Causing CG, Bonni S, Zhu H, Thomsen GH, Wrana JL (2000) Smad7 binds to Smurf2 to form an E3 ubiquitin ligase that targets the TGFβ receptor for degradation. *Mol Cell* 6: 1365–1375

134 Ebisawa T, Fukuchi M, Murakami G, Chiba T, Tanaka K, Imamura T, Miyazono K (2001) Smurf1 interacts with transforming growth factor-β type I receptor through Smad7 and induces receptor degradation. *J Biol Chem* 276: 12477–12480

135 Hata A, Lagna G, Massagué J, Hemmati-Brivanlou A (1998) Smad6 inhibits BMP/Smad1 signaling by specifically competing with the Smad4 tumor suppressor. *Genes Dev* 12: 186–197
136 Bai S, Shi X, Yang X, Cao X (2000) Smad6 as a transcriptional corepressor. *J Biol Chem* 275: 8267–8270
137 Kimura N, Matsuo R, Shibuya H, Nakashima K, Taga T (2000) BMP2-induced apoptosis is mediated by activation of the TAK1-p38 kinase pathway that is negatively regulated by Smad6. *J Biol Chem* 275: 17647–17652
138 Luo K, Stroschein SL, Wang W, Chen D, Martens E, Zhou S, Zhou Q (1999) The Ski oncoprotein interacts with the Smad proteins to repress TGFβ signaling. *Genes Dev* 13: 2196–2206
139 Wang W, Mariani FV, Harland RM, Luo K (2000) Ski represses bone morphogenic protein signaling in *Xenopus* and mammalian cells *Proc Natl Acad Sci USA* 97: 14394–14399
140 Yamaguchi K, Shirakabe K, Shibuya H, Irie K, Oishi I, Ueno N, Taniguchi T, Nishida E, Matsumoto K (1995) Identification of a member of the MAPKKK family as a potential mediator of TGF-β signal transduction. *Science* 270: 2008–2011
141 Yamaguchi K, Nagai S, Ninomiya-Tsuji J, Nishita M, Tamai K, Irie K, Ueno N, Nishida E, Shibuya H, Matsumoto K (1999) XIAP, a cellular member of the inhibitor of apoptosis protein family, links the receptors to TAB1-TAK1 in the BMP signaling pathway. *EMBO J* 18: 179–187
142 Sano Y, Harada J, Tashiro S, Gotoh-Mandeville R, Maekawa T, Ishii S (1999) ATF-2 is a common nuclear target of Smad and TAK1 pathways in transforming growth factor-β signaling. *J Biol Chem* 274: 8949–8957
143 Nakamura K, Shirai T, Morishita S, Uchida S, Saeki-Miura K, Makishima F (1999) p38 mitogen-activated protein kinase functionally contributes to chondrogenesis induced by growth/differentiation factor-5 in ATDC5 cells. *Exp Cell Res* 250: 351–363
144 Ahrens M, Ankenbauer T, Schroder D, Hollnagel A, Mayer H, Gross G (1993) Expression of human bone morphogenetic proteins-2 or -4 in murine mesenchymal progenitor C3H10T1/2 cells induces differentiation into distinct mesenchymal cell lineages. *DNA Cell Biol* 12: 871–880
145 Katagiri T, Yamaguchi A, Komaki M, Abe E, Takahashi N, Ikeda T, Rosen V, Wozney JM, Fujisawa-Sehara A, Suda T (1994) Bone morphogenetic protein-2 converts the differentiation pathway of C2C12 myoblasts into the osteoblast lineage. *J Cell Biol* 127: 1755–1766
146 Maliakal JC, Asahina I, Hauschka PV, Sampath TK (1994) Osteogenic protein-1 (BMP-7) inhibits cell proliferation and stimulates the expression of markers characteristic of osteoblast phenotype in rat osteosarcoma (17/2.8) cells. *Growth Factors* 11: 227–234
147 Lee KS, Kim HJ, Li QL, Chi XZ, Ueta C, Komori T, Wozney JM, Kim EG, Choi JY, Ryoo HM et al (2000) Runx2 is a common target of transforming growth factor β1 and bone morphogenetic protein 2, and cooperation between Runx2 and Smad5 induces

osteoblast-specific gene expression in the pluripotent mesenchymal precursor cell line C2C12. *Mol Cell Biol* 20: 8783–8792

148 Shi X, Yang X, Chen D, Chang Z, Cao X (1999) Smad1 interacts with homeobox DNA-binding proteins in bone morphogenetic protein signaling. *J Biol Chem* 274: 13711–13717

149 Yang X, Ji X, Shi X, Cao X (2000) Smad1 domains interacting with Hoxc-8 induce osteoblast differentiation. *J Biol Chem* 275: 1065–1072

150 Hullinger TG, Pan Q, Viswanathan HL, Somerman MJ (2001) TGFβ and BMP-2 activation of the OPN promoter: roles of Smad- and Hox-binding elements. *Exp Cell Res* 262: 69–74

151 Wan M, Shi X, Feng X, Cao X (2001) Transcriptional mechanisms of BMP-induced osteoprotegrin gene expression. *J Biol Chem* 276: 10119–10125

152 Nakanishi T, Kimura Y, Tamura T, Ichikawa H, Yamaai Y, Sugimoto T, Takigawa M (1997) Cloning of a mRNA preferentially expressed in chondrocytes by differential display-PCR from a human chondrocytic cell line that is identical with connective tissue growth factor (CTGF) mRNA. *Biochem Biophys Res Commun* 234: 206–210

153 Nishida T, Nakanishi T, Asano M, Shimo T, Takigawa M (2000) Effects of CTGF/Hcs24, a hypertrophic chondrocyte-specific gene product, on the proliferation and differentiation of osteoblastic cells *in vitro*. *J Cell Physiol* 184: 197–206

154 Afrakhte M, Morén A, Jossan S, Itoh S, Sampath K, Westermark B, Heldin C-H, Heldin N-E, ten Dijke P (1998) Induction of inhibitory Smad6 and Smad7 mRNA by TGF-β family members. *Biochem Biophys Res Commun* 249: 505–511

155 Takase M, Imamura T, Sampath TK, Takeda K, Ichijo H, Miyazono K, Kawabata M. (1998) Induction of Smad6 mRNA by bone morphogenetic proteins. *Biochem Biophys Res Commun* 244: 26–29

156 Chalaux E, Lopez-Rovira T, Rosa JL, Bartrons R, Ventura F (1998) JunB is involved in the inhibition of myogenic differentiation by bone morphogenetic protein-2. *J Biol Chem* 273: 537–543

157 Hollnagel A, Oehlmann V, Heymer J, Ruther U, Nordheim A (1999) Id genes are direct targets of bone morphogenetic protein induction in embryonic stem cells. *J Biol Chem* 274: 19838–19845

158 Ogata T, Wozney JM, Benezra R, Noda M (1993) Bone morphogenetic protein 2 transiently enhances expression of a gene, Id (inhibitor of differentiation), encoding a helix-loop-helix molecule in osteoblast-like cells. *Proc Natl Acad Sci USA* 90: 9219–9222

159 Moldes M, Lasnier F, Feve B, Pairault J, Djian P (1997) Id3 prevents differentiation of preadipose cells. *Mol Cell Biol* 17: 1796–1804

160 Jen Y, Weintraub H, Benezra R (1992) Overexpression of Id protein inhibits the muscle differentiation program: *in vivo* association of Id with E2A proteins. *Genes Dev* 6: 1466–1479

161 Melnikova IN, Christy BA (1996) Muscle cell differentiation is inhibited by the helix-loop-helix protein Id3. *Cell Growth Differ* 7: 1067–1079

162 Miyama K, Yamada G, Yamamoto TS, Takagi C, Miyado K, Sakai M, Ueno N, Shibuya

H (1999) A BMP-inducible gene, dlx5, regulates osteoblast differentiation and mesoderm induction. *Dev Biol* 208: 123–133
163 Acampora D, Merlo GR, Paleari L, Zerega B, Postiglione MP, Mantero S, Bober E, Barbieri O, Simeone A, Levi G (1999) Craniofacial, vestibular and bone defects in mice lacking the Distal-less-related gene Dlx5. *Development* 126: 3795–3809
164 Sirard C, Kim S, Mirtsos C, Tadich P, Hoodless PA, Itie A, Maxson R, Wrana JL, Mak TW (2000) Targeted disruption in murine cells reveals variable requirement for Smad4 in transforming growth factor β-related signaling. *J Biol Chem* 275: 2063–2070
165 Satokata I, Ma L, Ohshima H, Bei M, Woo I, Nishizawa K, Maeda T, Takano Y, Uchiyama M, Heaney S et al (2000) Msx2 deficiency in mice causes pleiotropic defects in bone growth and ectodermal organ formation. *Nat Genet* 24: 391–395
166 Service RF (2000) Tissue engineers build new bone. *Science* 289: 1498–1500
167 Franceschi RT, Wang D, Krebsbach PH, Rutherford RB (2000) Gene therapy for bone formation: *in vitro* and *in vivo* osteogenic activity of an adenovirus expressing BMP7. *J Cell Biochem* 78: 476–486
168 Fang J, Zhu YY, Smiley E, Bonadio J, Rouleau JP, Goldstein SA, McCauley LK, Davidson BL, Roessler BJ (1996) Stimulation of new bone formation by direct transfer of osteogenic plasmid genes. *Proc Natl Acad Sci USA* 93: 5753–5758
169 Zhang H, Bradley A (1996) Mice deficient for BMP2 are nonviable and have defects in amnion/chorion and cardiac development. *Development* 122: 2977–2986
170 Solloway MJ, Dudley AT, Bikoff EK, Lyons KM, Hogan BL, Robertson EJ (1998) Mice lacking Bmp6 function. *Dev Genet* 22: 321–339
171 Luo G, Hofmann C, Bronckers AL, Sohocki M, Bradley A, Karsenty G (1995) BMP-7 is an inducer of nephrogenesis, and is also required for eye development and skeletal patterning. *Genes Dev* 9: 2808–2820
172 Dudley AT, Lyons KM, Robertson EJ (1995) A requirement for bone morphogenetic protein-7 during development of the mammalian kidney and eye. *Genes Dev* 9: 2795–2807
173 Solloway MJ, Robertson EJ (1999) Early embryonic lethality in Bmp5;Bmp7 double mutant mice suggests functional redundancy within the 60A subgroup. *Development* 126: 1753–1768
174 Zhao GQ, Deng K, Labosky PA, Liaw L, Hogan BL (1996) The gene encoding bone morphogenetic protein 8B is required for the initiation and maintenance of spermatogenesis in the mouse. *Genes Dev* 10: 1657–1669
175 Galloway SM, McNatty KP, Cambridge LM, Laitinen MP, Juengel JL, Jokiranta TS, McLaren RJ, Luiro K, Dodds KG, Montgomery GW et al (2000) Mutations in an oocyte-derived growth factor gene (BMP15) cause increased ovulation rate and infertility in a dosage-sensitive manner. *Nat Genet* 25: 279–283
176 Sirard C, de la Pompa JL, Elia A, Itie A, Mirtsos C, Cheung A, Hahn S, Wakeham A., Schwartz L, Kern SE et al (1998) The tumor suppressor gene Smad4/Dpc4 is required for gastrulation and later for anterior development of the mouse embryo. *Genes Dev* 12: 107–119

177 Chang H, Huylebroeck D, Verschueren K, Guo Q, Matzuk MM, Zwijsen A (1999) Smad5 knockout mice die at mid-gestation due to multiple embryonic and extraembryonic defects. *Development* 126: 1631–1642

178 Ducy P, Desbois C, Boyce B, Pinero G, Story B, Dunstan C, Smith E, Bonadio J, Goldstein S, Gundberg C et al (1996) Increased bone formation in osteocalcin-deficient mice. *Nature* 382: 448–452

179 Liaw L, Birk DE, Ballas CB, Whitsitt JS, Davidson JM, Hogan BL (1998) Altered wound healing in mice lacking a functional osteopontin gene (spp1). *J Clin Invest* 101: 1468–1478

180 Yoshitake H, Rittling SR, Denhardt DT, Noda M (1999) Osteopontin-deficient mice are resistant to ovariectomy-induced bone resorption. *Proc Natl Acad Sci USA* 96: 8156–8160

181 Willing MC, Pruchno CJ, Atkinson M, Byers PH (1992) Osteogenesis imperfecta type I is commonly due to a COL1A1 null allele of type I collagen. *Am J Hum Genet* 51: 508–515

182 Chen D, Harris MA, Rossini G, Dunstan CR, Dallas SL, Feng JQ, Mundy GR, Harris SE (1997) Bone morphogenetic protein 2 (BMP-2) enhances BMP-3, BMP-4, and bone cell differentiation marker gene expression during the induction of mineralized bone matrix formation in cultures of fetal rat calvarial osteoblasts. *Calcif Tissue Int* 60: 283–290

183 Fedde KN, Blair L, Silverstein J, Coburn SP, Ryan LM, Weinstein RS, Waymire K, Narisawa S, Millan JL, MacGregor GR et al (1999) Alkaline phosphatase knock-out mice recapitulate the metabolic and skeletal defects of infantile hypophosphatasia. *J Bone Miner Res* 14: 2015–2026

184 Galvin KM, Donovan MJ, Lynch CA, Meyer RI, Paul RJ, Lorenz JN, Fairchild-Huntress V, Dixon KL, Dunmore JH, Gimbrone MA et al (2000) A role for smad6 in development and homeostasis of the cardiovascular system. *Nat Genet* 24: 171–174

185 Lyden D, Young AZ, Zagzag D, Yan W, Gerald W, O'Reilly R, Bader BL, Hynes RO, Zhuang Y, Manova K et al (1999) Id1 and Id3 are required for neurogenesis, angiogenesis and vascularization of tumour xenografts. *Nature* 401: 670–677

186 Schorpp-Kistner M, Wang ZQ, Angel P, Wagner EF (1999) JunB is essential for mammalian placentation. *EMBO J* 18: 934–948

187 Ducy P, Zhang R, Geoffroy V, Ridall AL, Karsenty G (1997) Osf2/Cbfa1: a transcriptional activator of osteoblast differentiation. *Cell* 89: 747–754

# Deciphering the binding code of BMP-receptor interaction

*Joachim Nickel, Matthias Dreyer and Walter Sebald*

Physiologische Chemie II, Biozentrum der Universität Würzburg, Am Hubland, 97074 Würzburg, Germany

## Introduction

BMPs and other members of the TGF-β superfamily are powerful secreted signalling proteins that determine development and homoeostasis of many organs and tissues [1, 2]. These comprise bone, cartilage and teeth as well as heart, kidney, muscle, skin, hair, reproductive tract, and several others. Despite the diversity of the biological functions, all ligands and receptors in this superfamily show on a molecular level many similarities in structure and function [3, 4].

The three-dimensional structure even of distantly related factors reveals an astonishingly similar backbone fold in most parts of the protein [5-9], although the amino acid sequences of the mature parts show only 30% identity among the most distant members and 70–90% only within special subgroups. Generally, type I and type II receptors with a cytoplasmic serine/threonine kinase domain are necessary for transmembrane signalling. SMAD proteins are special cytoplasmic signalling proteins for the TGF-β-like factors and their receptors [10].

More than 30 different TGF-β-like proteins known today in men and mice comprise BMPs, GDFs, TGF-βs, activins/inhibins, and others. The designation bone morphogenetic protein, "BMP," originally indicated that the protein induces ectopic bone or cartilage formation when analyzed in a Reddi-Sampath assay *in vivo* [11]. But many, especially the recently discovered so-called BMPs, most likely do not function during physiological bone formation or repair. Probably, the few type I and type II receptors (BRIA, -IB, ARI, -IB, BRII, ARII) established or discussed to participate in BMP signalling are promiscuous and can interact with more than one or multiple BMPs in experimental setups. The specificity and the affinities of these interactions, however, remain to be defined and quantified. It seems also important to explore if combinations of BMPs or BMP heterodimers are more efficient than homodimers and individual factors.

BMP-2 and other BMPs and TGF-β-like factors are notorious for interacting with a variety of proteins and molecules in addition to the type I and II receptor chains [12]. These additional proteins inhibit or modify the activity of the BMPs

Bone Morphogenetic Proteins, edited by Slobodan Vukicevic and Kuber T. Sampath

[12]. Some of these proteins interact specifically with one BMP or with a BMP subgroup, whereas others show a broader specificity. Several of these proteins, like noggin, chordin and bambi have been shown to block receptor binding. Only the BMP-2 epitope for glycosaminoglycan binding (heparin-binding epitope) has been studied in some detail [13].

The present review describes data on established three-dimensional structures of TGF-β-like proteins as well as on the structures of the receptor ectodomain of ARII [14] and of BRIA in complex with BMP-2 [15]. The structural data provide the framework to characterize the functional binding epitopes of BMPs and TGF-β like proteins for the type I and type II receptor chains as well as for heparinic sites.

## Subfamilies of TGF-β-like factors according to similarities of amino acid sequences

Sequences of 34 mammalian TGF-β-like factors are compiled in Figure 1. The mature proteins comprise (1) the "cystine-knot" domain [16, 17] starting with the first conserved cysteine and (2) a N-terminal segment upstream of that first conserved cysteine (arrow in Fig. 1). The "cystine-knot" domain is the functionally most important part, since it constitutes the binding epitopes for the type I and II receptors. The N-terminal segment, as detailed in the next section, seems to exert various functions in different proteins.

All TGF-β-like factors are dimers, usually homodimers (but see e.g. the heterodimeric inhibin). Six cysteines of the mature monomers form a typical pattern of three disulfide bonds (DSB) called "cystine knot". The seventh cysteine at position 78 (see Fig. 1) forms a disulfide bond between the monomers of nearly all the proteins. The most distant members, GDNF (glial derived neurotrophic factor), Mis (Müllerian duct inhibiting substance) and inhibin A share amino acid sequence identities in only 16–24%, 18–31% and 22–29% of their positions with the BMPs. Subgroups comprising one to three related factors can be discriminated. About 92% identical sequence positions occur between the closely related BMP-2s (BMP-2 and BMP-4), or GDF-8s (BMP-11/GDF-8). Identities of 75–80% exist among the TGF-βs (β1, β2 , and β3). TGF-βs and BMP-2s or BMP-7s share 30–35% identical positions (BMP-17 and BMP-18 are considered to be the same gene product in the SwissProt data bank).

Some subgroups are characterized by additional common properties (Fig. 2). An extra DSB exists in the activins, GDF-8s, GDF-15s and TGF-βs fixing the N-terminal segment to the cystine-knot. An interchain DSB linking the two monomers occurs in all known factors with the exception of GDF-3 and the BMP-15s. A typical pair of tryptophane residues is localized in the first finger loop of all TGF-β-like factors with the exception of the most distant members GDNF and MIS. These tryptophanes are separated by one residue (WxW group) in the GDF-8s and the TGF-βs and by two residues (WxxW group) in all other factors. Members of the same subgroup have usually the same interspaced residues. Finally, N-Glycosylation as deduced from the occurrence of the NxS/T sequence is also subgroup specific. A single NxS/T potential N-glycosylation site is located in the central α-helix of the BMP-2s, BMP-8s, and GDF-3. Two additional NxS/T sites exist in the N-terminal peptide of the BMP-7s. Many of the proteins cannot be N-glycosylated in the mature part. It is unclear if TGF-β-like factors can become O-glycosylated.

## N-terminal segment

The sequence preceding the cystine-knot domain is highly variable both in length and in amino acid composition (see Fig. 1). The length varies between six to seven residues in the BMP-10s and 37 residues in the BMP-7s and BMP-8s. (The reported N-terminal sequence of mGDF-7 is unusual in containing an uninterrupted stretch of 20 glycines.) Similarities in the sequence and size of these peptides exist, if at all, among members of the same subgroup. Nevertheless, the N-terminal segment is of functional importance. The additional DSB fixing this segment to the cystine-knot domain (C15 in Fig. 1) in several subgroups has been mentioned above and may be important for receptor binding of these proteins (see below). In the proteins that do not contain this additional DSB the N-terminal segment probably floats around freely and can interact with other proteins or molecules.

The BMP-2 provides a heparin-binding site in the N-terminal sequence. Binding of BMPs to the extracellular matrix and heparinic sites of glycosaminoglycans has already been inferred from cell culture experiments and the strong binding of BMP-2 and other BMPs to heparin-sepharose. Proteolytic cleavage abolishes binding to the extracellular matrix [18]. Substitution of the N-terminal segment of BMP-2 by a dummy sequence of the same size results in a BMP-2 variant that no longer binds to heparin at 150 mM NaCl. This variant has, however, a 5–10-fold higher biological activity in an embryonic chicken limb bud assay and a decreased activity *in vivo* in an ectopic bone formation assay [13].

A conspicuous feature of the N-terminal BMP-2 segment are two triplets of basic residues providing a high density of positive charges in their side chains (see Fig. 1). In total, 7 basic residues are present and no acidic negatively charged ones. These basic triplets are also present in BMP-4. BMP-2 variants containing one or two additional basic triplets in their N-terminal segment have been generated. They bind to higher levels and with a decreased dissociation rate to heparin (= higher affinity) (Fig. 3). This leads to a reduced biological activity in cell culture were heparinic sites compete with the receptors for the ligand. However, it seems to have a positive effect on ectopic bone formation *in vivo*, were it may stay longer to the application site and, therefore, may be effective at lower concentration and lead to a denser bone of higher quality (second generation BMPs).

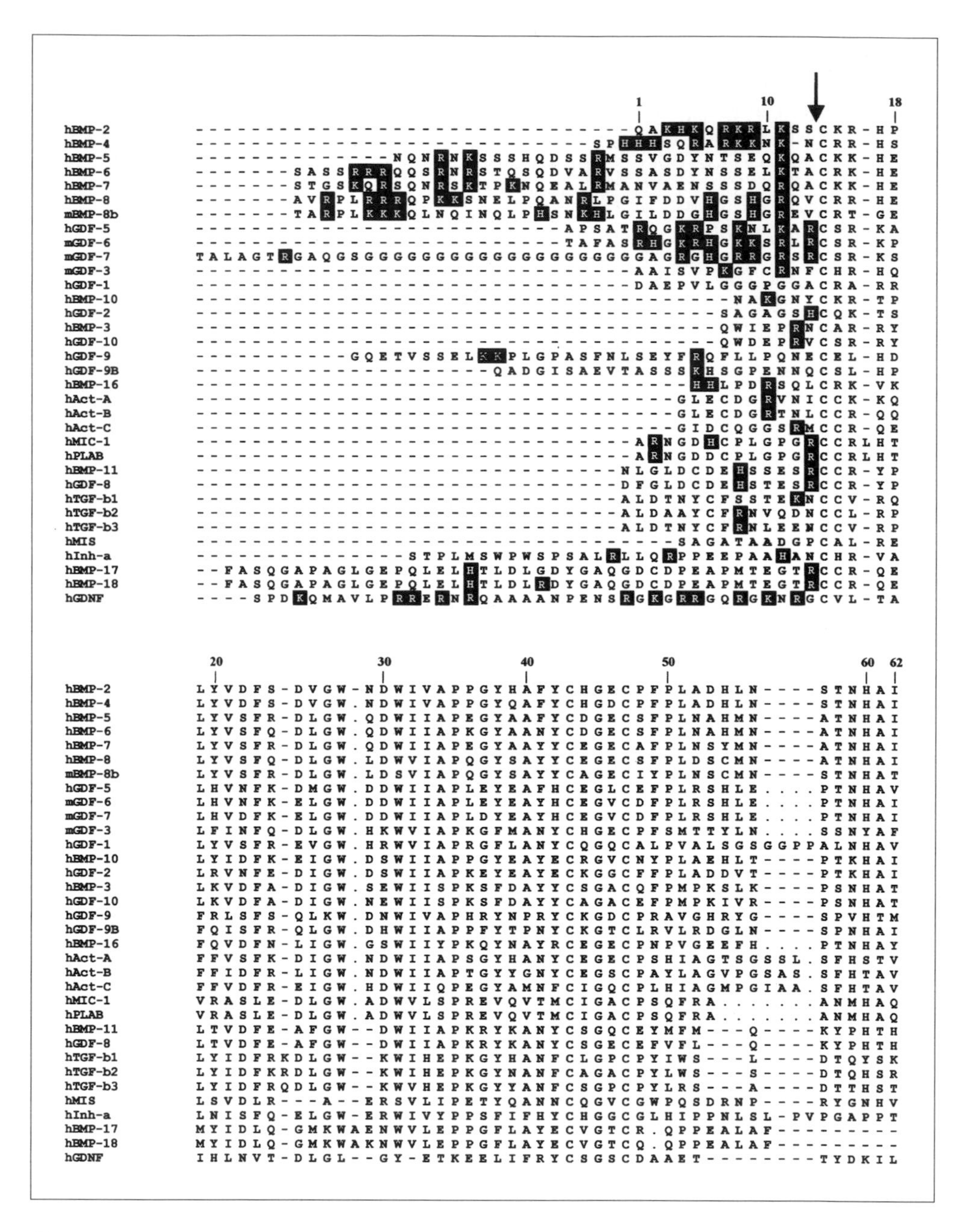

*Figure 1*
*Amino acid sequence alignment of the mature proteins of the TGF-β superfamily. Sequences from SwissProt and GenBank data bases were aligned first with the MultAlin program and then manually adjusted on the basis of a structure-based sequence alignment [9] using the*

```
                 63            70                      80                  90                  100            107
                 |             |                       |                   |                   |              |
hBMP-2           V Q T L V N S V - N S K - I P K A C C V P T E L S A I S M L Y L D E N E K V V L K - - - N Y Q D M V
hBMP-4           V Q T L V N S V - N S S - I P K A C C V P T E L S A I S M L Y L D E Y D K V V L K - - - N Y Q E M V
hBMP-5           V Q T L V H L M - F P D H V P K P C C A P T K L N A I S V L Y F D D S S N V I L K - - - K Y R N M V
hBMP-6           V Q T L V H L M - N P E Y V P K P C C A P T K L N A I S V L Y F D D N S N V I L K - - - K Y R N M V
hBMP-7           V Q T L V H F I - N P E T V P K P C C A P T Q L N A I S V L Y F D D S S N V I L K - - - K Y R N M V
hBMP-8           L Q S L V H L M - K P N A V P K A C C A P T K L S A T S V L Y Y D S S N N V I L R - - - K H R N M V
mBMP-8b          M Q A L V H L M - K P D I I P K V C C V P T E L S A I S L L Y Y D R N N N V I L R - - - R E R N M V
hGDF-5           I Q T L M N S M - D P E S T P P T C C V P T R L S P I S I L F I D S A N N V V Y K - - - Q Y E D M V
mGDF-6           I Q T L M N S M - D P G S T P P S C C V P T K L T P I S I L Y I D A G N N V V Y K - - - Q Y E D M V
mGDF-7           I Q T L L N S M - A P D A A P A S C C V P A R L S P I S I L Y I D A A N N V V Y K - - - Q Y E D M V
mGDF-3           M Q A L M H M A - D P K - V P K A V C V P T K L S P I S M L Y Q D S D K N V I L R - - - H Y E D M V
hGDF-1           L R A L M H A A - A P G A A D L P C C V P A R L S P I S V L F F D N S D N V V L R - - - Q Y E D M V
hBMP-10          I Q A L V H L K - N S Q K A S K A C C V P T K L E P I S I L Y L D K - G V V T Y K - - F K Y E G M A
hGDF-2           V Q T L V H L K - F P T K V G K A C C V P T K L S P I S V L Y K D D M G V P T L K - - Y H Y E G M S
hBMP-3           I Q S I V R A V G V V P G I P E P C C V P E K M S S L S I L F F D E N K N V V L K - - - V Y P N M T
hGDF-10          I Q S I V R A V G I I P G I P E P C C V P D K M N S L G V L F L D E N R N V V L K - - - V Y P N M S
hGDF-9           V Q N I I Y E K - L D S S V P R P S C V P A K Y S P L S V L T I E P D G S I A Y K - - - E Y E D M I
hGDF-9B          I Q N L I N Q L - V D Q S V P R P S C V P Y K Y V P I S V L M I E A N G S I L Y K - - - E Y E G M I
hBMP-16          I Q S L L K R Y - Q P H R V P S T C C A P V K T K P L S M L Y V D N G R - V L L D - - - H H K D M I
hAct-A           I N H Y R M R G H S P F A N L K S C C V P T K L R P M S M L Y Y D D G Q N I I K K - - - D I Q N M I
hAct-B           V N Q Y R M R G L N P - G T V N S C C I P T K L S T M S M L Y F D D E Y N I V K R - - - D V P N M I
hAct-C           L N L L K A N T A A G T T G G G S C C V P T A R R P L S L L Y Y D R D S N I V K T - - - D I P D M V
hMIC-1           I K T S L H R L - K P D T V P A P C C V P A S Y N P M V L I Q K T D T G V S L Q T - - - - Y D D L L
hPLAB            I K T S L H R L - K P D T E P A P C C V P A S Y N P M V L I Q K T D T G V S L Q T - - - - Y D D L L
hBMP-11          - - - L V Q Q A - N P R G S A G P C C T P T K M S P I N M L Y F N D K Q Q I I Y G - - - K I P G M V
hGDF-8           - - - L V H Q A - N P R G S A G P C C T P T K M S P I N M L Y F N G K E Q I I Y G - - - K I P A M V
hTGF-b1          V L A L Y N Q H - N P G A S A A P C C V P Q A L E P L P I V Y Y V G R K P K V E - - - - Q L S N M I
hTGF-b2          V L S L Y N T I - N P E A S A S P C C V S Q D L E P L T I L Y Y I G K T P K I E - - - - Q L S N M I
hTGF-b3          V L G L Y N T L - N P E A S A S P C C V P Q D L E P L T I L Y Y V G R T P K V E - - - - Q L S N M V
hMIS             V L L L K M Q A R G A A L A R P P C C V P T A Y A G K L L I S L S E E R I S A H - - - - H V P N M V
hInh-a           P A Q P Y S L L P G A - - - - Q P C C A A L P G T M R P L H V R T T S D G G Y S F K Y E T V P N L L
hBMP-17          - - - - - - - - - K W P F L G P R Q C I A S E T A S L P M I V S I K E G G R T R P Q V V S L P N M R
hBMP-18          - - - - - - - - - N W P F L G P R Q C I A S E T A S L P M I V S I K E G G R T R P Q V V S L P N M R
hGDNF            K N L S R N R R L V S D K V G Q A C C R P I A F D D D L S F L D D N L V Y H I L R - - - - - - K H S
```

```
                 108 110     114
                 |   |       |
hBMP-2           V E G C G C R - - - - - - - - - - - -
hBMP-4           V E G C G C R - - - - - - - - - - - -
hBMP-5           V R S C G C H - - - - - - - - - - - -
hBMP-6           V R A C G C H - - - - - - - - - - - -
hBMP-7           V R A C G C H - - - - - - - - - - - -
hBMP-8           V K A C G C H - - - - - - - - - - - -
mBMP-8b          V Q A C G C H - - - - - - - - - - - -
hGDF-5           V E S C G C R - - - - - - - - - - - -
mGDF-6           V E S C G C R - - - - - - - - - - - -
mGDF-7           V E A C G C R - - - - - - - - - - - -
mGDF-3           V D E C G C G - - - - - - - - - - - -
hGDF-1           V D E C G C R - - - - - - - - - - - -
hBMP-10          V S E C G C R - - - - - - - - - - - -
hGDF-2           V A E C G C R - - - - - - - - - - - -
hBMP-3           V E S C A C R - - - - - - - - - - - -
hGDF-10          V D T C A C R - - - - - - - - - - - -
hGDF-9           A T K C T C R - - - - - - - - - - - -
hGDF-9B          A E S C T C R - - - - - - - - - - - -
hBMP-16          V E E C G C L - - - - - - - - - - - -
hAct-A           V E E C G C S - - - - - - - - - - - -
hAct-B           V E E C G C A - - - - - - - - - - - -
hAct-C           V E A C G C S - - - - - - - - - - - -
hMIC-1           A K D C H C I - - - - - - - - - - - -
hPLAB            A K D C H C I - - - - - - - - - - - -
hBMP-11          V D R C G C S - - - - - - - - - - - -
hGDF-8           V D R C G C S - - - - - - - - - - - -
hTGF-b1          V R S C K C S - - - - - - - - - - - -
hTGF-b2          V K S C K C S - - - - - - - - - - - -
hTGF-b3          V K S C K C S - - - - - - - - - - - -
hMIS             A T E C G C R - - - - - - - - - - - -
hInh-a           T Q H C A C I - - - - - - - - - - - -
hBMP-17          V Q K C S C A S D G A L V P R R L Q P
hBMP-18          V Q K C S C A S D G A L V P R R L Q P
hGDNF            A K R C G C I - - - - - - - - - - - -
```

*Jalview program. The cystine-knot part of the proteins starts after the arrow with the cysteine at position 14. The basic residues (R, K, H) in the N-terminal segment are represented by white letters on a black background. Numbering is according to the BMP-2 sequence. Usually, the mature proteins are postulated to start after a RXXR furin cleavage site.*

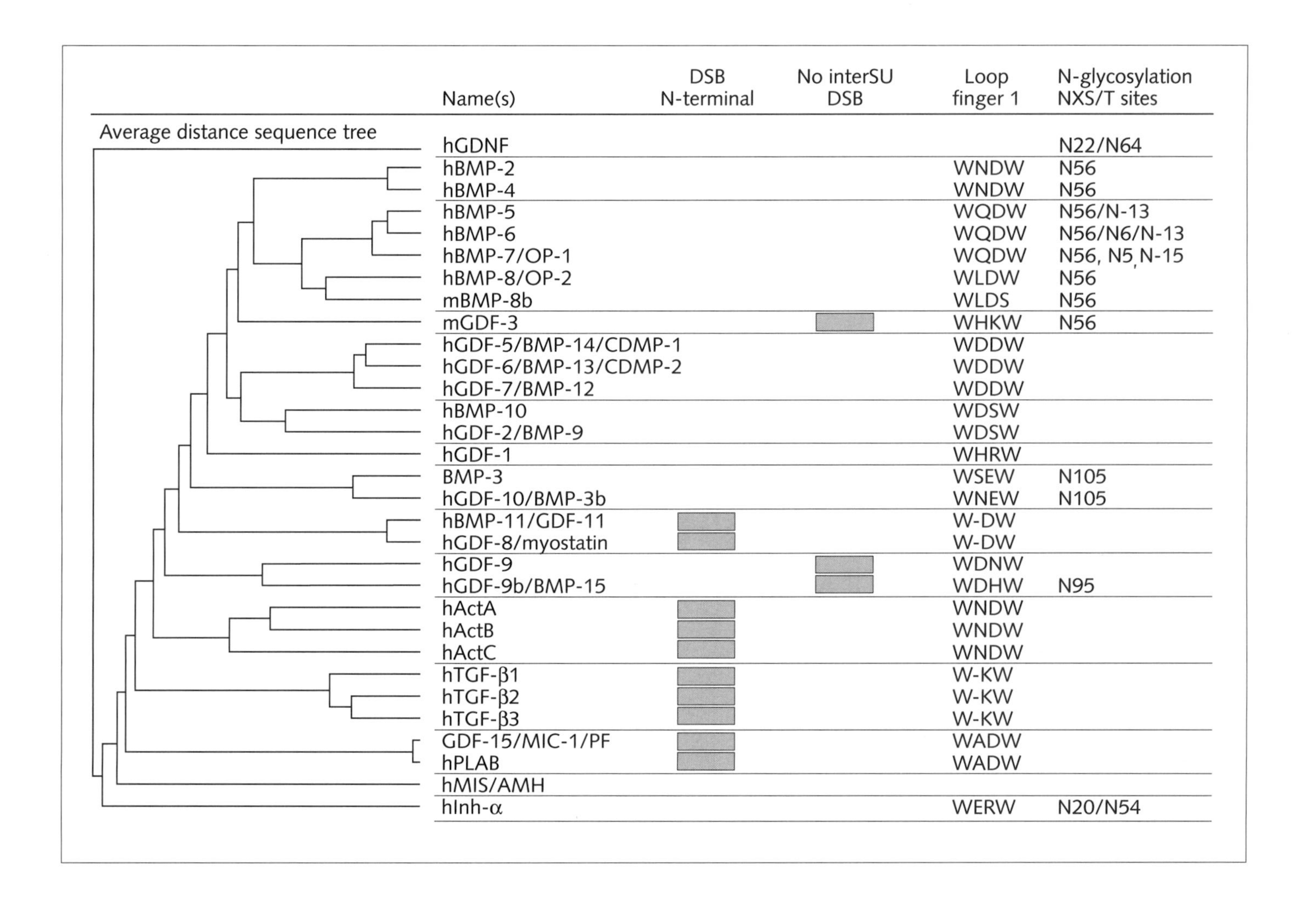
Average distance sequence tree
Name(s)
DSB N-terminal
No interSU DSB
Loop finger 1
N-glycosylation NXS/T sites
hGDNF
N22/N64
hBMP-2
WNDW
N56
hBMP-4
WNDW
N56
hBMP-5
WQDW
N56/N-13
hBMP-6
WQDW
N56/N6/N-13
hBMP-7/OP-1
WQDW
N56, N5, N-15
hBMP-8/OP-2
WLDW
N56
mBMP-8b
WLDS
N56
mGDF-3
WHKW
N56
hGDF-5/BMP-14/CDMP-1
WDDW
hGDF-6/BMP-13/CDMP-2
WDDW
hGDF-7/BMP-12
WDDW
hBMP-10
WDSW
hGDF-2/BMP-9
WDSW
hGDF-1
WHRW
BMP-3
WSEW
N105
hGDF-10/BMP-3b
WNEW
N105
hBMP-11/GDF-11
W-DW
hGDF-8/myostatin
W-DW
hGDF-9
WDNW
hGDF-9b/BMP-15
WDHW
N95
hActA
WNDW
hActB
WNDW
hActC
WNDW
hTGF-β1
W-KW
hTGF-β2
W-KW
hTGF-β3
W-KW
GDF-15/MIC-1/PF
WADW
hPLAB
WADW
hMIS/AMH
hInh-α
WERW
N20/N54

The GDF-5s also contain four to five basic residues and no acidic ones, but only GDF-6 contains a basic triplet. The long N-terminal segments of the BMP-5s and BMP-8s all contain three negatively charged side chains and between five and nine basic ones. The first 14 residues show similarities to the BMP-2 N-terminal segment. Thus, these factors potentially bind also to the negatively charged heparinic sites of the extracellular matrix. But this has not been analyzed in detail so far. Corresponding charge patterns cannot be seen in other TGF-β-like factors with the possible exception of GDNF. Thus, it seems that these proteins have no heparin-binding epitopes in their N-terminal segment. Discontinuous heparin binding epitopes may be present in the folded proteins. For example, the TGF-βs expose patches of six to seven basic amino acid side chains on the surface of the native dimer.

Collagen-binding epitopes have been fused to the N-terminus of TGF-β2 and collagen-binding could be established for the fusion protein but not for the wild type TGF-β2 [19].

## Primary sequences of type I and type II receptors extracellular domains

The extracellular domains of the TGF-β/Act/BMP receptors are likewise small in containing only between 96 (Alk-1) and 143 (TRII) amino acid residues. The binding domain established by crystal structure analysis [14, 20] shows 95 residues for the type II activin receptor ARII and 89 for the type I BMP receptor BRIA. A structure-based sequence alignment of the binding domains of known human type I and type II receptors is presented in Figure 4. The alignment does not include the C-terminal peptide connecting the binding domain to the membrane spanning segment of the receptor. These short connecting peptides consist of six to 15 residues in the type I or II receptors (Fig. 5).

Only a few positions are occupied by identical residues (C38, C59, C77, C102 and N108) in all proteins. The location of two DSBs seems to be diagnostic for spec-

*Figure 2*
*Similarities and typical features of TGF-β like proteins. The average distance sequence tree of the cystine-knot domain was constructed by the Jalview program. The abbreviations are: bone morphogenetic protein (BMP), growth and differentiation factors (GDF), osteogenic protein (OP), chondrocyte derived morphogenetic protein (CDMP), activin (Act), transforming growth factor (TGF), macrophage inhibitory cytokine (MIC), prostate factor (PF), placental TGF-β (PLAB), Muellerian duct inhibiting substance (MIS), Antimuellerian hormone (AMH), inhibin (Inh). The N-terminal disulfide bond (DSB) involves C15 (see Fig. 1). The inter-subunit disulfide bond (SU DSB) is probably missing, since C78 is absent. The finger 1 loop is the large L1 loop (see Fig. 6). Putative N-glycosylation sites are numbered according to the BMP-2 sequence in Figure 1.*

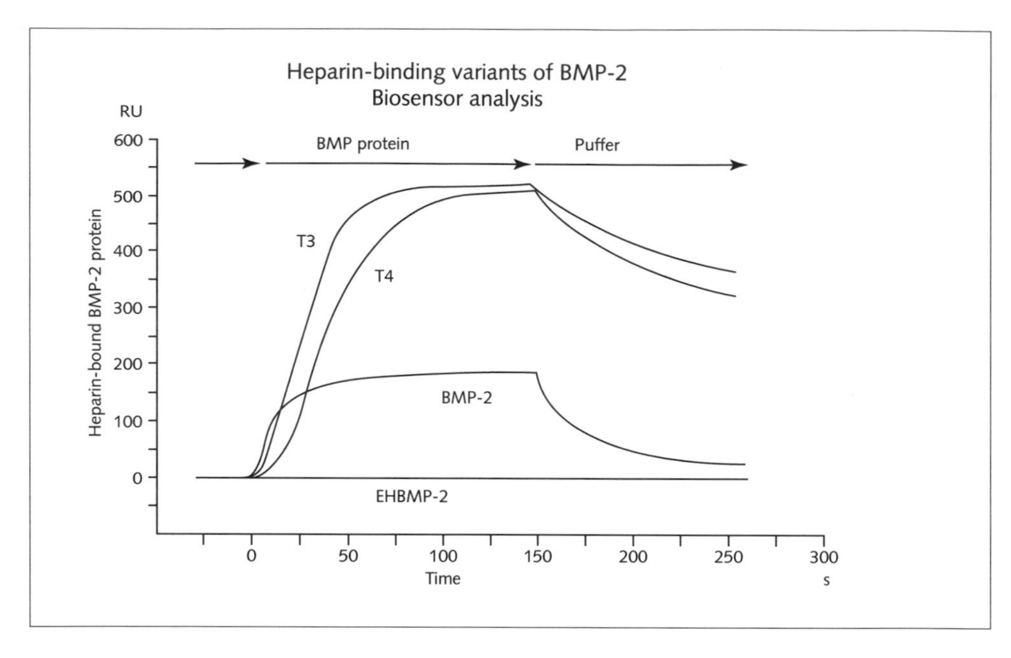


*Figure 3*
*The N-terminal segment of BMP-2 determines binding of BMP-2 to heparinic sites. Heparin was covalently attached to a biosensor matrix and binding of BMP-2 proteins was analyzed by plasmon resonance as described [13]. EHBMP-2 is a BMP-2 variant where the N-terminal segment has been exchanged by a peptide of the same size but without basic triplets. T3 and T4 are BMP-2 variants with insertions of one, respectively two additional triplets of basic residues.*

ifying a type I (DSB2) or a type II (DSB5) receptor. The first half-cystine in DSB3 and DSB4 is located at different positions in type I and type II receptors. A tryptophane residue is found in all type II receptors at position 63.

The average distance tree in Figure 5 demonstrates subgroups within the type I and the type II branches. The subgroup comprising BRIA and BRIB shows 47/89 = 53% identical positions in the binding domain, ARII and ARIIB 61/96 = 64%. BRII exhibits 33/96 = 34% sequence identities with ARIIB and 24/96 = 25% with ARII. The type I receptors have a low 25% sequence identity between the BRIA and BRIB or the ARI and ARIB binding domains.

The identity is even lower between pairs of type I and type II proteins, e.g. 15/96 or 15/89 = 16–17% between the binding domains of BRIA or BRIB and AR-II. But nevertheless, the backbone fold of these proteins is comparable for the core of the structure (see below). Therefore, the whole group of binding proteins have been assigned to one structural group containing the “three-finger-toxin” fold.

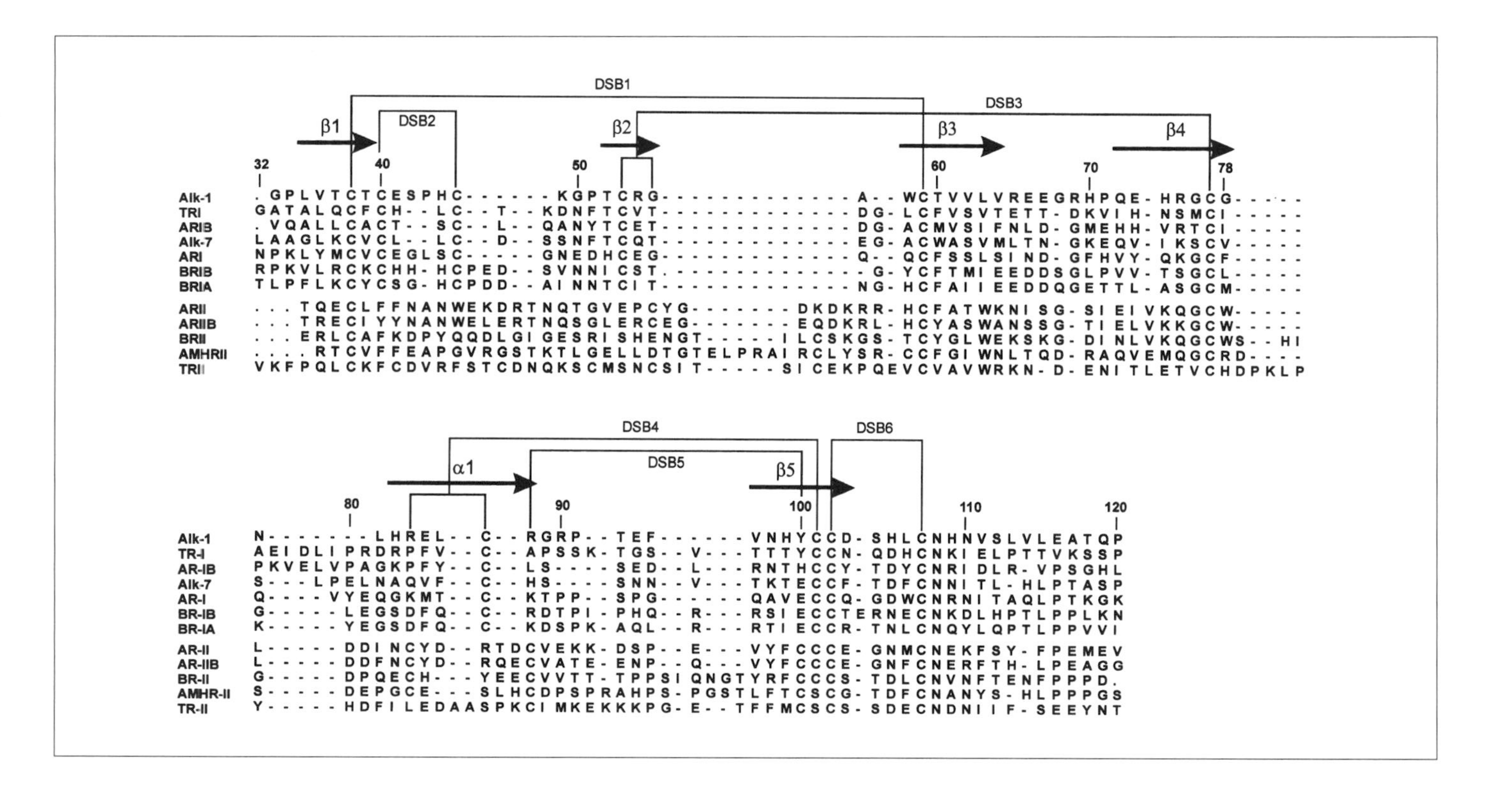


*Figure 4*

*Sequence alignment of the binding domains of type I and type II receptors of the TGF-β family. A structure-based alignment of the ARII and BRIA sequences was performed on the basis of the known crystal structures. The sequences of the type I and type II receptors were aligned with MultAlin and adjusted manually. Numbering of the positions, disulfide bonds (DSB), β-strands β1–β5, and helix α1 are drawn according to the BRIA protein. Abbreviations are: Activin-like kinase (Alk-1, Alk-7), activin receptor (ARI, ARIB, ARII, ARIIB), BMP receptor (BRIA, BRIB, BRII), TGF-β receptor (TRI, TRII), and anti-Muellerian-hormone receptor (AMHRII).*

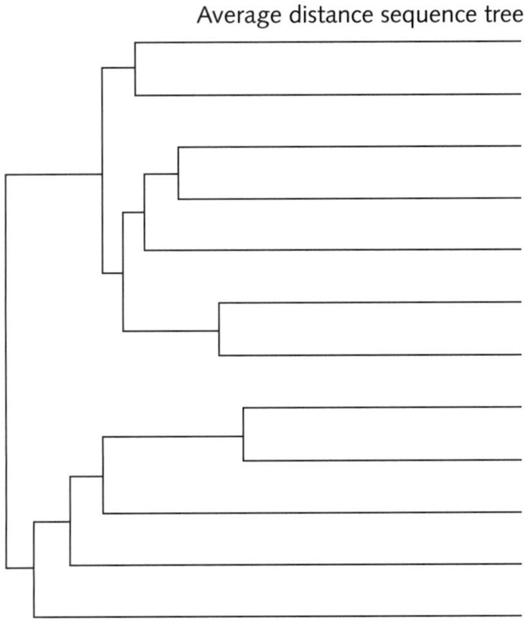

Average distance sequence tree

| Receptor | Binding domain | C-terminal segment | N-terminal segment | N-glycosylation NxS/T sites |
|---|---|---|---|---|
| | Number of residues | | | |
| Alk-1 | 79 | 10 | 7 | N110 |
| ARI | 83 | 11 | 8 | N110 |
| TRI | 89 | 6 | 5 | N50 |
| Alk-7 | 83 | | | N50/N95/N109 |
| ARIB | 85 | 13 | 5 | N50 |
| BRIB | 89 | 11 | 4 | |
| BRIA | 89 | 9 | 31 | N50 |
| ARII | 95 | 13 | 7 | N48/N65 |
| ARIIB | 96 | 13 | 7 | N48/N65 |
| BRII | 105 | 15 | 4 | N56/N96/N110 |
| AMHRII | 196 | 15 | 4 | N64/N110 |
| TRII | 101 | 11 | 31 | N54/N71 |

*Figure 5*
*Average distance tree of the binding domains of type I and type II receptors of the TGF-β family. The tree was constructed by Jalview on the basis of the alignment in Figure 4. The number of residues is specified for the binding domain, the C-terminal and the N-terminal segment. The numbering of the putative N-glycosylation sites is according to the BRIA sequence.*

The putative N-glycosylation sites tend to be conserved. Type I as well as type II receptors contain the N110 site. The segment around β-strand β2 (N48, N50, N54, N56) might be also a prefered glycosylation site in both receptor types. BMP receptor BRIB is probably not N-glycosylated in the extracellular binding domain.

## Three-dimensional structures of TGF-β-like factors

The backbone fold of the BMPs, as exemplified by BMP-7 [8] and BMP-2 [9], is very similar to that of the TGF-βs [5–7]. This is strikingly documented by the fact that the three-dimensional structure of BMP-2 could be solved by molecular replacement of the TGF-β2 crystal structure. Interestingly, even GDNF [21], a most distant member of the TGF-β family using even a different type of receptors, exhibits a comparable monomer fold and dimer assembly as the TGF-βs and BMPs.

A "left-hand" model of BMP-2 is depicted in Figure 6. Two β-sheets each composed of two interrupted β-strands represent two "fingers". Finger 1 has at its tip a large loop L1. Finger 2 has in the middle a crossing of the two strands. The central α-helix α3 represents the "palm" and is inserted between strands β5 and β6. The cystine-knot is formed by DSB1, 2, and 3 at the base of the fingers. (In the TGF-βs, the N-terminal peptide folds as a helix and is disulfid-bonded to a cysteine at the start of strand β1 from the same monomer.) The dimeric protein is assembled from the two monomers in such a way that the left hand is rotated around a two-fold axis perpendicular to the β-strands and in plane of the β-sheets, so that the N-terminal ends are oriented to the same side of the protein. In most members of the TGF-β family, there is an extra disulfide bond (DSB4) connecting the two monomers (see Fig. 2).

Differences between the backbone fold of the BMPs and TGF-βs exist especially in finger loops L1 and L4, in the orientation of the central α3 helix, and most pronouncedly in the pre-helix loop L2 and in the N-terminal segment. As described below the finger loops L1 and L4 of BMP-2 are only peripherally involved in receptor binding. However, the α3 helix and the pre-helix loop L2 occur at the center of the epitope for type I receptor interaction. The α-helical N-terminus of the TGF-βs is disulfide-bonded to the cystine-knot *via* a special cysteine at the start of strand β1 and may be important for receptor binding in this group of ligands.

The crystal structures of both BMP-2 and TGF-β3 show a bound organic molecule located in the hydrophobic finger-helix cavity. The pentandiol in BMP-2 establishes also a hydrogen bond to N59. The tetrahydrofuran in TGF-β3 forms a hydrogen bond to W28. Remarkably, in the BMP-2/BRIAec complex the phenyl ring of receptor F85 occupies these cavities.

In the three-dimensional structures of BMP-2 and BMP-7, high temperature factors are found for the backbone atoms of finger loops L1 and L4 suggesting a high mobility of these segments. The pre-helix loop L2 seems to be mobile in the free

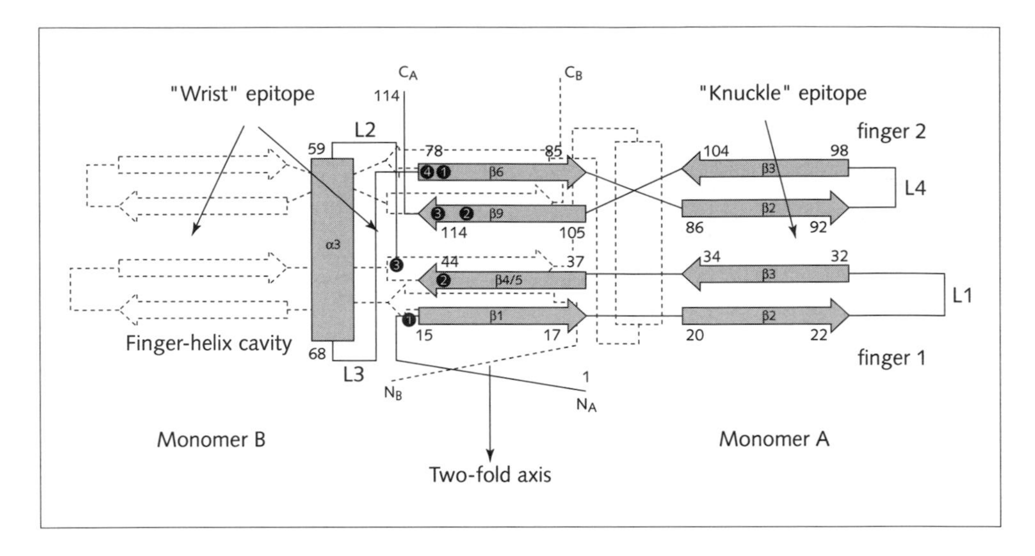


*Figure 6*
*Diagram of the "hand-like" BMP-2 structure. The folding of monomer A of BMP-2 is drawn with shaded β-strands β1 to β9 and helix α3. Monomer B is drawn with broken lines. N- and C-terminal ends are indicated for monomers A and B. Loops L1 to L4, disulfide bonds DSB1 to DSB4 (white numbers in black circles), amino acid positions, and fingers 1 and 2 are marked for monomer A only. One of the "wrist" epitopes for $BRIA_{ec}$ binding (comprising the "finger-helix cavities") and one of the "knuckle" epitopes for $BRII_{ec}$ binding are also indicated.*

BMP-2, but not in the BMP-2/$BRIA_{ec}$ complex (see below). Whereas L2 is involved in receptor binding, this mobility most likely influences the binding affinity and possibly determines specificity for different BMP ligands.

## Three-dimensional structure of ARII and BRIA

To date, a crystal structure of the free ectodomain of the ARII receptor $ARII_{ec}$ [14] and the complex between the ectodomain of the BRIA receptor $BRIA_{ec}$ and BMP-2 [15] have been elucidated. $ARII_{ec}$ has been expressed in *P. pastoris* and enzymatically deglycosylated. The BRIAec protein was expressed in *E. coli*.

The $ARII_{ec}$ protein consists of seven β-strands, that form a two-stranded (β1 and β2), a three-stranded (β4, β3 and β6), and another two-stranded (β5 and β7) β-sheet as depicted in the diagram in Figure 7. Strands β1/β2, β3/β4 as well as β5/β6 and their interconnecting loops α1/L1, L3, L5 represent three finger-like structures, similar as the three-finger-toxin fold of neuro- and cardiotoxins and fasciculin. The pro-

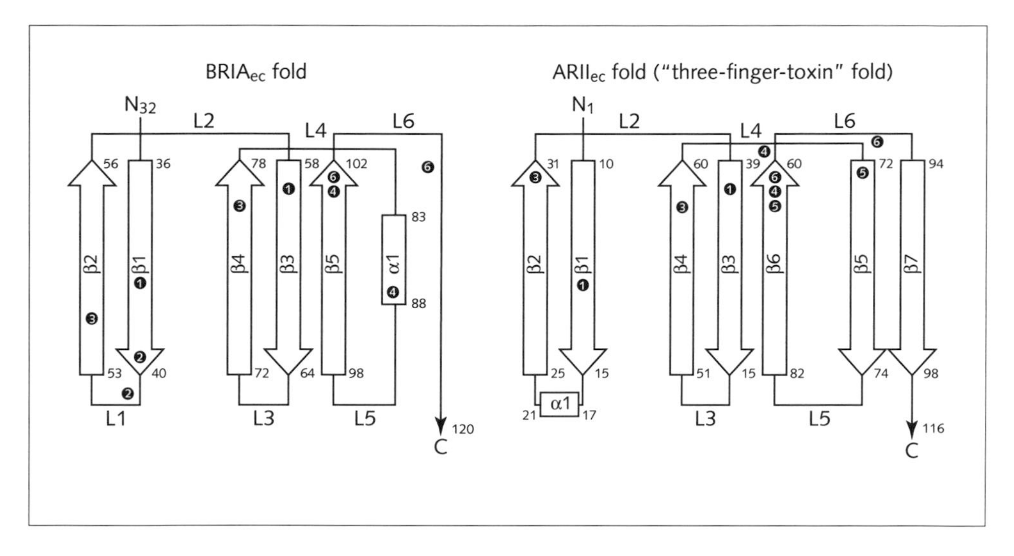


*Figure 7*
*Diagram of the $BRIA_{ec}$ fold and the $ARII_{ec}$ fold ("three-finger-toxin" fold). The β-strands β1 to β5 of BRIA and β1 to β7 of ARII, the α helix, the loops L1 to L6, as well as numbers indicating amino acid positions are shown for both receptor proteins. Disulfide bonds DSB1, 2, 3 ,4 , and 6 for $BRIA_{ec}$ and DSB1, 3, 4, 5, and 6 for $ARII_{ec}$ are depicted as white numbers in black circles.*

tein is stabilized by five DSBs that are conserved in all type II receptors. Three DSBs (DSB1, 3 and 6) are present at comparable positions in $BRIA_{ec}$ (see Fig. 4). The binding epitope of $ARII_{ec}$ is not completely known. Mutational analysis indicates that binding affinity for activin A and inhibin A is disrupted after substituting F42, W60 or F83 by alanine [22].

The three-dimensional structure of the ectodomain of the type I receptor BRIA ($BRIA_{ec}$) was deduced from the crystal structure of the $BRIA_{ec}$ /BMP-2 complex. The $BRIA_{ec}$ fold shows five β-strands that form a two-stranded (β1 and β2) and a three-stranded (β4, β3, and β5) sheet. The DSB 1, 3 and 6 and the backbone of the two β-sheets are similar to the corresponding regions of the type II receptor ARII. These elements of $BRIA_{ec}$ can be superimposed to those of $ARII_{ec}$ [15].

The discriminating element between the type I and type II ectodomain is (1) the long over-hand segment connecting strands β4 and β5, (2) the attachment of the C-terminal peptide, and (3) the orientation and structure of loop L1. In $BRIA_{ec}$ the long loop adopts an α-helical structure that is fixed by DSB4 to β5 at the border of the concave side of the protein (the finger 3 is not present); the C-terminal segment forms an extended peptide that runs from top to bottom over the convex back of the protein; The β1/β2 loop L1 is linked by a type I specific DSB to the sheet. In

$ARII_{ec}$ the long over-hand segment between loops L4 and L5 forms a new β5 strand that together with a C-terminal β7 strand forms a small β-sheet. A reorientation of DSB4 and a new DSB5 fasten this two-stranded β-sheet at the back of the protein. The loop L1 forms a short α helix.

The $BRIA_{ec}$ fold exits probably also in other type I ectodomains, since the type I specific DSBs and other sequence features are found in all subtypes. The three-finger-toxin fold of $ARII_{ec}$ seems to occur in further type II receptors, as deduced from the common DSB pattern and sequence similarities. Thus it is tempting to discuss the binding epitopes established for $BRIA_{ec}$ and $ARII_{ec}$ in the context of the receptor subtype families.

## The ligand-binding epitope of BRIA

The 24 contact residues of BRIA for the BMP-2 ligand are located on the concave or "palm" side of the hand-like receptor protein (Fig. 8). These residues constitute three binding clusters: (1) a "groove", (2) a "block" and (3) a "knob". The hydrophobic bottom of the groove is constituted by side chains and backbone elements of strands β4 (C77, M78), β3 (F60, I62), and β5 (I99). The left wall of the groove formed by loop L1 (D46, P45, H43) and the right wall formed by β2 (T55), loop L4 (K79, E81) and helix α1 (Q86) are assembled predominantly by polar and charged side chains. The groove is open at the upper end but closed at the lower end by a block formed by the side chains of loop L3 (E64) and of loop L5 (R97, D89, Q94, A93, S90, and K92). In the middle of the right wall the hydrophobic side chain of F85 from the α1 helix protrudes like a knob. It is encircled by the charged or polar side chains of Q86, E81, D84, K88, and D89, derived from helix α1 and the adjoining loop regions.

In the complex with BMP-2, the percentual loss of accessible surface area (Fig. 9A) is above 80% for the central "groove" residues T55, F60, I62, C77, and M78, as well as for "knob" residues F85, Q86, and S90. Among these residues hydrophobic side chains predominate. More than 50% of the accessible surface is buried in the complex for the more peripheral residues H43, P45, K79, G82, D84, D89, R97 and I99. These residues to the most part have charged side chains.

## The BMP-2 epitope 1 for $BRIA_{ec}$ binding ("wrist epitope")

The complementary epitope of BMP-2 for BRIA binding comprises 24 residues. This so-called "wrist epitope" is constituted by both monomers [23]. Monomers A and B contribute 16, and 8 residues to this epitope respectively (see Fig. 6). Three binding clusters can be discriminated, (1) a hydrophobic "hole" (corresponding to the "finger-helix cavity" in [9]), (2) a "rim", and (3) an extended "pre-helix loop". The

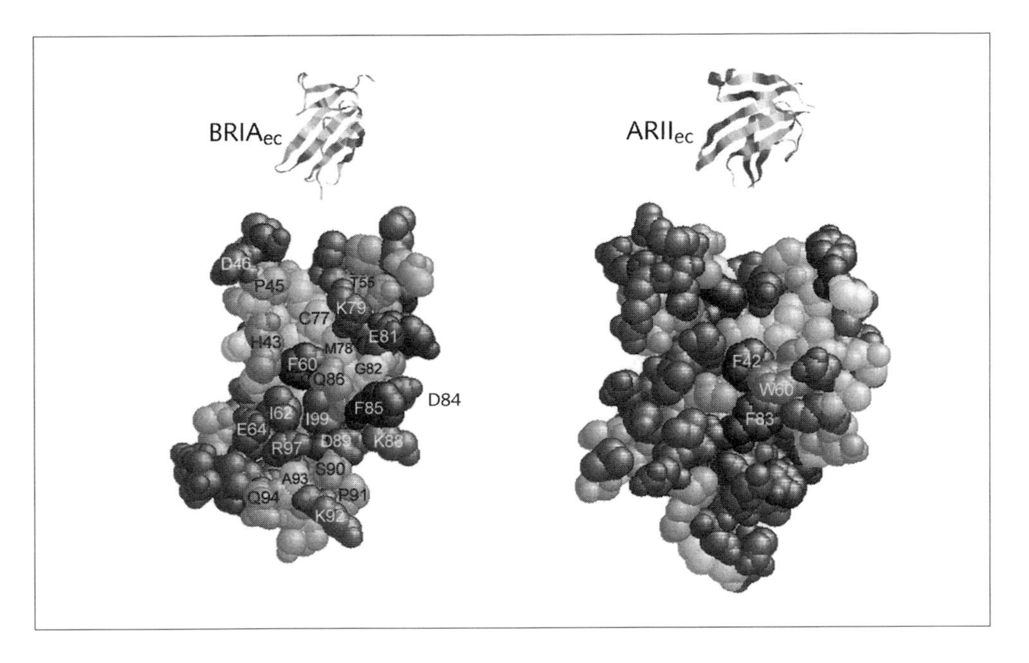


*Figure 8*
*BRIA residues buried in the BMP-2 contact and ARII side chains involved in activin/inhibin binding. Space-filling models of both proteins were drawn with RasMol. The 24 residues of $BRIA_{ec}$ that bury more than 15% of their accessible surface area in the complex with BMP-2 are indicated. The three functional residues of $ARII_{ec}$ were identified by mutational analysis. Both proteins are shown in roughly the same orientation as indicated by the two ribbon models in the upper part. Recently, alanine scanning mutational analysis of $BRIA_{ec}$ has demonstrated [32] that binding affinity for BMP-4 was reduced about 15-fold in receptor F85A and I62A variants.*

hydrophobic hole contains at its bottom $M89_B$ (β7), $M106_B$ (β8) and $V63_A$ (α3). The walls of the hole are formed by $W28_B$ and $W31_B$ in loop L1, $K101_B$ and $Y103_B$ in β8 as well as $N59_A$, $I62_A$ and $L66_A$ in helix α3. The extended pre-helix loop (L2) comprises residues $P48_A$, $F49_A$, $P50_A$, $L51_A$, $A52_A$, $D53_A$, $H54_A$ and $N56_A$. (3) Peripheral receptor contacts are also established at the lower rim of BMP-2 by residues $V26_B$ and $G27_B$ in loop L1 as well as $S69_A$ and $V70_A$ in loop L3 ($K15_A$ in β1 forms an ion pair with D46 in receptor L1).

Many residues of BMP-2 bury more than 80% of their accessible surface in the contact with $BRIA_{ec}$. Among these are constituents of the hydrophobic hole (W28, W31, N59, I62, V63, L66, M89, Y103, and M106), as well as residues of the extended pre-helix loop (F49, P50, L51, A52, and D53). The "rim" residues V26, G27, S69, and V70 bury 60 to 80% of their surface area in the complex.

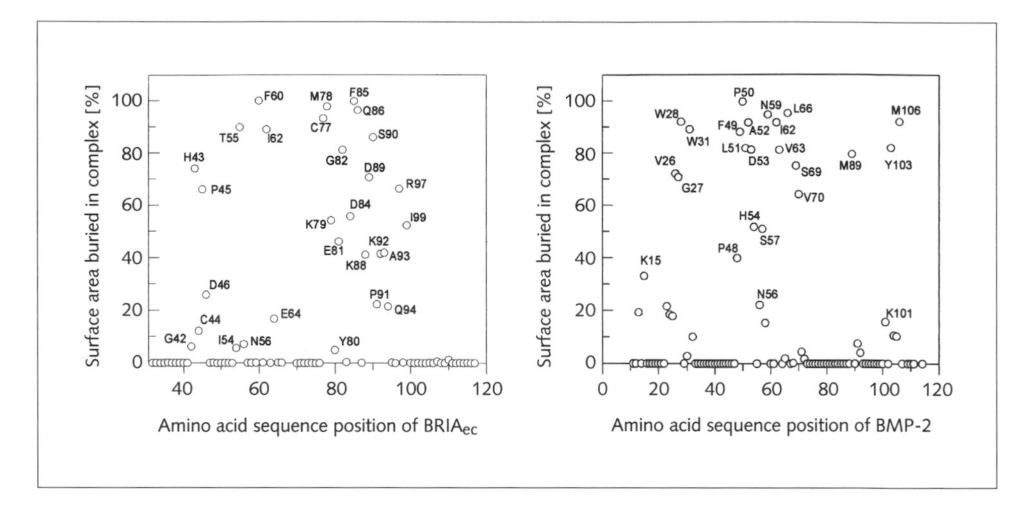


*Figure 9*

*The contact residues of BRIA$_{ec}$ and BMP-2. The accessible surface areas buried in the complex were calculated as percent of the accessible surface areas of the free proteins.*

## Hydrogen bonds in the BMP-2/BRIA$_{ec}$ contact

In the BMP2/BRIA$_{ec}$ contact 11 hydrogen bonds can be identified (Fig. 10). Four hydrogen bonds encircle the "knob-into the hole" element. These are formed between BMP-2 N59 and recE81, Y103 and recD84, W28 and recD89 as well as L51and recQ86. Remarkably, the receptor provides all acceptor atoms and BMP-2 all donator atoms for these hydrogen bonds (see table in Fig. 10). Four bonds occur in the "groove-loop" contact between BMP-2 D53 and recT55, D53 and recC77, D53 and recK79, as well as H54 and recH43. Three bonds are found in the "block-rim" cluster between BMP-2 V26 and recS90, S69 and recQ94, as well as S69 and recR97. Four main chain atoms of BRIA$_{ec}$ and four of BMP-2 are engaged in these hydrogen bonds. Two of them are main chain/main chain bonds (recC77/D53 and recS90/V26). An ionic interaction exits only for recK79 and D53.

## Mutational analysis of the BMP-2 epitope 1 ("wrist epitope")

A systematic mutational analysis of BMP-2 employing alanine and charged side-chain substitutions yielded a functional epitope for BRIA$_{ec}$ binding [23]. The location of this functional epitope is in agreement with the structural epitope identified by X-ray analysis described above. Pre-helix loop side chains of F49, P50, A52 and H54 were identified as binding determinants, as well as side chains of W31, I62,

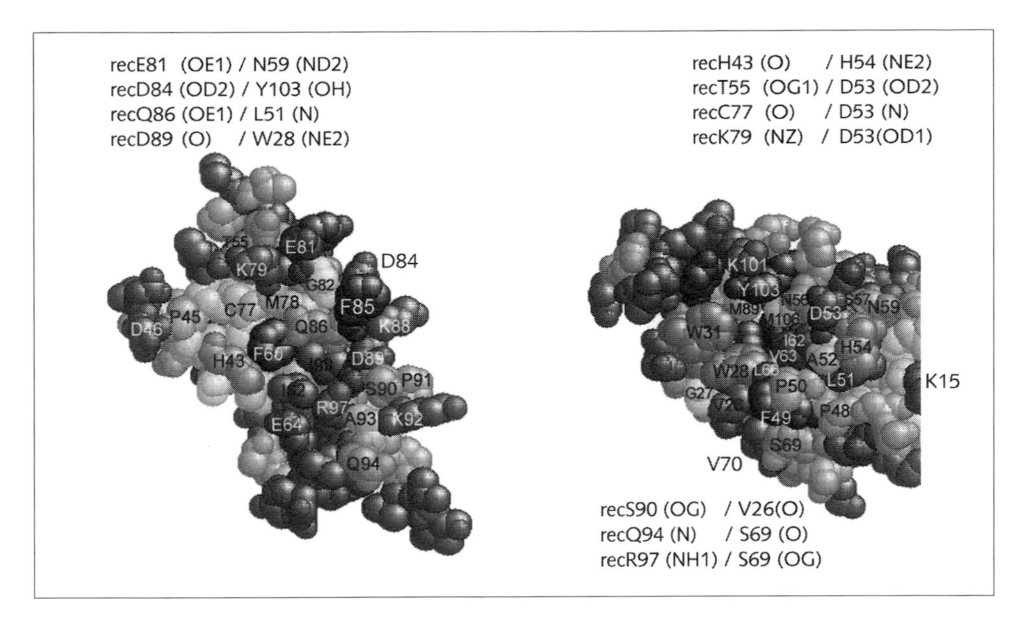


*Figure 10*
*The BMP-2/BRIA$_{ec}$ complex: contact residues and hydrogen bonds. The open-book view of the space-filling models depicts the location of the contact residues in BRIA$_{ec}$ and BMP-2 (only one half of the BMP-2 molecule is shown). The hydrogen bonds in the "knob/hole"-, the "peptide/groove"-, and the "bank/rim"-cluster are drawn around the model, with first the receptor atom and after the diagonal stroke the BMP-2 atom.*

L66, K101 and Y103 from the hydrophobic hole. S69 and V26 when substituted also yielded BMP-2 variants with reduced binding affinity. Thus, all three binding clusters of interacting residues contribute to the binding affinity.

A more detailed inspection, however, shows several interesting and unexpected properties of the mutant BMP-2s. (1) No hot spot of binding energy, i.e. no main binding determinants leading to a massive decrease in affinity after substitution, could be identified. The decrease in binding affinity was at most 10-fold after alanine substitutions and at most 30-fold after charged side chain substitutions. This could simply indicate that the most informative mutants were not analyzed; for example, the W28A BMP-2 could not be refolded and isolated. It is a distinct possibility, however, that hydrogen bonds involving main chain atoms are of functional importance. The contribution of such interactions to the binding free energy cannot be directly addressed by a mutational analysis. The small increase of affinity, however, in the BMP-2 D53A mutant protein might be caused by a stabilization of the recC77(O)/D53(N) hydrogen bond. (2) The mutant BMP-2s affected at V26, F49, P50, and H53 (and also at K101 and Y103) exhibit a slow-down of the association rate constant for the complex formation with $BRIA_{ec}$. This property is highly unusual, since in many mutant proteins studied so far the dissociation rate constant has been found to be increased. The on-rate effect may be related to the observation, that the pre-helix loop where several of these proteins are altered shows high temperature factors of the backbone atoms. This suggests an increased structural flexibility of this segment already in the wild type protein that may influence the probability of a productive complex formation. Amino acid substitutions in or near to this segment might further enhance the flexibility.

The BMP-2 mutants altered in side chains of the $BRIA_{ec}$ contact exhibited altered biological activity. They showed increased ED50 values during a C2C12 cell-based alkaline phophatase induction assay. Mutant proteins with decreased affinity for the $BRIA_{ec}$ were active only at higher concentrations. As described below, several mutant BMP-2s substituted in the epitope for type II receptor binding ($BRII_{ec}$, $ARII_{ec}$) showed a different phenotype. They were BMP-2 antagonists with partial or no agonist activities and competed with wild type BMP-2 for $BRIA_{ec}$ binding.

## BMP-2 epitope 2 for type II receptor binding ("knuckle epitope")

BMP-2 contains a second epitope for type II receptor binding [23]. Epitope 2 has a low affinity for receptor interaction. With immobilized $BRII_{ec}$ equilibrium binding of BMP-2 with a dissociation constant Kd of 100 nM is measured on a Biacore system. All BMP-2 mutants with a specific alteration in $BRII_{ec}$ binding have substitutions in side-chains clustering together at the back of fingers 1 and 2 of one monomer ("knuckle" epitope). The BMP-2 mutant A34D showed the highest decrease in $BRII_{ec}$ binding (50-fold) and in parallel undetectable levels (more than 100-fold reduction) of biological activity in the C2C12 ALP induction assay. The A34D mutant inhibited the activity of wild type BMP-2 with an $IC_{50}$ (20–50 nM) similar to the $ED_{50}$ (10–20 nM) of BMP-2 in this assay. This indicates that A34D represents a complete high-affinity antagonist. Other substitutions in epitope 2 yielded partial agonistic/antagonistic proteins. This finding supports the view that BMP-2 binds *in situ* with high affinity to the type I receptor (BRIA), and that the type II receptor chain (BRII or ARII) subsequently associates with the low affinity site of the bound BMP-2 in the membrane.

## Cooperation of multiple binding epitopes and receptor chains

Each of the BMP-2 knuckle epitopes is localized close to one of the wrist epitopes. Thus, trans-activation of the cytoplasmic parts of the receptor chains should be possible. Another symmetry-related pair of the two epitopes is localized some distance apart at the other pole of BMP-2, and it is unknown in how far the multiple epitopes cooperate. This is especially interesting since homodimeric type I and homodimeric type II receptor chains have been identified in whole cells [24]. Possibly, two BMP-2 proteins are cross-linked by one homodimeric receptor chain. This could lead to higher aggregates of BMP-2-activated receptor complexes. It is also unclear how multiple binding sites might influence the affinity of the receptor for BMP-2; we do not know if the affinity measured between BMP-2 and the receptor ectodomain on the biosensor results from a 1:1 or a 1:2 interaction. This question is relevant for *in situ* ligand binding, since cooperation of two low-affinity sites could result in a high affinity binding.

## Topology of the BMP-2 receptor complex in the membrane

The ectodomain $BRIA_{ec}$ of type I receptor BRIA is small. The binding domain consists of only 89 amino acid residues and it is connected to the transmembrane domain by a short 9-residue peptide segment. The ectodomain of the type II receptor BRII is slightly larger than $BRIA_{ec}$. The binding domain of $BRII_{ec}$, according to the sequence alignment in Figure 4, has 105 residues and the connecting segment has 15 residues. The result is that the binding domains of both type I and type II receptors are located near to the membrane surface with the connecting peptides allowing some freedom of mobility. Considering the three-dimensional structure of the BMP-2/$BRIA_{ec}$ complex as well as the topology of the two pairs of binding epitopes, the BMP-2 ligand seems to be bound to the receptor chains with the twofold axis of rotation perpendicular to the plane of the membrane, as depicted in Figure 11.

If four receptor chains are attached to the BMP-2 protein, it is likely that the cytoplasmic domains of the type I and type II receptors interact due to the juxtaposition of the extracellular binding epitopes. An interaction of the cytoplasmic domains of chains from different pairs seems to be topologically possible only for heterotypic interaction, but not simultaneously for the two possible homodimers. These considerations are intriguing considering that the TGF-β receptors exist in the membrane as stable homodimers [25–27].

## Affinity and topology of type I and type II receptor binding

We would like to propose that in all TGF-β like proteins epitope 1 ("wrist" epitope) binds only type I receptors irrespective of whether it is a high-affinity BMP/GDF (BRIA, BRIB) receptor or a low-affinity TGF-β/activin receptor chain (TRI, ARI, ARIB). This implicates that epitope 2 ("knuckle" epitope) binds only type II receptor chains irrespective of whether this is a low-affinity BMP/GDF receptor chain

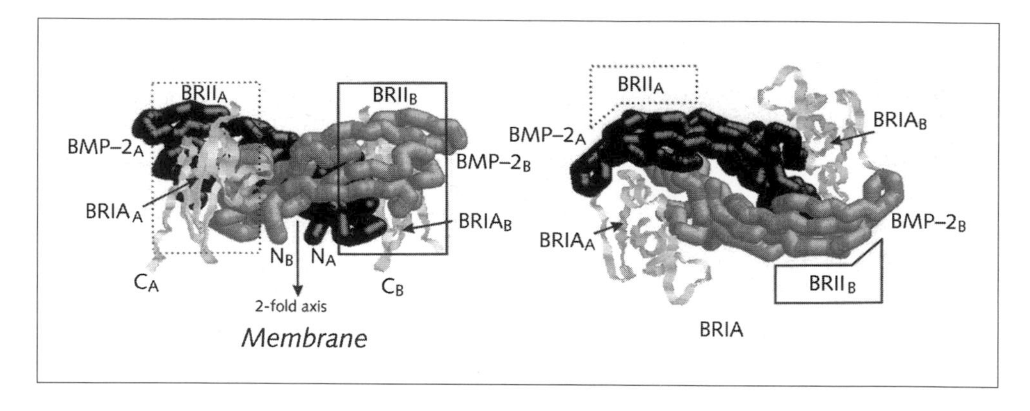


*Figure 11*
*The topology of the BMP-2/BRIA$_{ec}$ complex in relation to the membrane. The ribbon diagram of the complex on the left side is drawn with the two-fold axis perpendicular to the membrane. On the right side the top view of the complex is shown. The N-terminal ends of both monomers of BMP-2 and the C-terminal end of BRIA$_{ec}$ are indicated. The probable location of two BRII$_{ec}$ as deduced form mutational analysis of the "knuckle" epitope is depicted as rectangles.*

(BRII, ARII, ARIIB) or a high-affinity TGF-β/activin receptor chain (TRII, ARII, ARIIB). This hypothesis rests on the following observations:

The binding of BMP-2 to the BRIA receptor seems to be critically dependent on the presence of the receptor helix α1 with the protruding F85 knob fitting into the hole of epitope 1 of BMP-2. This helix including the knob motif is not present in the type II receptor ARII as discussed above. An α1 helix with a phenylalanine or another large hydrophobic side chain, however, seems to be present at the knob position in all type I receptors (with the exception of Alk-1).

## TGF-β like proteins: similarities in the binding epitopes

The wrist epitope (epitope 1) shows a remarkable high similarity of side chains in the "hydrophobic-hole" motif in many subgroups of the TGF-β family (Fig. 12). The two tryptophanes in the large finger 1 loop are a hallmark of the whole family with the exception of the most distant members GDNF and MIS. At the positions corresponding to BMP-2 V63, M89 and M106, large hydrophobic side chains (V, L, I, M) always occur. Other relevant positions are occupied by identical or isofunctional side chains in most members. Exceptions in the TGF-βs are the positions 62, 101, 103. Nevertheless, this binding motif seems to be functioning throughout the TGF-β family.

| Position | 15 | 26 | 27 | 28 | 31 | 48 | 49 | 50 | 51 | 52 | 53 | 54 | 56 | 57 | 59 | 62 | 63 | 66 | 69 | 70 | 89 | 101 | 103 | 106 |
|---|---|---|---|---|---|---|---|---|---|---|---|---|---|---|---|---|---|---|---|---|---|---|---|---|
| Monomer | A | B | B | B | B | A | A | A | A | A | A | A | A | A | A | A | A | A | A | A | B | B | B | B |
| Cluster | nd | r | r | h | h | p | p | p | p | p | p | p | p | p | p | h | h | h | r | r | h | h | h | h |
| WBB | | X | | X | | | | | X | | X | X | | | X | | | | X | | | | X | |
| | | 0 | | NE2 | | | | | N | | N | NE2 | | | ND2 | | | | 0 | | | | OH | |
| | | | | | | | | | | | 0D1 | | | | | | | | OG | | | | | |
| | | | | | | | | | | | 0D2 | | | | | | | | | | | | | |
| | | | | | | | | | | | | | | | | | | | | | | | | |
| hBMP-2 | K | V | G | W | W | P | F | P | L | A | D | H | N | S | N | I | V | L | S | V | M | K | Y | M |
| hBMP-4 | R | V | G | W | W | P | F | P | L | A | D | H | N | S | N | I | V | L | S | V | M | K | Y | M |
| | | | | | | | | | | | | | | | | | | | | | | | | |
| hBMP-5 | K | L | G | W | W | S | F | P | L | N | A | H | N | A | N | I | V | L | L | M | V | K | Y | M |
| hBMP-6 | R | L | G | W | W | S | F | P | L | N | A | H | N | A | N | I | V | L | L | M | V | K | Y | M |
| hBMP-7 | K | L | G | W | W | A | F | P | L | N | S | Y | N | A | N | I | V | L | F | I | V | K | Y | M |
| | | | | | | | | | | | | | | | | | | | | | | | | |
| hGDF-5 | S | M | G | W | W | E | F | P | L | R | S | H | E | P | N | V | I | L | S | M | I | K | Y | M |
| mGDF-6 | S | L | G | W | W | D | F | P | L | R | S | H | E | P | N | I | I | L | S | M | I | K | Y | M |
| mGDF-7 | S | L | G | W | W | S | F | P | L | R | S | H | E | P | N | I | I | L | S | M | I | K | Y | M |
| | | | | | | | | | | | | | | | | | | | | | | | | |
| hTGF-β1 | C | L | G | W | W | P | Y | I | W | S | | | L | D | Q | K | V | L | Q | H | I | Q | L | M |
| hTGF-β2 | C | L | G | W | W | P | Y | L | W | S | | | S | D | Q | R | V | L | T | I | I | Q | L | M |
| hTGF-β3 | C | L | G | W | W | P | Y | L | R | S | | | A | D | T | T | V | L | T | L | I | Q | L | M |

*Figure 12*
*Alignment of putative contact residues for type I receptor binding in the BMP-2s, BMP-7s, GDF-5s and TGF-βs. The positions of the 24 residues identified in BMP-2 as well as their occurrence in monomers A and B, in "hole"-, "peptide"-, or "rim"-clusters, as well as contributions to hydrogen bonds are shown in the upper part. Shaded areas indicate positions occupied by non-identical or non-isofunctional residues.*

The second binding element, the "pre-helix loop," varies considerably in length between the subfamilies. In addition, differences in charge, polarity or size are found at six out of eight positions. Remarkably, the residues at positions 53 and 54 of BMP-2 that form hydrogen bonds to the receptor seem to be deleted in the TGF-βs. Only the positions corresponding to BMP-2 F49 and P50 show similarities. This suggests that the "pre-helix loop" and, possibly, also the "rim" element determine the specificity and affinity of interaction with the type I receptor. It remains to be established if these elements contribute to binding in the postulated low-affinity interaction between, for example TGF-β2 and TRI.

On the putative receptor epitopes, BRIA F85 has an identical counterpart at the corresponding positions of BRIB, ARIB and TRI or residues with a large hydrophobic side chain (V, M) in ARI and Alk-7. Thus, a "hydrophobic knob" seems to be present in these type I receptors. No helix and, accordingly, no knob-motif can be discriminated at the type II receptor ARII and probably not on other type II receptor proteins (see Figs. 7 and 8). Other side chains of the contact are not clearly similar among all the type I proteins. However, in the BRIA/BRIB subgroup the epitopes are occupied by identical or isofunctional residues at 83% (20/24) of the positions. This is higher than in the rest of the protein (63.6% = 56/88), indicating the promiscuity of the BMP receptors in BMP-2 ligand binding. Differences at some positions (see, for example, G/K79, H/Q94) may be related to the observations that BMP-2 binds to $BRIA_{ec}$ with a ca 10-fold higher affinity than to $BRIB_{ec}$, and that BMP-7 and GDF-5 bind preferentially to BRIB.

Remarkably, cross-linking experiments in transfected COS cells have shown that BMP-7 binds to ARI and BRIB, whereas GDF-5 binds preferentially BRIB and BMP-2s bind preferentially BRIA [28–31]. ARI has a putative epitope containing only 33% (8/24) identical or isofunctional positions compared to those of BRIA or BRIB. The ligand-receptor specificities identified by cross-linking experiments, however, have not always been demonstrated to be functionally relevant. How such a low similarity of binding epitopes could be compatible with the reported binding of a common ligand represents an intriguing problem.

Type II receptor ectodomains represent a separate group of proteins whose binding domains seem to be slightly larger (95–106 residues) than the ectodomain of the type I receptors (79–89 residues). For none of the type II receptors a ligand contact has been structurally defined. For ARII functional residues F42, W60 and F83 have been established by mutational analysis [22], corresponding to BRIA positions of F60, M78 and I99. As seen in the contact alignment (Fig. 13) identical or isofunctional residues occur at the corresponding positions of ARIIB and BRII, but not those of TRII and AMHRII. In line with these similarities all three type II receptors, BRII, ARII and ARIIB, have been found to function as low-affinity chains for BMP-2s, BMP-7s and GDF-5s. Interestingly, the ARII and ARIIB receptors function additionally as high-affinity chains for the activins.

## Reading the binding code

The final goal of the structural and functional analysis of ligand-receptor interactions in the TGF-β family is the understanding of specificity and affinity of binding. In principle, it might be feasible to read the binding code for these interactions. Toward this end more three-dimensional structures of ligand-receptor complexes (structural epitopes) as well as a mutational/interaction analysis of more epitopes (functional epitopes) have to be executed.

Of special interest is the question of how and where the high affinity epitope of the TGF-βs, activins, GDF-8s and others is assembled. In light of our present knowledge it seems possible that the knuckle epitope is converted into a high-affinity epitope by the acquisition of binding residues at the N-terminal helix immobilized *via*

| Position | 43 | 45 | 46 | 55 | 60 | 62 | 64 | 77 | 78 | 79 | 81 | 82 | 84 | 85 | 86 | 88 | 89 | 90 | 91 | 92 | 93 | 94 | 97 | 99 |
|---|---|---|---|---|---|---|---|---|---|---|---|---|---|---|---|---|---|---|---|---|---|---|---|---|
| Cluster | g | g | g | g | g | g | b | g | g | g | K | K | K | K | K/g | K | K | b | b | b | b | b | b | g |
| WBB | X | | | X | | | | X | | X | X | | X | | X | | X | X | | | | X | X | |
| | O | | | OG1 | | | | O | | NZ | OE1 | | OD2 | | OE1 | | O | OG | | | | N | NH1 | |
| | | | | | | | | | | | | | | | | | | | | | | | | |
| Alk-1 | H | | | G | T | V | V | C | G | N | | L | R | E | L | R | G | R | P | | T | E | V | H |
| TR-I | L | | | T | F | S | T | C | I | A | R | D | P | F | V | A | P | S | S | K | T | G | T | T |
| AR-IB | S | | | T | M | S | F | C | I | P | A | G | P | F | Y | L | S | | | | S | E | R | T |
| Alk-7 | L | | | T | W | S | M | C | V | S | L | N | Q | V | F | H | S | | | | S | N | T | T |
| AR-I | S | | | G | F | S | S | C | F | Q | E | Q | K | M | T | K | T | P | P | | S | P | Q | V |
| BR-IB | H | P | E | T | F | M | E | C | L | G | E | G | D | F | Q | R | D | T | P | I | P | H | R | I |
| BR-IA | H | P | D | T | F | I | E | C | M | K | E | G | D | F | Q | K | D | S | P | K | A | Q | R | I |
| AR-II | N | E | K | C | f | T | K | C | W | L | D | I | C | Y | D | C | V | E | K | K | D | S | V | F |
| AR-IIB | N | E | L | C | Y | S | A | C | W | L | D | F | C | Y | D | C | V | A | T | E | E | N | V | F |
| BR-II | Y | Q | D | E | Y | L | E | C | W | G | P | Q | C | H | | C | V | V | T | T | T | P | Y | F |
| AMHR-II | P | V | R | D | F | I | N | C | R | S | E | P | C | E | | C | D | P | S | P | A | H | L | T |
| TR-II | R | S | T | C | V | V | R | C | H | Y | D | F | L | E | D | C | I | M | K | E | K | K | F | M |

*Figure 13*
*Alignment of putative contact residues for ligand binding in type I and type II receptor ectodomains. The positions of the 24 contact residues for BMP-2 identified in $BRIA_{ec}$ as well as their occurrence in the "groove"-, "knob"-, or "block"-cluster, and their participation in hydrogen bonds are shown in the upper part. The structure based sequence alignment of the sequences of $BRIA_{ec}$ and $ARII_{ec}$ was performed as in Figure 4.*

a specific DSB at the border of the knuckle epitope. The wrist epitope can be visualized as converted into a low-affinity epitope by a decrease in binding affinity of the highly divergent pre-helix loop segment.

Finally, it has to be established to what extent the multiple binding epitopes at the BMP surface cooperate, or if they are independent.

## *Acknowledgement*
This work was supported by the Deutsche Forschungsgemeinschaft (SFB487, TPB1 und TPB2) and by the Fond der Chemischen Industrie. The authors thank Christian Söder and Alexandra Will for excellent technical assistance and Petra Knaus and Thomas Kirsch for helpful discussions.

## References

1 Hogan BL (1996) Bone morphogenetic proteins in development. *Curr Opin Genet Dev* 6: 432–438

2 Reddi AH (1998) Role of morphogenetic proteins in skeletal tissue engineering and regeneration. *Nat Biotechnol* 16: 247–252
3 Heldin CH, Miyazono K, ten Dijke P (1997) TGF-beta signalling from cell membrane to nucleus through SMAD proteins. *Nature* 390: 465–471
4 Massague J (1998) TGF-beta signal transduction. *Annu Rev Biochem* 67: 753–791
5 Schlunegger MP, Grutter MG (1992) An unusual feature revealed by the crystal structure at 2.2 A resolution of human transforming growth factor-beta 2. *Nature* 358: 430–434
6 Daopin S, Piez KA, Ogawa Y, Davies DR (1992) Crystal structure of transforming growth factor-beta 2: an unusual fold for the superfamily [see comments]. *Science* 257: 369–373
7 Mittl PR, Priestle JP, Cox DA, McMaster G, Cerletti N, Grutter MG (1996) The crystal structure of TGF-beta 3 and comparison to TGF-beta 2: implications for receptor binding. *Protein Sci* 5: 1261–1271
8 Griffith DL, Keck PC, Sampath TK, Rueger DC, Carlson WD (1996) Three-dimensional structure of recombinant human osteogenic protein 1: structural paradigm for the transforming growth factor beta superfamily. *Proc Natl Acad Sci USA* 93: 878–883
9 Scheufler C, Sebald W, Hulsmeyer M (1999) Crystal structure of human bone morphogenetic protein-2 at 2.7 A resolution. *J Mol Biol* 287: 103–115
10 Piek E, Heldin CH, Ten Dijke P (1999) Specificity, diversity, and regulation in TGF-beta superfamily signaling. *Faseb J* 13: 2105–2124
11 Reddi AH (1997) Bone morphogenetic proteins: an unconventional approach to isolation of first mammalian morphogens. *Cytokine Growth Factor Rev* 8, 11–20
12 Massague J, Chen YG (2000) Controlling TGF-beta signaling. *Genes Dev* 14, 627–644
13 Ruppert R, Hoffmann E, Sebald W (1996) Human bone morphogenetic protein 2 contains a heparin-binding site which modifies its biological activity. *Eur J Biochem* 237: 295–302
14 Greenwald J, Fischer, WH, Vale WW, Choe S (1999) Three-finger toxin fold for the extracellular ligand-binding domain of the type II activin receptor serine kinase. *Nat Struct Biol* 6: 18–22
15 Kirsch T, Sebald W, Dreyer MK (2000) Crystal structure of the BMP-2–BRIA ectodomain complex [see comments]. *Nat Struct Biol* 7: 492–496
16 Murray Rust J, McDonald NQ, Blundell TL, Hosang M, Oefner C, Winkler, Bradshaw RA (1993) Topological similarities in TGF-beta 2, PDGF-BB and NGF define a superfamily of polypeptide growth factors. *Structure* 1: 153–159
17 Sun PD, Davies DR (1995) The cystine-knot growth-factor superfamily. *Annu Rev Biophys Biomol Struct* 24: 269–291
18 Koenig BB, Cook JS, Wolsing DH, Ting J, Tiesman JP, Correa PE, Olson CA, Pecquet AL, Ventura F, Grant RA et al (1994) Characterization and cloning of a receptor for BMP-2 and BMP-4 from NIH 3T3 cells. *Mol Cell Biol* 14: 5961–5974
19 Andrades JA, Han B, Becerra J, Sorgente N, Hall FL, Nimni ME (1999) A recombinant human TGF-beta1 fusion protein with collagen-binding domain promotes migration,

growth, and differentiation of bone marrow mesenchymal cells. *Exp Cell Res* 250: 485–498

20 Greenwald J, Le V, Corrigan A, Fischer W, Komives E, Vale W, Choe S (1998) Characterization of the extracellular ligand-binding domain of the type II activin receptor. *Biochemistry* 37: 16711–16718

21 Eigenbrot C, Gerber N (1997) X-ray structure of glial cell-derived neurotrophic factor at 1.9 A resolution and implications for receptor binding [letter]. *Nat Struct Biol* 4, 435–438

22 Gray PC, Greenwald J, Blount AL, Kunitake KS, Donaldson CJ, Choe S, Vale W (2000) Identification of a binding site on the type II activin receptor for activin and inhibin. *J Biol Chem* 275: 3206–3212

23 Kirsch T, Nickel J, Sebald W (2000) BMP-2 antagonists emerge from alterations in the low-affinity binding epitope for receptor BMPR-II. *EMBO J* 19: 3314–3324

24 Gilboa L, Nohe A, Geissendorfer T, Sebald W, Henis YI, Knaus P (2000) Bone morphogenetic protein receptor complexes on the surface of live cells: a new oligomerization mode for serine/threonine kinase receptors. *Mol Biol Cell* 11: 1023–1035

25 Luo K, Lodish HF (1997) Positive and negative regulation of type II TGF-beta receptor signal transduction by autophosphorylation on multiple serine residues. *EMBO J* 16: 1970–1981

26 Weis Garcia F, Massague J (1996) Complementation between kinase-defective and activation-defective TGF-beta receptors reveals a novel form of receptor cooperativity essential for signaling. *EMBO J* 15: 276–289

27 Gilboa L, Wells RG, Lodish HF, Henis YI (1998) Oligomeric structure of type I and type II transforming growth factor beta receptors: homodimers form in the ER and persist at the plasma membrane. *J Cell Biol* 140: 767–777

28 ten Dijke P, Yamashita H, Sampath TK, Reddi AH, Estevez M, Riddle DL, Ichijo H, Heldin CH, Miyazono K (1994) Identification of type I receptors for osteogenic protein-1 and bone morphogenetic protein-4. *J Biol Chem* 269: 16985–16988

29 Liu F, Ventura F, Doody J, Massague J (1995) Human type II receptor for bone morphogenic proteins (BMPs): extension of the two-kinase receptor model to the BMPs. *Mol Cell Biol* 15: 3479–3486

30 Rosenzweig BL, Imamura T, Okadome T, Cox GN, Yamashita H, ten Dijke P, Heldin CH, Miyazono K (1995) Cloning and characterization of a human type II receptor for bone morphogenetic proteins. *Proc Natl Acad Sci USA* 92: 7632–7636

31 Nishitoh H, Ichijo H, Kimura M, Matsumoto T, Makishima F, Yamaguchi A, Yamashita H, Enomoto S, Miyazono K (1996) Identification of type I and type II serine/threonine kinase receptors for growth/differentiation factor-5. *J Biol Chem* 271: 21345–21352

32 Hatta T, Konishi H, Katoh E, Natsume T, Ueno N, Kobayashi Y, Yamazaki T (2000) Identification of the ligand-binding site of the BMP type IA receptor for BMP-4. *Biopolymers* 55: 399–406

# Biology of bone morphogenetic proteins

*Snjezana Martinovic*[1], *Fran Borovecki*[1], *Kuber T. Sampath*[2] *and Slobodan Vukicevic*[1]

[1]Department of Anatomy, School of Medicine, University of Zagreb, Zagreb 10 000 Croatia;
[2]Genzyme Corporation, One Mountain Road, Framingham, MA 01701-9322, USA

## TGF-β superfamily of proteins

Morphogens are signaling molecules that provide positional information to developing tissues and control conformation and histologic architecture of tissues by regulating specific gene expression. The morphogenic feature of BMPs was first described by Marshall Urist in 1965, when he discovered that demineralized bone matrix induced bone formation at extraskeletal sites [1] (see the chapter by Rueger). Since then, the molecules responsible for this phenomenon were isolated, cloned and identified as members of the TGF-β superfamily [2–9]. These signaling molecules were identified in many species, suggesting that they evolved from a group of ancestral genes with their functions refined to meet the needs of particular species. Among 17 proteins so far identified as BMPs, eight have been found to be involved in regulating bone formation and repair. The process of ectopic bone formation is similar to the endochondral bone formation seen during embryonic skeletal development, and the capability of forming new bone is shared by no other growth factor. BMPs are pleiotropic regulators that act at all the important steps in the cascade of events that form new bone: chemotaxis of progenitor cells, mitosis, differentiation and proliferation of chondrocytes and osteoblasts [9, 10]. BMPs also stimulate extracellular matrix formation [9, 11–16] and bind to specific matrix molecules [8, 17–19, 20–22] affecting bone remodeling. Many studies on the cellular activities of BMPs indicate that, as expected from their activities in animal systems, they essentially act as differentiation factors, causing induction and increased expression of multiple differentiated phenotypes in mesenchymal cells [23–30]. Besides skeleton, BMPs play a role in the development of other organ and tissue systems that form *via* mesenchymal-epithelial interactions and possibly function to deliver or interpret positional information in a wide variety of organisms [31–33].

The fundamental importance of BMPs can be inferred from the broad spectrum of species from which very similar BMP molecules have been isolated. The *Drosophila* BMP homologue, *decapentalegic* (*dpp*) protein [34], responsible for proper dorsal-ventral patterning of the early embryo is closely related to mam-

Bone Morphogenetic Proteins, edited by Slobodan Vukicevic and Kuber T. Sampath

malian BMP-2 and BMP-4 (75% homogeneity). Indeed, *Drosophila dpp* protein can induce cartilage and bone when implanted in mammals 700 million years distant [35], and mammalian BMP-4 can rescue defects caused by *dpp* mutations [36]. The function of Vg1, the factor whose messenger RNA is found in the vegetal hemisphere of the *Xenopus* oocyte, is less certain, although it is indicated it may also be involved in embryonic development [37]. Another subgroup represented by BMP-5, BMP-6, BMP-7 and the last discovered member of this subgroup, BMP-8 [38], is closely related to *60A*, a protein expressed in the early *Drosophila* embryo and responsible for the development of the gastrointestinal tract [39, 40]. BMP-3 itself represents another subgroup, although separated from the above mentioned two groups of proteins, it is the next most closely related TGF-β superfamily member. TGF-βs are clearly separated from the BMP family, namely TGF-β1, β2 and β3 in humans, TGF-β4 in chicken, and TGF-β5 in *Xenopus* show an average of only about 37% amino acid identity in the seven-cysteine region to BMP family members. Müllerian inhibiting substance and inhibin α are the most distantly related members of the TGF-β superfamily (see also the chapter by Paralkar et al.).

## BMPs in development

### *Xenopus laevis*: a model for exploring the developmental role of BMPs and BMP antagonists

Several critical observations regarding the role of BMPs and their secreted antagonists in early vertebrate embryogenesis have been made in the South African clawed frog, *Xenopus laevis* [31]. *Xenopus* BMP family members identified at early blastulae/gastrulae stage with different expression pattern are ADMP, BMP-2, BMP-4, BMP-7, Vg1 and GDF6 [41, 42]. BMP-4 has been shown as the major ventralizing and potent mesoderm-inducing signal during the gastrulation phase of *Xenopus* development [31, 43]. Prior to the onset of overt gastrulation, the molecular signals from the Spemann organizer including BMP-4 specify the dorso-ventral pattern of the early embryo in a very precise and dose-dependent manner.

Target genes that are transcriptionally regulated in response to BMP signaling in early *Xenopus* embryos are numerous and include transcription factors *xvent1*, *xvent2*, *xmsx1*, *mix1*, *xhox3*, *xfd1'* and *xmyf5*, signaling molecules *xSmad8* and *xwnt8*, as well as *xbmp4* itself [44–49]. Mostly, the expression of these genes is stimulated by BMP-4 and inhibited by BMP-4 inhibitors. Current evidence suggests that BMP-related molecules are required in organizer patterning, mesoderm induction in the marginal zone during blastula stages and subsequently, in specifying dorsal-ventral fates, repressing the development of dorsal tissues such as the neural tube, notochord and muscle [41, 47, 50]. The *Xenopus* embryo expresses a number of genes encoding BMP inhibitors, like chordin [51], noggin [52] and follistatin

[53], which act as dorsal de-repressors and also regulate cell fate during normal early development. These BMP antagonists are able to directly bind potent ventralizing factors, like BMP-2 and BMP-4, with high affinity and BMP-7 with low affinity, preventing association with their respective receptors, thereby rendering them inactive and establishing a morphogen gradient of BMP activity [54].

Overexpression studies in the early zebrafish embryo demonstrate that chordin and noggin have the same dorsalizing properties as their *Xenopus* homologues [55]. The null mutation in the zebrafish chordin gene disrupts the development of dorsal tissues, but noggin and follistatin are excluded from the zebrafish organizer [56]. Follistatin was originally identified because of its high affinity for activin, but it also has affinity for BMP-4 and BMP-7 [53, 57, 58]. Most recently, additional related proteins named cerberus, DAN, Gremlin and BAMBI have been shown to antagonize BMP signaling in *Xenopus* embryos [59, 60]. In striking contrast to noggin or chordin [51, 61], BAMBI is strictly coexpressed with BMP-4 during early *Xenopus* embryogenesis [60], thus being a member of the BMP-4 synexpression group. Besides *Xenopus*, the existence of an evolutionarily conserved BMP-4 synexpression group has been documented in mammals [62].

Overlapping expression and continuous presence of BMPs and their inhibitory proteins throughout *Xenopus* development suggest that similar mechanisms may exists at later developmental stages of *Xenopus* embryo. The existence of multiple inhibitory binding proteins in regulating BMP signaling has not been understood yet, but they can serve different functions within the BMP signaling pathway. For example, chordin may function as storage for BMP-4 since the proteolitically cleaved chordin has a low affinity for BMP-4, releasing the active BMP-4 [63]. On the other hand, follistatin may target BMP molecules for degradation, regulating their availability in cellular microenvironment [64], and may be required to clear activins and BMPs from the cellular environment. Interaction between BMPs and their binding proteins enables the inhibition of BMP signaling which has proved to be an important mechanism regulating cell fate decisions in early development. Because of the high conservation during evolution, these mechanisms probably influence the development of many other organisms. However, recent investigations suggest the possible involvement of new binding proteins providing a permissive signal that allows high BMP signaling in the embryo [65].

## BMPs as signals in organ development

During the development of multicellular organisms the formation of complex patterns relies on specific cell-cell signaling events. For tissues to become spatially organized and cells to become committed to specialized fates it is absolutely crucial for proper development that the underlying signaling systems receive and route information correctly. Recently, a wealth of genetic and biochemical experimental data has been collected about evolutionary conserved signaling families, such as the Dpp/BMPs, Wnts and hedgehogs, in flies, worms, and vertebrates. These signaling molecules form a crucial group of regulators of induction and patterning of embryonic germ layers in metazoa.

The BMP expression pattern as well as the analysis of spontaneously mutated or genetically depleted animals have demonstrated a much broader range of their function (see *Chromosomal localization and developmental function* in this chapter). These activities are mainly localized at sites of epithelial-mesenchymal interactions, including but not restricted to the skeleton [20, 66–74]. BMPs also influence the craniofacial development and initiation of tooth buds [75–80] and play a role in maintenance of vascular smooth muscle cells as well as in specification of cardiogenic mesoderm and early development of the heart [81–84]. They are essential for migration and/or fusion of the heart primordial and cardiomyocyte differentiation [85], even contributing to the left-right asymmetry of the heart [86]. Other signals taking part in those events, like activin or TGF-β, seem to be regulated both spatially and temporally by interplay between BMPs and their antagonists [83, 87].

The existence of the functional BMP system in the rat ovary, replete with ligand, receptor, and novel cellular functions suggests their involvement in morphogenesis of the reproductive system. It has been shown that BMPs differentially regulate FSH-dependent steroidogenesis during the normal rat estrous cycle [88]. Moreover, PDF, another member of the BMP family, could be regulated by androgens in the prostate [89], thus emphasizing the role of BMP family members in the reproductive system [89, 90] (see the chapter by Paralkar et al).

Evolutionary relationships between the amphibian, avian, and mammalian digestive systems revealed a common embryonic expression of BMPs, suggesting their prime importance as mesenchymal signals involved in the formation of stomach glands [91, 92] with possible protective role in maintenance of the adult intestinal epithelium [93].

## BMPs: Chromosomal localization and developmental function using gene disruption and overexpression

Little is known about the structure of BMP genes. BMP-2 and BMP-4 genes, for example, show high similarity to the *Drosophila dpp* gene, with conserved position of a single intron within the coding region [94], and the BMP-7 gene is structurally related to murine *Vgr-1* gene [4]. BMP genes have been linked to specific chromosomes in mouse as well as in the human genome (Tab. 1). The chromosomal localization of BMPs suggests close linkage to several morphogenetic developmental anomalies [95, 96]. The roles of individual BMPs have been studied through identification of mutated genes in classic mouse mutants or through conventional gene-targeting approaches, gene disruption and overexpression of genes encoding mem-

*Table 1 - Chromosomal localizations of the BMPs*

| | **Human chromosome/ disease** | **Mouse chromosome/ mutation** | **References** |
|---|---|---|---|
| BMP-2 | 20p12 / Holt-Oram syndrome | 2 / tight skin syndrome (*tsk*) | [94, 95, 102–106, 111, 115] |
| BMP-3 | 4p14.8-q21 / Dentinogenesis imperfecta II | 5 / increased bone volume | [95, 105, 107, 108] |
| BMP-3B | 10 | | |
| BMP-4 | 14q22-23 / Holt-Oram syndrom | 14 / pugnose (*pn*), no mesoderm formation | [94, 109–112, 114, 115, 137] |
| BMP-5 | 6 | 9 / Short-ear (*se*) | [116–119, 125, 138] |
| BMP-6 | 6 | 13 / congenital hydrocephalus | [119–121] |
| BMP-7 | 20 / Holt-Oram syndrome | 2 / impaired kidney and eye development | [4, 69, 98, 99 115, 119, 137, 138] |
| BMP-8A | – | 4 / germ-cell deficiency | [122] |
| BMP-8B | – | 4 / germ-cell deficiency | [123] |
| CDMP-1 (GDF-5, | 20 / Grebe syndrome, Hunter-Thompson disease | 2 / brachypodism | [124–131] |
| BMP-14) | 2 | – | [131] |
| CDMP-2 (GDF-6, | | | |
| BMP-13) | 3 | – / improper development of dorsal spinal cord | [131, 132] |
| CDMP-3 (GDF-7, BMP-15) | | | |
| GDF-8 | | – / increased skeletal muscle mass | [133] |
| GDF-9 | | – / infertility, impaired folliculogenesis | [134, 135] |
| GDF-10 | | 14 / none | 136 |

bers of the BMP family, BMP receptors and SMAD proteins (see also the chapter by Korchynsky and ten Dijke, and the chapter by Luyten et al.). Collectively, these studies confirmed that BMPs have significant roles in the development of the skeleton, nervous system, eye, kidney and heart [68–70, 97–99].

However, gene disruption experiments did not always deduct the total extent of the BMP function. Namely, some homozygous knock-out animals were embryonic lethal, which prevented the disclosure of the true impact of the disruption. In TGF-β knock-out mice the function was also masked by the fact that the maternal protein in heterozygous mothers crossed the placental barrier at early stages of development, resulting in the maternal rescue of offspring [100]. Whether BMPs circulating in biological fluids [101] of heterozygous mice can also cross the placental barrier and mask the true developmental role will be discussed in the chapter by Borovecki et al. This merely indicates that gene disruption will not necessarily result in a protein deficiency.

Disruption of the gene encoding BMP-2 expresses the most highly malformed phenotype. Homozygous mice are embryonic lethal between E7.0 and E10.5 [102]. This is caused by the persistence of the proamniotic canal, a transient embryonic structure, the preservation of which leads to malformation of the amnion and the chorion. In mutant embryos the heart develops in the exocoelomic cavity or does not develop at all. Delay in allantois development, open neural tubes and overall slower growth of these embryos is also observed. The defects are consistent with previously detected patterns of expression of BMP-2 in mouse extraembryonic mesoderm and promyocardium [103]. The mutation of BMP-2 gene localized on mouse chromosome 2 showed that it is a candidate gene for the tight skin (tsk) mutant (Tab. 1). These animals show increased bone, cartilage and tendon growth with excessive collagen deposition in the subcutaneous connective tissue. On the contrary, overexpression of BMP-2 in the developing embryo of *Xenopus laevis* leads to ventralization, through inhibition of dorsalizing factors, such as β-tubulin and α-actin [104]. In chick embryos, BMP-2 is expressed in mesenchyme surrounding early cartilage condensations in the developing limb.

In humans, BMP-2 gene is assigned to chromosome 20 and it is positively linked to Holt-Oram syndrome [105, 106], characterized by defects in cardiac and skeletal development resulting in septal and upper limb deformities.

The mouse BMP-3 gene is localized on chromosome 5, but the human homologue has been assigned to chromosome 4 (between p14 and q21). Interestingly, dentinogenesis imperfecta type II, a disease of tooth development has been associated with human chromosome 4 (Tab. 1). BMP-3, also called osteogenin, is the most ample member of the BMP family in demineralized bone, accounting for more than 60% of the total amount of BMPs [3], suggesting an important role in the skeletal homeostasis [68]. A recent study on homozygous BMP-3 deficient mice showed that mutants (Fig. 1B), although possessing a normal skeletal phenotype, have increased bone density with total trabecular bone volume twice that of the wild-type animals [107]. The increased bone density is not a consequence of the reduced osteoclast number or increased number of osteoblasts. As a negative regulator of bone density *in vivo*, BMP-3 might effect the regulation of osteoclast function and osteoblast proliferation and/or differentiation [107]. Experiments using bone loading cham-

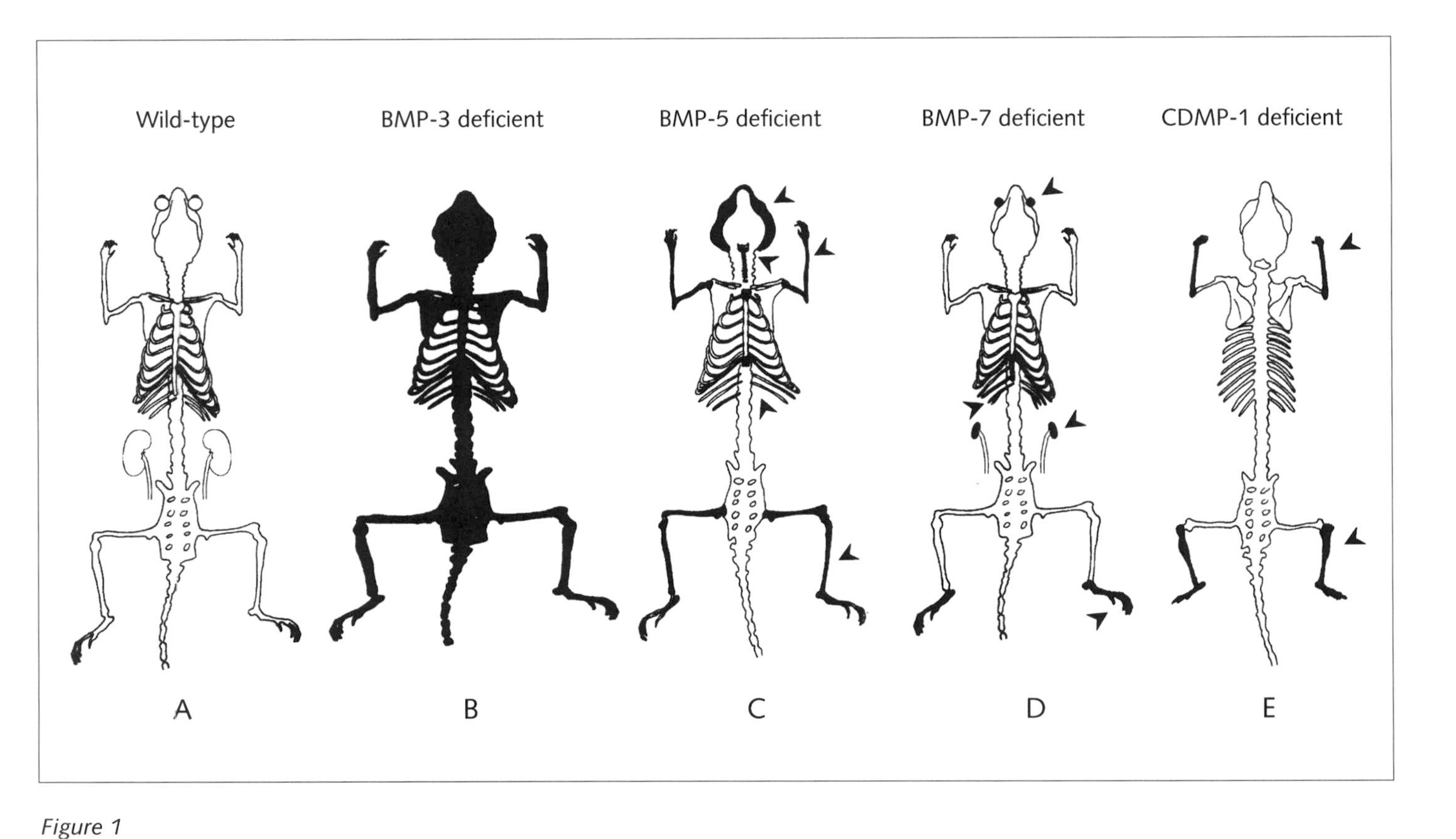


*Figure 1*
*Schematic depiction of a normal mouse (A) and BMP-3, -5, -7 and CDMP-1 (GDF-5) deficient mice (B). BMP-3 deficient mice show increased bone mineral density throughout the whole skeleton, while BMP-5 deficient mice exhibit deformities of the appendicular skeleton, thorax and auricular cartilage (arrows) (C). (D) Mice lacking a functional BMP-7 gene have impaired development of the eye, malformed ribs and feet and undeveloped kidneys (arrows). (E) GDF-5 deficient mice have a shortened appendicular skeleton.*

bers in rats have shown that mechanical stimuli decrease expression of BMP-3 allowing the formation of cartilage and bone [108], which is in line with its role as a negative regulator of osteogenesis.

The BMP-4 gene is localized to chromosome 14 both in mouse and human genome. It may be a candidate for the pugnose mutation (*pn*) in mice, characterized by abnormalities in skull bone development, and has a possible association with Holt-Oram syndrome described in humans (Tab. 1). Another BMP-4-related gene has been assigned to mouse X chromosome, but as human homologue of this gene has not been found, the mouse sequence might be a pseudogene. Inactivation by homologous recombination of the BMP-4 gene leads to anomalies in extraskeletal tissues and embryonic lethality between E6.5 and E9.5, and a variable phenotype in homozygous animals. A majority of mutant embryos show highly impaired mesodermal differentiation [109]. Some homozygous mutants develop to the head fold or beating heart/early somite stage, or beyond, and are developmentally retarded with disorganized posterior structures and a reduction in the extraembryonic mesoderm, including blood islands. Heterozygous BMP-4 mutant mice exhibit craniofacial malformation, microphthalmia and preaxial polydactyly. The plethora and diversity of abnormalities observed indicate that BMP-4 is crucial for normal gastrulation and mesoderm formation. This is also corroborated by previous findings that BMP-4 is needed for differentiation and proliferation of the posterior mesoderm, from which the extraembryonic mesoderm of the amnion, allantois and yolk-sack, as well as the ventral-lateral mesoderm develops [110]. BMP-4 is normally expressed in the perichondrium of the developing cartilage elements. Overexpression of BMP-2 and BMP-4 produced by using retroviral vectors caused enlarged and malformed cartilage elements and joint fusions by increasing the matrix production and number of chondrocytes [111]. The formation of the periosteum was considerably delayed. An overexpression of BMP-4 has been found in lymphocytes and fibroblast-like cells derived from fibroproliferative lesions in patients with *fibrodysplasia ossificans progressiva* (FOP), a rare human autosomal-dominant disorder characterized by progressive heterotopic ossification and congenital malformation of the big toes [112]. Given the osteogenic capability of BMPs, any of BMP genes could be a candidate for FOP. But, overexpression of the BMP-4 gene has been found in lymphocytes of patients with FOP, suggesting the disease could result from an error in the regulation of this gene [112]. Normal lymphocytes do not produce BMPs, but express ALK-3, a BMP specific receptor [113]. Therefore, in patients with FOP, lymphocytes capable of expressing BMP-4 are presumably recruited to the connective tissue from the bloodstream after soft tissue injury. Increased doses of BMP-4 in the connective tissue may lead to fibroproliferative lesions. Gastric cancer cells also show increased expression of BMP-4 mRNA. These cells can be classified as poorly and well differentiated. Poorly differentiated types show greater tendency towards bone metastasis and patients with this type of cancer have a decreased life expectancy. Expression of BMP mRNA has been examined in seven different gastric cancer cell lines and results have shown increased expression of BMP-4 [114].

Salivary pleomorphic adenomas, which are often associated with ectopic cartilaginous tissue formation, have also been examined in regards to expression of different members of the BMP family. A marked increase in expression of BMP-2, BMP-4 and BMP-7 mRNA has been found. However, chondroid formation and expression of the type II collagen was most frequently observed in pleomorphic adenomas overexpressing BMP-2 mRNA. BMP-2 was also detected in modified myoepithelia cells around the chondroid tissue and basement membranes [115].

The BMP-5 gene is localized on human chromosome 6, and the phenotype resulted from its mutation in mouse has been studied for over 40 years [116, 117]. The mutation of the BMP-5 gene alters size, shape and number of many different skeletal elements with greatly reduced size of the external ear, named a short ear mutation. The short ear mouse displays numerous skeletal abnormalities (Fig. 1C), such as reduction in body size, absence of the xyphoid process, reduction of ventral processes of the cervical vertebrae, deletion of one pair of ribs and, the most prominent change of all, a reduced size of the auricle [118]. Mutant adult animals also have a reduced capacity to repair rib fractures. Short ear mice also develop a number of other extraskeletal abnormalities, like hydronephrosis, as well as misplacement of gonads, lung cysts, liver granulomas and neuromuscular tail kinks. BMP-5 is expressed in the mesenchyme of the affected skeleton elements and in the periosteum. It is also expressed in liver, lung, bladder and intestine [116]. The expression pattern corresponds to the localization of the affected tissues and organs.

The BMP-6 gene is present on human chromosome 6 with no reported disease association, and on mouse chromosome 13, possibly near the congenital hydrocephalus (ch) locus, which is associated with abnormalities in the growth and differentiation of the skeletal system and kidney [119]. However, mice with targeted null mutation at the BMP-6 locus are viable and fertile, and show no obvious difference in the skeleton to the wild-type animals. Upon closer examination of skeletogenesis in late pregnancy, delayed ossification of the developing sternum is observed [120]. As other members of the BMP family overlap with the BMP-6 expression, especially BMP-2, this apparent lack of defects in mutant mice could be the result of the functional redundancy. BMP-6 is expressed during the development of the epidermis, coinciding with the commencement of stratification. It declines 1 week after the birth. To study the effects of increased expression of BMP-6 in the epidermis, transgenic mice with inherent overexpression of BMP-6 in suprabasal layers of the intrafollicular epidermis were created [121]. The pattern of transgene expression influences the effects on proliferation and differentiation to a large extent. Consistent and strong expression of BMP-6 leads to lessened cell proliferation in the embryonic and perinatal epidermis, but had hardly any effect on differentiation. Weaker and irregular expression induces hyperproliferation and parakeratosis in the adult epidermis and disturbed differentiation. Histologically, the later findings show high similarity to psoriasis.

The gene for BMP-7 is localized to chromosome 2 in mouse and chromosome 20 in human genome. In humans, both chromosomes 2 and 20 have been implicated in Holt-Oram syndrome, so that BMP-2, BMP-4 or BMP-7 might be involved. Deletion in the mature domain of the BMP-7 coding gene produced no apparent malformations in heterozygous animals. However, crosses between these heterozygotes produce a very distinctive phenotype in a quarter of neonates. Mice are smaller in size, have polydactyly in the hindlimbs, exhibit abnormally formed thoracic skeleton and have either anophthalmia or microphthalmia (Tab. 1). Most importantly, these animals die of uremia within 24 h of birth due to small dysgenic kidneys with hydroureters (Fig. 1D). The kidneys have no identifiable metanephric mesenchyme and no evidence of glomeruli formation in the cortical region [98, 99].

Mice lacking BMP-8A exert normal phenotype throughout embryonic and postnatal development. However, in 47% of homozygous mutants, germ-cell degeneration occurs. A small proportion of homozygous mutants also show degeneration of the epididymal epithelium. BMP-8A thus plays a pivotal role in spermatogenesis and regulation of epididymal function [122]. Targeted mutation of the BMP-8B gene also leads to germ-cell deficiency and sterility. This occurs because of impaired proliferation and differentiation of germ cells as well as premature apoptosis of spermatocytes [123].

## GDFs: Chromosomal localization and developmental function

Genes encoding cartilage derived morphogenetic proteins, CDMP-1 (GDF-5), CDMP-2 (GDF-6) and CDMP-3 (GDF-7) are localized on human chromosomes 20, 2 and 3, respectively. However, the *brachypodism* mouse phenotype has been studied long before the discovery of BMPs/CDMPs (GDFs). The most prominent feature of these animals is reduction in length of the appendicular skeleton. The axial skeleton is largely unaffected. The defects in the limbs affect metacarpals and metatarsals, along with altered patterning segments in the digits of the limbs. *Brachypodism* is a direct result of three independent mutations in the GDF-5 gene [124, 125]. GDF-5 is expressed during joint formation *in vivo* [126, 127] and malformations in *brachypodism* mouse could be due to impaired chondrogenesis. However, ear, sternum, rib or vertebral morphology is not affected (Fig. 1E). The only known human mutation in a gene encoding a member of the TGF-β superfamily described is the mutation of CDMP-1 gene (*cdmp-1*), a human homologue of GDF-5 [128]. *Cdmp-1* mutations have been implicated in two recessive chondrodysplasias: the Hunter-Thompson chondrodysplasia [129] and the chondrodysplasia Grebe type [130]. The Hunter-Thompson chondrodysplasia is caused by insertion of 22 bp in the mature region of the *cdmp-1* gene, while the cause of the chondrodysplasia Grebe type seems to be a single replacement of cysteine by tyrosine in a mature TGF-β domain of the *cdmp-1* gene. In both cases, the appendicular skeleton is severely shortened, while the axial skeleton remains largely intact (see the chapter by Luyten et al.). It has been shown that recombinant GDF-5, 6 and 7 proteins implanted subcutaneously in a bone collagen carrier induce tendon and ligament structures in the subcutaneous bone induction assay in rats [131].

GDF-7 is selectively expressed in the cells of the roof plate in the developing central nervous system [132]. GDF-7 null mutant embryos lack a specific class of neurons, which are important for dorsal spinal cord development. GDF-7 could play a crucial role in the assignment of neuronal identity within the mammalian CNS (see the chapter by Lein et al.).

GDF-8 deficient animals with induced mutation in mice and spontaneous mutation in double-muscled Belgian blue and Piedmontese cattle exert an extensive increase in skeletal muscle mass (see the chapter by Paralkar et al). The weight of individual muscles in mutants is increased two- or three-fold when compared to wild-type animals. This suggests a role of GDF-8 as a negative regulator of the skeletal muscle growth [133].

GDF-9 is a member of the BMP family important in the development and maintenance of the reproductive system in mice. It is expressed at high levels in the mammalian oocyte and mice lacking GDF-9 are infertile. This occurs because of impaired folliculogenesis [134, 135].

GDF-10 is expressed during development in the craniofacial region and the vertebral column of the skeleton. During adult life it is highly expressed in the brain and in the uterus [136]. Mice carrying null mutation for the GDF-10 gene, however, do not show any obvious abnormality in the development, confirming that gene knock-out experiments do not necessarily have functional consequences.

## Double deficiencies in genes encoding BMPs

Absence of malformations observed in some mutants lacking functional BMP encoding genes, and simultaneous expression of several BMP members in different tissues have pointed to a possible functional redundancy in the role of these proteins. Therefore, several phenotypes have been investigated in which the function of two genes encoding different members of the BMP family has been disrupted.

Doubly heterozygous BMP-4 and BMP-7 mice develop defects in the rib cage and distal part of the limbs [137]. These two morphogens seem to act in cooperation in the mesenchymal condensations of the affected skeletal regions, possibly through regulation of apoptosis.

BMP-5/-6 double mutants show sternal defects similar to those found in BMP-6 single mutants. However, these defects tend to be slightly exacerbated in the double mutant.

Mice with simultaneous deficiency in BMP-5 and BMP-7 show the most severe phenotype. Coexpression of both morphogens seems to be pivotal for development of allantois, heart, branchial arches, somites and the forebrain since mutant embryos die at E10.5 with extensive defects of the aforementioned tissues [138].

Null mutants with simultaneous deficiency in BMP-5 and GDF-5/CDMP-1 exert defects, which cannot be observed in either of the single mutants. Disruption of the sternebrae within the sternum and abnormal formation of fibrocartilaginous joints between the sternebrae and the ribs are the most prominent of those defects [125].

## Disruptions in the genes encoding BMP antagonists

Heterozygous noggin deficient mice possess normal phenotypes. Skeletal structures in homozygous animals however exhibit abnormalities. The defects are especially striking in the vertebrae, ribs and limbs, with the severity of axial defects increasing caudally [139]. The skull and cervical vertebrae are basically normal, but thoracic vertebrae are fused. They also fail to close dorsally. Ribs are reduced in number and have abnormal morphology. The appendicular skeleton in mutant animals is also shortened. All these processes seem to arise from a lack of noggin leading to increased BMP activity after the chondrogenesis has started. A majority of the joints is also fused. Elbows and digits are fused and have cartilaginous spurs as a result of a failure to specify the joint. Unregulated expression of GDF-5/CDMP-1 in the joint regions seems to play a pivotal role in those processes. Absence of local regulation of BMP members, especially BMP-6, which is expressed in the hypertrophic zone of cartilage in the joints, most probably also plays a role in impaired articular development (see the chapter by Luyten et al.).

Malformations, as described, both skeletal and extraskeletal, are numerous, but studies of localization of different BMPs imply that deficient phenotypes should be more severe. This apparent discrepancy is, most likely, caused by mechanisms, which are still not fully understood. Firstly, BMPs overlap, both in localization and function. Only at localizations in which one BMP is predominant, like BMP-7 in the kidney mesenchyme, will the deficiency of that morphogen lead to impaired development and function. Secondly, maternal morphogens might also play an important role in early embryonic development, disguising or totally eliminating deficiencies, which might lead to irregular or impaired development. This has been shown to be the case in TGF-β deficient mice, and is probably in the root of variations of phenotypes in BMP-4 deficient animals. Early mesenchyme induction in BMP-7 deficient animals could also be linked to the maternal BMP-7 circulating in the bloodstream of BMP-7 deficient embryos [101] (see the chapter by Borovecki et al.). This indicates that genetic and functional evidence, when determining the role of a certain morphogen, often differs greatly. The genetic findings, which mainly derive from studies in cell cultures and on gene deficient animals, although valuable, are not always confirmed when put to a test in a physiological surrounding. It is only through the combination of genetic and functional data that one can reveal the complex web of interactions, which weave the delicate balance of a gene function.

## Appendicular skeleton

BMPs have multiple functions in development of the appendicular skeleton, specifically in the establishment of the anteroposterior axis and morphogenesis of the limb, and formation of articular joints [32, 140] (see the chapter by Luyten et al.). Anteroposterior patterning of the vertebrate limb is achieved by sequential long-range and short-range sonic hedgehog signaling (Shh), allowing continued proximodistal specification of limb elements [141, 142]. Those signals act initially long range to prime the region of the limb competent to form digits and thus control digit number. Later, Shh acts short range to induce expression of BMPs, whose morphogenetic action specifies digit identity. In the final stages of limb morphogenesis the undifferentiated cells of the distal growing tip of the limb can follow two distinct fates, chondrogenesis and apoptotic cell death, forming the digits and the interdigital regions. It seems that both processes are controlled by BMPs in an interactive loop with noggin, GDFs, TGF-βs, FGFs and hedgehog signaling [143–146]. Moreover, patterning along the dorso-ventral axis of the embryo is regulated by a gradient of secreted morphogens of the BMP-4/*Dpp* family. This gradient is formed by the opposing activities of BMP-sequestering proteins and BMP-releasing metalloproteases. Coordinated regulation of the activities of BMPs and their inhibitors is essential for skeletal development since loss-of-function experiments show that both BMPs and BMP inhibitory signals, such as noggin, are required to establish proper formation of skeletal tissues [147–149]. At early embryonic stages, BMP-2 and 4 can be detected at the apical ectodermal ridge and posterior mesenchymal condensations of limb buds [150]. BMP-7 appears to have a more diffuse distribution [72, 151]. Later, numerous BMPs are expressed in the perichondrium surrounding long bones, ribs, vertebrae, and craniofacial bones. BMP-6 seems to be expressed at a later stage of embryonic bone development when chondrocytes undergo hypertrophic maturation [152, 153].

Besides antero-posterior and dorso-ventral patterning of the embryo, it has recently been shown that BMP-4, its ligand and downstrcam Smad1 protein are transiently expressed on the right side of the Hensen's node of the chick embryo when left-right polarity is being established [154]. Furthermore, a key role for BMP-4 in this process is suggested by maintaining sonic hedgehog asymmetry [155].

## Joints

Several studies indicate that members of the bone morphogenetic protein family promote cartilage formation, and it seems that they are required at two steps of limb chondrogenesis: formation of prechondrogenic condensations and their differentiation into chondrocytes [28].

Development of the joints is also influenced by several BMPs [127]. BMP-2 and BMP-4 induce apoptosis in the undifferentiated limb mesenchyme. Perichondrial expression of BMP-7 follows a proximal-to-distal sequence and is characterized by interruptions in the regions of joint formation, so that BMP-7 may inhibit joint formation while stimulating radial growth and differentiation of developing limb cartilage. BMP-2 may be involved in determining the joint shape. Cartilage-derived morphogenetic proteins (CDMP-1/GDF-5/BMP-14 and CDMP-2/GDF-6/BMP-13) show strong expression at the sites of joint development and weak expression in the perichondrium (see the chapter by Luyten et al.). CDMP-1 in combination with Wnt-14 is crucial for joint positioning and early events in joint formation [156].

## Axial skeleton

Besides complex activities of BMPs in the morphogenesis of the appendicular skeleton, there are also reports on their roles in early somitogenesis and proper development of the axial skeletal elements, such as vertebrae, ribs and scapula. These structures develop from the embryonic somatic mesoderm through interactions with neural tube/notochord and skin ectoderm. BMPs seem to play important roles in these tissue interactions, since perturbation of BMP signaling in somitogenesis resulted in vertebral and rib malformations [147]. Again, they act in concert with other growth factors involved in the formation of the sclerotome, in particular the secreted sonic hedgehog (Shh). Shh signals are required only transiently and act to change the competence of target cells to respond to BMPs. The later stages of this process specifically depend on BMP signaling, which acts to trigger the chondrogenic differentiation [157].

## Teeth

The expression of six different BMPs have been described in developing teeth suggesting roles during several stages of morphogenesis, including initiation of tooth development [75, 78, 158], morphogenesis of the epithelium and mandibular mesenchyme [79, 159], differentiation of dentin and enamel forming cells, and deposition of extracellular matrices. BMPs have also been expressed during closure of the

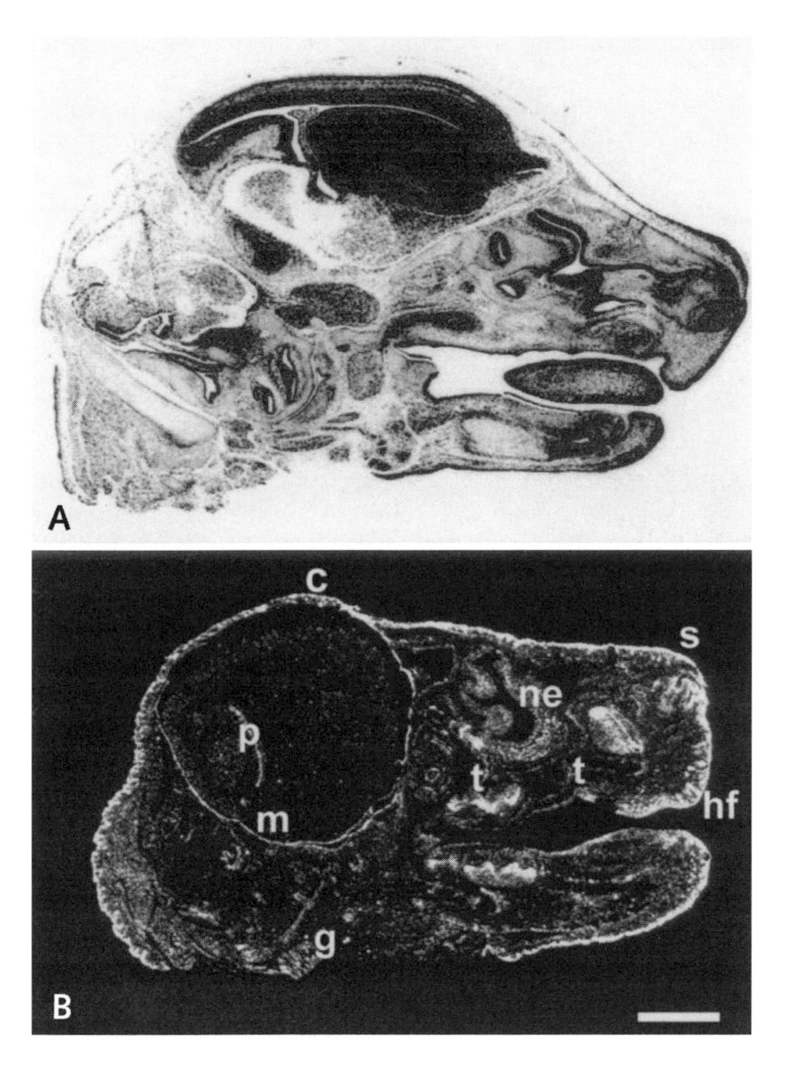


*Figure 2*
*BMP-7 transcripts are found in several craniofacial structures of a developing rat embryo. (A) Brightfield image of the sagital cranial section of a 17.5-day rat embryo. (B) A corresponding darkfield section indicating transcripts in calvarial bone (c), chorioid plexus (p), meninges (m), nasal epithelium (ne), molar and incisor teeth (t), skin (s), hair follicles (hf), salivary gland (g). Bar, 350 μm.*

sutures of calvarial bones suggesting roles in the calvarial bone development and confirming their role in regulating cell communication during the craniofacial development [160–162] (Fig. 2).

## Bone and bone marrow, cartilage and muscle differentiation

Functions of individual BMPs have been extensively studied *in vitro* using a number of different cultured cell lines, with results generally indicating complex effects that depend on the cell type and culture conditions [163]. For example, in a line of mouse mesodermal progenitor cells, low doses of BMP-2 induced differentiation into adipocytes and high concentrations produced chondrocytes and bone cells [24]. Multipotent cell types respond to different BMPs by increasing either differentiation or proliferation, and similar effects were found in osteoblastic cell lines. The effect of BMPs on cell proliferation is different: proliferation of osteosarcoma cells is stimulated by BMP-2 and BMP-7, while proliferation of osteoblasts is stimulated by BMP-7, but inhibited by BMP-2 and BMP-3 [66, 164–166]. BMP-7 also promotes growth and maturation of chick sternal chondrocytes [167] (Fig. 3) *via* binding to type X collagen promotor [168], but primary mammalian articular chondrocytes do not undergo hypertrophy in similar culturing conditions [169]. Differentiation of osteoblastic and preosteoblastic cells is stimulated by the addition of BMP-3 [170], although they express mRNA and protein for other BMPs during differentiation process *in vitro* [171]. The expression of BMPs could be modulated by exogenously added growth factors like dexametasone or estradiol. The data suggest that only one BMP is required and sufficient for differentiation of osteoprogenitor cells towards a more mature phenotype, and that the function of BMP-4 can be replaced by BMP-7, another member of this family [171] (Fig. 4).

Moreover, the important role for BMPs was observed in the maintenance of the vascular smooth muscle cell phenotype, hence vascularization is a prerequisite in the development and homeostasis of normal cartilaginous and bone tissue [172]. Besides stimulation of genes specific for the smooth muscle cell phenotype, the strong antiproliferative effect of BMP-7 on primary human aortic smooth muscle cells *in vitro* was observed suggesting that BMP-7 could prevent vascular proliferative disorders [81]. BMP-7 is also able to inhibit inflammatory cytokine-mediated ICAM production in smooth muscle cells *in vitro* as well as in peritubular renal capillaries *in vivo* [81, 173], thus confirming the important role in the maintenance of vascular integrity. The expression of BMP-2 has also been reported in various human blood vessels and vascular cell types, and direct effect on migration of human aortic vascular smooth muscle cells has been shown [174].

There is also accumulating evidence that BMPs are candidates for regulators of hemopoietic differentiation and function of mature blood cells in the adult life. The regulation of hemopoiesis is a complex process, which requires signaling between stromal cells, stem cells and progenitor cells. Recent studies have confirmed the effect of BMPs on highly primitive as well as highly differentiated hemopoietic cells *in vitro* [175, 176], but their involvement in the adipocytic differentiation pathway has also been suggested [177]. Primitive $CD34^{+}CD38^{-}$ cells could respond to exogenously added BMP-2, -4, and -7, which regulated their proliferation and differenti-

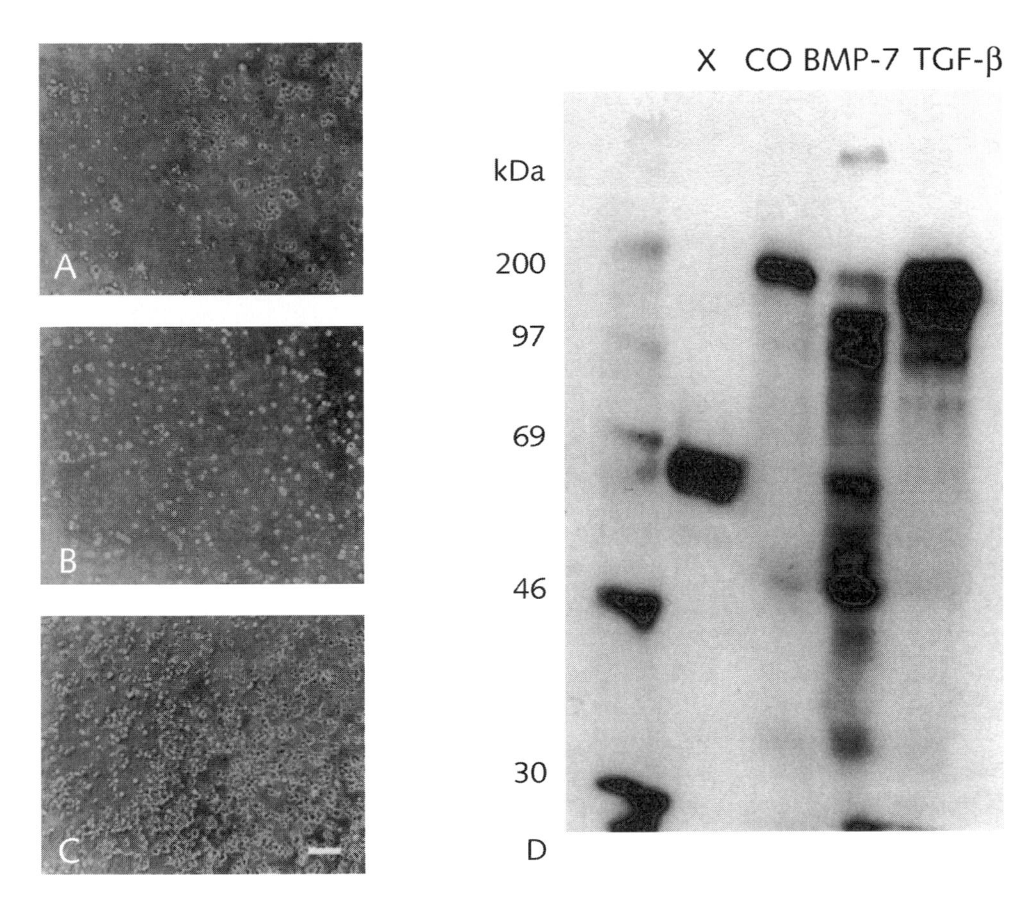


*Figure 3*
*BMP-7 induces clonal proliferation and maturation of day 15 chick sternal chondrocyte agarose cultures in serum-free medium. Cells were grown in agarose for 3 weeks in chemically defined medium at a density of $1 \times 10^5$ cells/well. Left panel, photomicrographs of living cultures treated with: (A) control; (B) TGF-β1 (10 ng/ml; (C) BMP-7 (50 ng/ml). Bar, 25 μm. Right panel, collagen biosynthesis gel, first lane shows the molecular mass standard; lane X, type X collagen (positive control); lane CO, control cells; lane BMP-7, cells treated with 50 ng/ml of BMP-7; lane TGF-β (10 ng/ml).*

ation with a direct effect on stem cell survival [175]. Another study confirmed the expectation that normal adult hemopoietic cell lines express BMP genes (BMP-2, -4, -6, -7) as well as other members of the TGF-β superfamily, with lineage-restricted patterns of expression [176, 178]. It has also been found that BMP-9 acts as a hemopoietic hormone [179]. The expression and presence of BMP molecules have also been reported in a normal adult tissue, which represents the absolute prerequi-

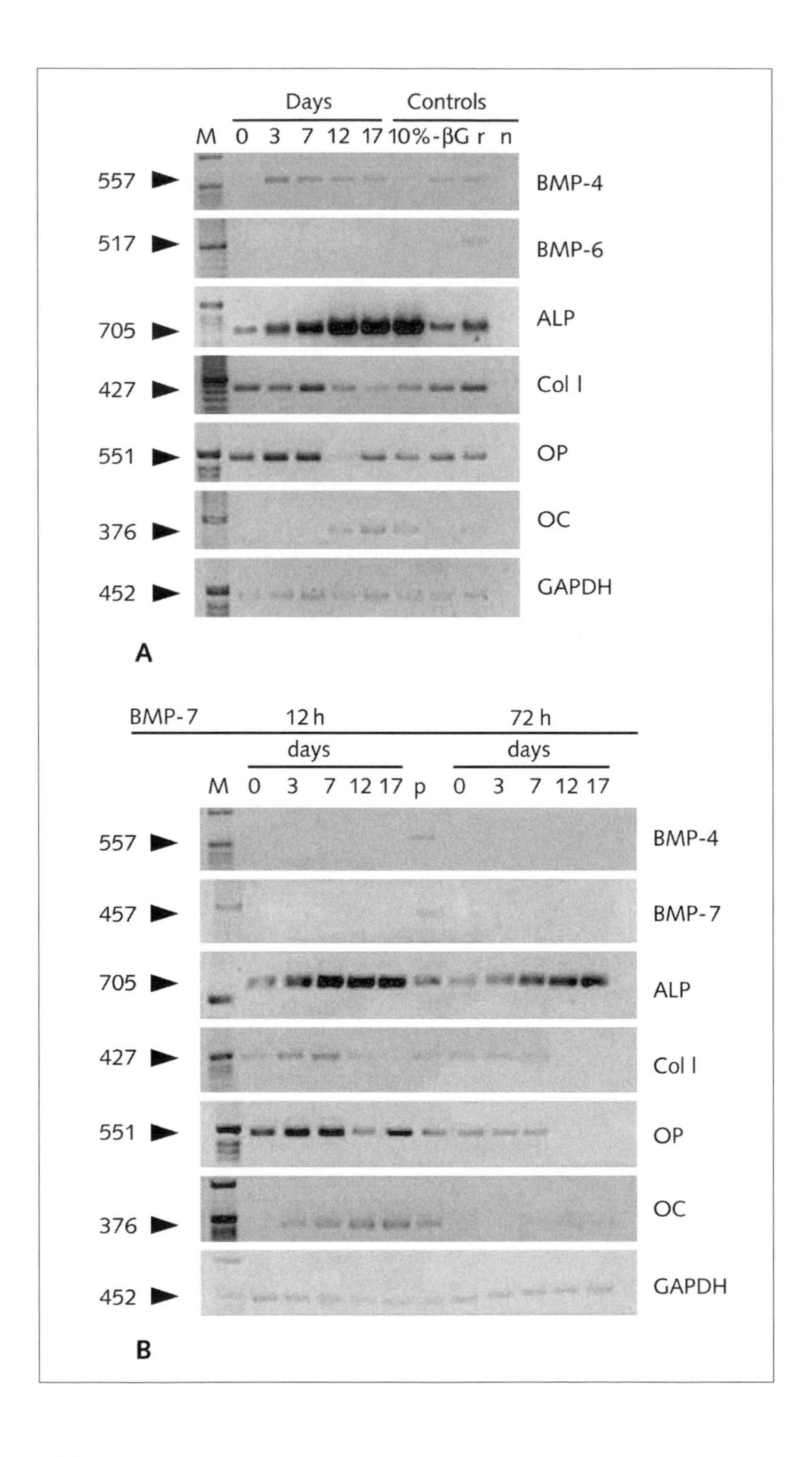

Days
Controls
M 0 3 7 12 17 10%-βG r n
557
517
705
427
551
376
452
BMP-4
BMP-6
ALP
Col I
OP
OC
GAPDH
A
BMP-7
12 h
72 h
days
days
M 0 3 7 12 17 p 0 3 7 12 17
557
457
705
427
551
376
452
BMP-4
BMP-7
ALP
Col I
OP
OC
GAPDH
B

site for maintenance of hemopoiesis *in vitro*. Normal human bone marrow stromal cells synthesize and produce BMP-3, BMP-4 and BMP-7 (Fig. 5) as well as type I receptors and receptor-related and common mediator Smad molecules, thus, implicating important roles in autocrine/paracrine mechanism regulating hemopoiesis [113] (Fig. 4).

## BMP applications: bone and beyond bone

BMPs are capable of restoring lost bone in the post-fetal life by recapitulating the cellular events that are involved in the formation of bone during the embryonic development [10]. The recently completed prospective randomized clinical study for the restoration of tibial nonunions in humans by recombinant BMP-7 containing collagenous devices [180] (see the chapter by Giltaij et al.) offers significant promise in the demonstration that the events responsible for tissue formation in the embryo can form strategies for therapeutic development in man. Clinical results on the use of recombinant human BMPs in orthopedic reconstruction and craniofacial repair strongly support their use in bone regeneration in humans (see the chapters by Blockhuis et al. and Terheyden et al.).

Identification and characterization of BMP-specific type I and II receptor complex and subsequent intracellular signaling *via* BMP-specific Smad intracellular proteins and identification of BMP responding elements in tissue specific target genes, provide a basis for endogenous up-regulation of BMPs in individuals with osteoporosis and various metabolic bone diseases [173] (see the chapter by Korchynsky and ten Dijke).

The demonstration that the application of recombinant BMPs is capable of regenerating a variety of tissues, like bone, cartilage, tendon, ligament, peridontium and dentin, kidney or central nervous system in various animal models suggests that the specific biological action of BMPs is determined by responding cells and the

*Figure 4*
*Expression of BMPs and osteoblast differentiation markers in osteoblastic cells during differentiation process* in vitro. *MC3T3-E1 cells were grown for 17 days in DMEM with 1% FCS; β-glycerophosphate (5 mM) and ascorbic acid (50 μg/ml). RNA was isolated at designated times and semi-quantitative RT-PCR performed using specific primers for BMP-2 to BMP-7, ALP (alkaline phosphatase), Col I (collagen type I), OP (osteopontin) and OC (osteocalcin). (A) Control cultures expressed BMP-4 mRNA and differentiation markers. (B) BMP-7 treated cells (20 ng/ml) expressed differentiation markers regardless complete inhibition of BMP expression, suggesting BMP-7 can replace the function of endogenous BMP-4. GAPDH, house-keeping gene; p, positive control lane; n, negative control lane.*

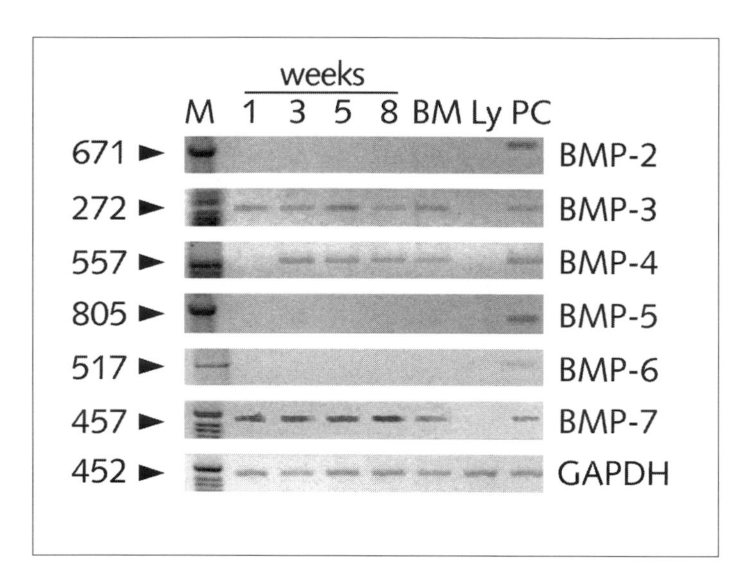


*Figure 5*
*BMP expression in stromal cells from human bone marrow long-term culture. Stromal cells were obtained from healthy donors by standard biopsy procedure and cultivated up to 8 weeks in appropriate conditions. Total RNA was extracted from freshly isolated (BM) or cultivated bone marrow samples (1, 3, 5, 8 weeks), and semi-quantitative RT-PCR performed using specific primers for BMP-2 to BMP-7. Stromal cells expressed mRNA for BMP-3, BMP-4 and BMP-7 throughout entire investigation period. GAPDH, house-keeping gene; Ly, periferal lymphocytes; PC, positive control.*

microenvironment available at the site of injury [173, 181, 182]. Studies on gain and loss of function indicate that in addition to their morphogenic role in the musculoskeletal system, BMPs serve as inductive signals for overall tissue development during embryogenesis, and suggest that they may have therapeutic utility in nervous, urogenital, cardiovascular, pulmonary and reproductive organ systems in the adult life. The role of BMPs in several other systems will be extensively discussed in other chapters of this book.

## Conclusion

Apart from the unique bone-inductive ability of BMPs, the last decade has brought a wealth of morphological, genetic and biochemical data emphasizing their essential function in developmental processes and overall morphogenesis of many distant species. Besides their profound role in bone tissue regeneration and maintenance,

BMPs act as differentiation factors, as well as physiological regulators in homeostasis of different tissues. Multiple therapeutic uses in a variety of clinical indications are foreseeable.

## References


1 Urist MR (1965) Bone: formation by autoinduction. *Science* 150: 893–899
2 Wozney JM, Rosen V, Celeste AJ, Mitsock LM, Whitters MJ, Kriz RW, Hewick RM, Wang EA (1988) Novel regulators of bone formation: molecular clones and activities. *Science* 242: 1528–1534
3 Luyten FP, Cunningham NS, Vukicevic S, Paralkar V, Ripamonti U, Reddi AH (1992) Advances in osteogenin and related bone morphogenetic proteins in bone induction and repair. *Acta Orthop Belg* 58 (Suppl 1): 263–267
4 Ozkaynak E, Rueger DC, Drier EA, Corbett C, Ridge RJ, Sampath TK, Oppermann H (1990) OP-1 cDNA encodes an osteogenic protein in the TGF-beta family. *EMBO J* 9: 2085–2093
5 Sampath TK, Maliakal JC, Hauschka PV, Jones WK, Sasak H, Tucker RF, White KH, Coughlin JE, Tucker MM, Pang RH et al (1992) Recombinant human osteogenic protein-1 (hOP-1) induces new bone formation *in vivo* with a specific activity comparable with natural bovine osteogenic protein and stimulates osteoblast proliferation and differentiation *in vitro*. *J Biol Chem* 267: 20352–20362
6 Celeste AJ, Iannazzi JA, Taylor RC, Hewick RM, Rosen V, Wang EA, Wozney JM (1990) Identification of transforming growth factor beta family members present in bone-inductive protein purified from bovine bone. *Proc Natl Acad Sci USA* 87: 9843–9847
7 Luyten FP, Cunningham NS, Ma S, Muthukumaran N, Hammonds RG, Nevins WB, Woods WI, Reddi AH (1989) Purification and partial amino acid sequence of osteogenin, a protein initiating bone differentiation. *J Biol Chem* 264: 13377–13380
8 Vukicevic S, Paralkar VM, Reddi AH (1993) Extracellular matrix and bone morphogenetic proteins in cartilage and bone development and repair. *Adv Mol Cell Biol* 6: 207–224
9 Reddi AH (1994) Bone and cartilage differentiation. *Curr Opin Genet Dev* 4: 737–744
10 Reddi AH, Huggins C (1972) Biochemical sequences in the transformation of normal fibroblasts in adolescent rats. *Proc Natl Acad Sci USA* 69: 1601–1605
11 Reddi AH (1995) Cartilage morphogenesis: role of bone and cartilage morphogenetic proteins, homeobox genes and extracellular matrix. *Matrix Biol* 14: 599–606
12 Reddi AH (1998) Role of morphogenetic proteins in skeletal tissue engineering and regeneration. *Nat Biotechnol* 16: 247–252
13 Reddi AH (2000) Morphogenetic messages are in the extracellular matrix: biotechnology from bench to bedside. *Biochem Soc Trans* 28: 345–349
14 Franceschi RT (1999) The developmental control of osteoblast-specific gene expression:

role of specific transcription factors and the extracellular matrix environment. *Crit Rev Oral Biol Med* 10: 40–57

15 Nishida Y, Knudson CB, Eger W, Kuettner KE, Knudson W (2000) Osteogenic protein 1 stimulates cells-associated matrix assembly by normal human articular chondrocytes: up-regulation of hyaluronan synthase, CD44, and aggrecan. *Arthritis Rheum* 43: 206–214

16 Nishida Y, Knudson CB, Kuettner KE, Knudson W (2000) Osteogenic protein-1 promotes the synthesis and retention of extracellular matrix within bovine articular cartilage and chondrocyte cultures. *Osteoarthritis Cartilage* 8: 127–136

17 Paralkar VM, Nandedkar AK, Pointer RH, Kleinman HK, Reddi AH (1990) Interaction of osteogenin, a heparin binding bone morphogenetic protein, with type IV collagen. *J Biol Chem* 265: 17281–4

18 Paralkar VM, Vukicevic S, Reddi AH (1991) Transforming growth factor beta type 1 binds to collagen IV of basement membrane matrix: implications for development. *Dev Biol* 143: 303–308

19 Paralkar VM, Weeks BS, Yu YM, Kleinman HK, Reddi AH (1992) Recombinant human bone morphogenetic protein 2B stimulates PC12 cell differentiation: potentiation and binding to type IV collagen. *J Cell Biol* 119: 1721–1728

20 Vukicevic S, Latin V, Chen P, Batorsky R, Reddi AH, Sampath TK (1994) Localization of osteogenic protein-1 (bone morphogenetic protein-7) during human embryonic development: high affinity binding to basement membranes. Biochem Biophys Res Commun 198: 693–700

21 Zhu Y, Oganesian A, Keene DR, Sandell LJ (1999) Type IIA procollagen containing the cysteine-rich amino propeptide is deposited in the extracellular matrix of prechondrogenic tissue and binds to TGF-beta1 and BMP-2. *J Cell Biol* 144: 1069–1080

22 Suzawa M, Takeuchi Y, Fukumoto S, Kato S, Ueno N, Miyazono K, Matsumoto T, Fujita T (1999) Extracellular matrix-associated bone morphogenetic proteins are essential for differentiation of murine osteoblastic cells *in vitro*. *Endocrinology* 140: 2125–2133

23 Katagiri T, Yamaguchi A, Ikeda T, Yoshiki S, Wozney JM, Rosen V, Wang EA, Tanaka H, Omura S, Suda T (1990) The non-osteogenic mouse pluripotent cell line, C3H10T1/2, is induced to differentiate into osteoblastic cells by recombinant human bone morphogenetic protein-2. *Biochem Biophys Res Commun* 172: 295–299

24 Wang EA, Israel DI, Kelly S, Luxenberg DP (1993) Bone morphogenetic protein-2 causes commitment and differentiation in C3H10T1/2 and 3T3 cells. *Growth Factors* 9: 57–71

25 Puleo DA (1997) Dependence of mesenchymal cell responses on duration of exposure to bone morphogenetic protein-2 *in vitro*. *J Cell Physiol* 173: 93–101

26 Ji X, Chen D, Xu C, Harris SE, Mundy GR, Yoneda T (2000) Patterns of gene expression associated with BMP-2-induced osteoblast and adipocyte differentiation of mesenchymal progenitor cell 3T3-F442A. *J Bone Miner Metab* 18: 132–139

27 Gazit D, Turgeman G, Kelley P, Wang E, Jalenak M, Zilberman Y, Moutsatsos I (1999)

Engineered pluripotent mesenchymal cells integrate and differentiate in regenerating bone: a novel cell-mediated gene therapy. *J Gene Med* 1: 121–133
28 Pizette S, Niswander L (2000) BMPs are required at two steps of limb chondrogenesis: formation of prechondrogenic condensations and their differentiation into chondrocytes. *Dev Biol* 219: 237–249
29 Reddi AH (1995) Bone morphogenetic proteins, bone marrow stromal cells, and mesenchymal stem cells. Maureen Owen revisited. *Clin Orthop* 313: 115–119
30 Reddi AH (2000) Morphogenesis and tissue engineering of bone and cartilage: inductive signals, stem cells, and biomimetic biomaterials. *Tissue Eng* 6: 351–359
31 Hogan BLM (1996) Bone morphogenetic proteins: multifunctional regulators of vertebrate development. *Genes Dev* 10: 1580–1594
32 Hogan BLM (1996) Bone morphogenetic proteins in development. *Curr Opin Gen Dev* 6: 432–438
33 Reddi AH (2000) Bone morphogenetic proteins and skeletal development: the kidney-bone connection. *Pediatr Nephrol* 14: 598–601
34 Padget RW, St Johnston RD, Gelbart WM (1987) A transcript from a *Drosophila* pattern gene predicts a protein homologous to the transforming growth factor-β family. *Nature (London)* 325: 81–84
35 Sampath TK, Rashka EK, Doctor JS, Tucker RF, Hoffmann FM (1993) *Drosophila* transforming growth factor superfamily proteins induce endochondral bone formation in mammals. *Proc Natl Acad Sci USA* 90: 6004–6008
36 Padget RW, Wozney JM, Gelbart WM (1993) Human BMP sequences can confer normal dorsal-ventral patterning in the *Drosophila* embryo. *Proc Natl Acad Sci USA* 90: 2905–2909
37 Weeks DL, Melton DA (1987) A maternal mRNA localized to the vegetal hemisphere in *Xenopus* eggs codes for a growth factor related to TGF-β. *Cell* 51: 861–867
38 Ozkaynak E, Schnegelsberg PN, Jin DF, Clifford GM, Warren FD, Drier EA, Oppermann H (1992) Osteogenic protein-2. A new member of the transforming growth factor-beta superfamily expressed early in embryogenesis. *J Biol Chem* 267: 25220–25227
39 Wharton KA, Thomsen GH, Gelbart WM (1991) *Drosophila 60A* gene, another transforming growth factor β family member, is closely related to human bone morphogenetic proteins. *Proc Natl Acad Sci USA* 88: 9214–9218
40 Doctor JS, Jackson PD, Rashka KE, Visalli M, Hoffmann FM (1992) Sequence, biochemical characterization and developmental expression of a new member of the TGF-β superfamily in *Drosophila melanogaster*. *Dev Biol* 151: 491–505
41 Dale L, Jones MC (1999) BMP signalling in early *Xenopus* development. *Bioesseys* 21: 751–760
42 Chang C, Hemmati-Brivanlou A (1999) *Xenopus* GDF6, a new antagonist of noggin and a partner of BMPs. *Development* 126: 3347–3357
43 Hemmati-Brivanlou A, Thomsen GH (1995) Ventral mesodermal patterning in *Xenopus* embryos: expression patterns and activities of BMP-2 and BMP-4. *Dev Genet* 17: 78–89

44 Lahder R, Mohun RJ, Smith JC, Snape AM (1996) Xom: a Xenopus homeobox gene that mediates the early effects of BMP-4. *Development* 122: 2385–2394
45 Suzuki A, Ueno N, Hemmati-Brivanlou A (1997) Xenopus msx1 mediates epidermal induction and neural inhibition by BMP-4. *Development* 124: 3037–3044
46 Kaufmann E, Paul H, Friedle H, Metz A, Scheucher M, Clement JH, Knochel W (1996) Antagonistic actions of activin A and BMP-2/4 control dorsal lip-specific activation of the early response gene XFD-1' in *Xenopus laevis* embryos. *EMBO J* 15: 6739–6749
47 Dosch R, Gawantka V, Delius H, Blumenstock C, Niehrs C (1997) BMP-4 acts as a morphogen in dorsoventral mesoderm patterning in *Xenopus*. *Development* 124: 2325–2334
48 Jones CM, Lyons KM, Lapan PM, Wright CVE, Hogan BLM (1992) DVR-4 (bone morphogenetic protein-4) as a posterior-ventralizing factor in *Xenopus* mesoderm induction. *Development* 115: 639–647
49 Jones CM, Smith JC (1998) Establishment of a BMP-4 morphogen gradient by long-range inhibition. *Dev Biol* 194: 12–17
50 Dosch R, Niehrs C (2000) Requirement for anti-dorsalizing morphogenetic protein in organizer patterning. *Mech Dev* 90: 195–203
51 Piccolo S, Sasai Y, Lu B, De Robertis EM (1996) Dorsoventral patterning in *Xenopus*: inhibition of ventral signals by direct binding of chordin to BMP-4. *Cell* 86: 589–598
52 Zimmerman LB, De Jesus-Escobar JM, Harland RM (1996) The Spemann organizer signal noggin binds and inactivates bone morphogenetic protein 4. *Cell* 86: 599–606
53 Iemura S, Yamamoto TS, Takagi C, Uchiyama H, Natsume T, Shimasaki S, Sugino H, Ueno N (1998) Direct binding of follistatin to a complex of bone-morphogenetic protein and its receptor inhibits ventral and epidermal cell fates in early *Xenopus* embryo. *Proc Natl Acad Sci USA* 95: 9337–9342
54 Capdevila J, Johnson RL (1998) Endogenous and ectopic expression of noggin suggests a conserved mechanism force regulation of BMP function during limb and somite patterning. *Dev Biol* 197: 205–217
55 Furthauer M, Thisse B, Thisse C (1999) Three different noggin genes antagonize the activity of bone morphogenetic proteins in the zebrafish embryo. *Dev Biol* 214: 181–196
56 Bauer H, Meier A, Hild M, Stachel S, Economides A, Hazelett D, Harland RM, Hammerschmidt M (1998) Follistatin and noggin are excluded from the zebrafish organizer. *Dev Biol* 204: 488–507
57 Hemmati-Brivanlou A, Kelly OG, Melton DA (1994) Follistatin, an antagonist of activin, is expressed in the Spemann organizer and displays direct neuralizing activity. *Cell* 77: 283–295
58 Fainsod A, Deissler K, Xelin R, Marom K, Epstein M, Pillemer G, Steinberisser H, Blum M (1997) The dorsalizing and neural inducing gene follistatin is an antagonist of BMP-4. *Mech Dev* 63: 39–50
59 Hsu DR, Economides AN, Wang X, Eimon PM, Harland RM (1998) The *Xenopus* dor-

salizing factor Gremlin identifies a novel family of secreted proteins that antagonize BMP activities. *Mol Cell* 1: 673–683

60 Onichtchouk D, Chen YG, Dosch R, Gawantka V, Delius H, Massague J, Niehrs C (1999) Silencing of TGF-beta signalling by the pseudoreceptor BAMBI. *Nature* 401: 480–485

61 Holley SA, Neul JL, Attisano L, Wrana JL, Sasai Y, O'Connor MB, De Robertis EM, Ferguson EL (1996) The *Xenopus* dorsalizing factor noggin ventralizes *Drosophila* embryos by preventing DPP from activating its receptor. *Cell* 86: 607–617

62 Grotewold L, Plum M, Dildrop R, Peters T, Ruther U (2001) Bambi is coexpressed with Bmp-4 during mouse embryogenesis. *Mech Dev* 100: 327–330

63 Piccolo S, Agius E, Lu B, Goodman S, Dale L, De Robertis EM (1997) Cleavage of chordin by the Xolloid metalloprotease suggests a role for proteolytic processing in the regulation of Spemann organizer activity. *Cell* 91: 407–416

64 Hashimoto O, Nakamura T, Shohi H, Shimasaki S, Hayashi Y, Sugino H (1997) A novel role of follistatin, an activin-binding protein, in the inhibition of activin action in rat pituitary cells. Endocytotic degradation of activin and its acceleration by follistatin associated with cell-surface heparan sulfate. *J Biol Chem* 272: 13835–13842

65 Oelgeschlager M, Larrain J, Geissert D, De Robertis EM (2000) The evolutionarily conserved BMP-binding protein twisted gastrulation promotes BMP signalling. *Nature* 405: 757–763

66 Vukicevic S, Luyten FP, Reddi AH (1990) Osteogenin inhibits proliferation and stimulates differentiation in mouse osteoblast-like cells (MC3T3-E1). *Biochem Biophys Res Commun* 166: 750–756

67 Vukicevic S, Paralkar VM, Cunningham NS, Gutkind JS, Reddi AH (1990) Autoradiographic localization of osteogenin binding sites in cartilage and bone during rat embryonic development. *Dev Biol* 40: 209–214

68 Vukicevic S, Helder MN, Luyten FP (1994) Developing human lung and kidney are major sites for synthesis of bone morphogenetic protein-3 (osteogenin). *J Histochem Cytochem* 42: 869–875

69 Vukicevic S, Kopp JB, Luyten FP, Sampath TK (1996) Induction of nephrogenic mesenchyme by osteogenic protein 1 (bone morphogenetic protein 7). *Proc Natl Acad Sci USA* 93: 9021–9026

70 Helder MN, Ozkaynak E, Sampath KT, Luyten FP, Latin V, Oppermann H, Vukicevic S (1995) Expression pattern of osteogenic protein-1 (bone morphogenetic protein-7) in human and mouse development. *J Histochem Cytochem* 43: 1035–1044

71 Takahashi H, Ikeda T (1996) Transcripts for two members of the transforming growth factor-beta superfamily BMP-3 and BMP-7 are expressed in developing rat embryos. *Dev Dyn* 207: 439–449

72 Lyons KM, Hogan BL, Robertson EJ (1995) Colocalization of BMP7 and BMP2 RNAs suggest that these factors cooperatively mediate tissue interactions during murine development. *Mech Dev* 50: 71–83

73 Lyons KM, Pelton RW, Hogan BLM (1989) Patterns of expression of murine Vgr-1 and

BMP 2a suggest that transforming growth factor-β-like genes coordinately regulate aspects of embryonic development. *Gen Development* 3: 1657–1668

74 Martinovic S, Latin V, Suchanek E, Stavljenic-Rukavina A, Sampath TK, Vukicevic S (1996) Osteogenic protein-1 is produced by human fetal trophoblasts *in vivo* and regulates the synthesis of chorionic gonadotropin and progesterone by trophoblasts *in vitro*. *Eur J Clin Chem Clin Biochem* 34: 103–109

75 Vainio S, Karavanova I, Jowett A, Thesleff I (1993) Identification of BMP-4 as a signal mediating secondary induction between epithelial and mesenchymal tissues during early tooth development. *Cell* 75: 45–58

76 Thesleff I (1995a) Homeobox genes and growth factors in regulation of craniofacial and tooth morphogenesis. *Acta Odontol Scand* 53: 129–134

77 Thesleff I, Vaahtokari A, Kettunen P, Aberg T (1995b) Epithelial-mesenchymal signaling during tooth development. *Connect Tissue Res* 32: 9–15

78 Aberg T, Wozney J, Thesleff I (1997) Expression patterns of bone morphogenetic proteins (Bmps) in the developing mouse tooth suggest roles in morphogenesis and cell differentiation. *Dev Dyn* 210: 383–396

79 Helder MN, Karg H, Bervoets TJM, Vukicevic S, Burger EH, D'Souza RN, Woltgens JHM, Karsenty G, Bronckers ALJJ (1998) Bone morphogenetic protein-7 (osteogenic protein-1, OP-1) and tooth development. *J Dent Res* 77: 545–554

80 Peters H, Balling R (1999) Teeth: where and how to make them. *Trends Genet* 15: 59–65

81 Dorai H, Vukicevic S, Sampath TK (2000) Bone morphogenetic protein-7 (osteogenic protein-1) inhibits smooth muscle cell proliferation and stimulates the expression of markers that are characteristic of SMC phenotype *in vitro*. *J Cell Physiol* 184: 37–45

82 Schultheiss TM, Burch JB, Lassar AB (1997) A role for bone morphogenetic proteins in the induction of cardiac myogenesis. *Genes Dev* 11: 451–462

83 Ladd AN, Yatskievych TA, Antin PB (1998) Regulation of avian cardiac myogenesis by activin/TGFbeta and bone morphogenetic proteins. *Dev Biol* 204: 407–419

84 Yamada M, Revelli JP, Eichele G, Barron M, Schwartz RJ (2000) Expression of chick Tbx-2, Tbx-3, and Tbx-5 genes during early heart development: evidence for BMP2 induction of Tbx2. *Dev Biol* 228: 95–105

85 Walters MJ, Wayman GA, Christian JL (2001) Bone morphogenetic protein function is required for terminal differentiation of the heart but not for early expression of cardiac marker genes. *Mech Dev* 100: 263–273

86 Branford WW, Essner JJ, Yost HJ (2000) Regulation of gut and heart left-right asymmetry by context-dependent interactions between xenopus lefty and BMP4 signaling. *Dev Biol* 223: 291–306

87 Yamagishi T, Nakajima Y, Miyazono K, Nakamura H (1999) Bone morphogenetic protein-2 acts synergistically with transforming growth factor-beta3 during endothelial-mesenchymal transformation in the developing chick heart. *J Cell Physiol* 180: 35–45

88 Shimasaki S, Zachow RJ, Li D, Kim H, Iemura S, Ueno N, Sampath K, Chang RJ, Erick-

son GF (1999) A functional bone morphogenetic protein system in the ovary. *Proc Natl Acad Sci USA* 96: 7282–7287

89 Paralkar VM, Vail AL, Grasser WA, Brown TA, Xu H, Vukicevic S, Ke HZ, Qi H, Owen TA, Thompson DD (1998) Cloning and characterization of a novel member of the transforming growth factor-beta/bone morphogenetic protein family. *J Biol Chem* 273: 13760–13767

90 Harris SE , Harris MA, Mahy P, Wozney J, Feng JQ, Mundy GR (1994) Expression of bone morphogenetic protein messenger RNAs by normal rat and human prostate cancer cells. *Prostate* 24: 204–211

91 Smith DM, Grasty RC, Theodosiou NA, Tabin CJ, Nascone-Yoder NM (2000) Evolutionary relationships between the amphibian, avian, and mammalian stomachs. *Evol Dev* 2: 348–359

92 Narita T, Saitoh K, Kameda T, Kuroiwa A, Mizutani M, Koike C, Iba H, Yasugi S (2000) BMPs are necessary for stomach gland formation in the chicken embryo: a study using virally induced BMP-2 and noggin expression. *Development* 127: 981–988

93 Maric I, Poljak L, Zoricic S, Bobinac D, Sampath TK, Maliakal J, Vukicevic S (2001) Systemic administration of BMP-7 accelerates healing of inflammatory bowel disease in rat (unpublished observations)

94 Feng JQ, Harris MA, Ghosh-Choudhury N, Feng M, Mundy GR, Harris SE (1994) Structure and sequence of mouse bone morphogenetic protein-2 gene (BMP-2): comparison of the structures and promoter regions of BMP-2 and BMP-4 genes. *Biochim Biophys Acta* 1218: 221–224

95 Dickinson ME, Kobrin MS, Silan CM, Kingsley DM, Justice MJ, Miller DA, Ceci JD; Lock LF, Lee A, Buchberg AM et al (1990) Chromosomal localization of seven members of the murine TGF-b superfamily suggest close linkage to several morphogenetic mutant loci. *Genomics* 6: 505–520

96 Ceci JD, Kingsley DM, Silan CM, Copeland NG, Jenkins NA (1990) An interspecific backcross linkage map of the proximal half of mouse chromosome 14. *Genomics* 87: 9843–9847

97 Ducy P, Karsenty G (2000) The family of bone morphogenetic proteins. *Kidney Int* 57: 2207–2214

98 Dudley AT, Lyons K, Robertson EJ (1995) A requirement for bone morphogenetic protein-7 during development of the mammalian kidney and eye. *Genes Dev* 9: 2795–2807

99 Luo G, Hofmann C, Bronckers AL, Sohocki M, Bradley A, Karsenty G (1995) BMP-7 is an inducer of nephrogenesis, and is also required for eye development and skeletal patterning. *Genes Dev* 9: 2808–2820

100 Letterio JJ, Geiser AG, Kulkarni AB, Roche NS, Sporn MB, Roberts AB (1994) Maternal rescue of transforming growth factor-β1 null mice. *Science* 264: 1936–1938

101 Borovecki F, Jelic M, Bosukonda D, Sampath K, Vukicevic S. Osteogenic protein-1 (bone morphogenetic protein-7) is available to the fetus through placental transfer during early stages of development. *Kidney Int; in press*

102 Hongbin Z, Bradley A (1996) Mice deficient for BMP-2 are nonviable and have defects in amnion/chorion and cardiac development. *Development* 122: 2977–2986

103 Lyons KM, Pelton RW, Hogan BLM (1990) Organogenesis and pattern formation in the mouse: RNA distribution patterns suggest a role for bone morphogenetic protein-2A (BMP-2A). *Development* 109: 833–844

104 Clement JH, Fettes P, Knochel S, Lef J, Knochel W (1995) Bone morphogenetic protein 2 in early development of Xenopus laevis. *Mech Dev* 52: 357–370

105 Tabas JA, Zasloff M, Wasmuth JJ, Emanuel BS, Altherr MR, McPherson JD, Wozney JM, Kaplan FS (1991) Bone morphogenetic protein: chromosomal localization of human genes for BMP1, BMP2A, and BMP3. *Genomics* 9: 283–289

106 Rao VV, Loffler C, Wozney JM, Hansmann I (1992) The gene for bone morphogenetic protein 2A (BMP2A) is localized to human chromosome 20p12 by radioactive and non-radioactive *in situ* hybridization. *Hum Genet* 90: 299–302

107 Daluiski A, Engstrand T, Bahamonde ME, Gamer LW, Agius E, Stevenson SL, Cox K, Rosen V, Lyons KM (2001) Bone morphogenetic protein-3 is a negative regulator of bone density. *Nat Genet* 27: 84–88

108 Aspenberg P, Basic N, Tagil M, Vukicevic S (2000) Reduced expression of BMP-3 due to mechanical loading: a link between mechanical stimuli and tissue differentiation. *Acta Orthop Scand* 71: 558–562

109 Winnier G, Blessing M, Labosky PA, Hogan BLM (1995) Bone morphogenetic protein-4 is required for mesoderm formation and patterning in the mouse. *Gen Dev* 9: 2105–2116

110 Lawson KA, Pedersen RA (1992) Clonal analysis of cell fate during gastrulation and early neurulation in the mouse. Postimplantation development in the mouse. *CIBA Found* 165: 3–26

111 Duprez D, Bell EJ, Richardson MK, Archer CW, Wolpert L, Bricker PM, Francis-West PH (1996) Overexpression of BMP-2 and BMP-4 alters the size and shape of developing skeletal elements in the chick limb. *Mech Dev* 57: 145–157

112 Shafritz AB, Shore EM, Gannon FH, Zasloff MA, Taub R, Muenke M, Kaplan FS (1996) Overexpression of an osteogenic morphogen in fybrodysplasia ossificans progressiva. *N Engl J Med* 335: 555–561

113 Martinovic S, Kisic V, Mazic S, Basic N, Jakic-Razumovic J, Batinic D, Labar B, Vukicevic S (2001) Expression of bone morphogenetic proteins in stromal cells from human bone marrow long-term culture (unpublished observations)

114 Katoh M, Terada M (1996) Overexpression of bone morphogenetic protein (BMP)-4 mRNA in gastric cancer cel lines of poorly differentiated type. *J Gastroenterol* 31: 137–139

115 Kusafuka K, Yamaguchi A, Kayano T, Fujiwara M, Takemura T (1998) Expression of bone morphogenetic proteins in salivary pleomorphic adenomas. *Virchows Arch* 432: 247–253

116 King JA, Marker PC, Seung KJ, Kingsley DM (1994) BMP5 and the molecular, skeletal, and soft-tissue alterations in short ear mice. *Dev Biol* 166: 112–122

117 Green MC (1968) Mechanism of the pleiotropic effects of the short-ear mutant gene in the mouse. *J Exp Zool* 167: 129–150
118 Kingsley DM, Bland AE, Grubber JM, Marker PC, Russell LB, Copeland NG, Jenkins NA (1992) The mouse short ear skeletal morphogenesis locus is associated with defects in a bone morphogenetic member of the TGFβ superfamily. *Cell* 71: 399–410
119 Hahn GV, Cohen RB, Wozney JM, Levitz CL, Shore EM, Zasloff MA, Kaplan FS (1992) A bone morphogenetic protein subfamily: chromosomal localization of human genes for BMP5, BMP6, and BMP7. *Genomics* 14: 759–762
120 Solloway MJ, Dudley AT, Bikoff EK, Lyons KM, Hogan BL, Robertson EJ (1998) Mice lacking Bmp6 function. *Dev Genet* 22: 321–339
121 Blessing M, Schrimacher P, Kaiser S (1996) Overexpression of bone morphogenetic protein-6 (BMP-6) in the epidermis of transgenic mice: inhibition or stimulation of proliferation depending on the pattern of transgene expression and formation of psoriatic lesions. *J Cell Biol* 135: 227–239
122 Zhao GQ, Deng K, Labosky PA, Liaw L, Hogan BL (1996) The gene encoding bone morphogenetic protein 8B is required for the initiation and maintenance of spermatogenesis in the mouse. *Genes Dev* 10: 1657–1669
123 Zhao GQ, Liaw L, Hogan BL (1998) Bone morphogenetic protein 8A plays a role in the maintenance of spermatogenesis and the integrity of the epididymis. *Development* 125: 1103–1112
124 Storm EE, Huynh TV, Copeland NG, Jenkins NA, Kingsley DM, Lee SJ (1994) Limb alterations in brachypodism mice due to mutations in a new member of the TGFβ-superfamily. *Nature* 368: 639–643
125 Storm EE, Kingsley DM (1996) Joint patterning defects caused by single and double mutations in members of the bone morphogenetic protein (BMP) family. *Development* 122: 3969–3979
126 Francis-West PH, Abdelfattah A, Chen P, Allen C, Parish J, Ladher R, Allen S, MacPherson S, Luyten FP, Archer CW (1999) Mechanisms of GDF-5 action during skeletal development. *Development* 126: 1305–1315
127 Francis-West PH, Parish J, Lee K, Archer CW (1999) BMP/GDF-signalling interactions during synovial joint development. *Cell Tissue Res* 296: 111–119
128 Chang SC, Hoang B, Thomas JT, Vukicevic S, Luyten FP, Ryba NJ, Kozak CA, Reddi AH, Moos M Jr (1994) Cartilage-derived morphogenetic proteins. New members of the transforming growth factor-beta superfamily predominantly expressed in long bones during human embryonic development. *J Biol Chem* 269: 28227–28234
129 Thomas JT, Lin K, Nandedkar M, Camargo M, Cervenka J, Luyten FP (1996) A human chondrodysplasia due to a mutation in a TGF-β superfamily member. *Nat Gen* 12: 315–8
130 Thomas JT, Kilpatrick MW, Lin K, Erlacher L, Lembessis P, Costa T, Tsipouras P, Luyten FP (1997) Disruption of human limb morphogenesis by a dominant negative mutation in CDMP1. *Nat Genet* 17: 58–64
131 Wolfman NM, Hattersley G, Cox K, Celeste AJ, Nelson R, Yamaji N, Dube JL, DiBla-

sio-Smith E, Nove J, Song JJ et al (1997) Ectopic induction of tendon and ligament in rats by growth and differentiation factors 5, 6 and 7, members of the TGF-beta gene family. *J Clin Invest* 100: 321–330
132 Lee KJ, Mendelsohn M, Jessell TM (1998) Neuronal patterning by BMPs: a requirement fir GDF7 in the generation of a discrete class of commissural interneurons in the mouse spinal cord. *Genes Dev* 12: 3394–3407
133 McPherron AC, Lawler AM, Lee SJ (1997) Regulation of skeletal muscle mass in mice by a new TGF-beta superfamily member. *Nature* 387: 83–90
134 Elvin JA, Changning Y, Wang P, Nishimori K, Matzuk MM (1999) Molecular characterization of the follicle defects in the growth differentiation factor 9-deficient ovary. *Mol Endocrin* 6: 1018–1035
135 Elvin JA, Yan C, Matzuk MM (2000) Oocyte-expressed TGF-β superfamily members in female fertility. *Mol Cell Endocrin* 159: 1–5
136 Zhao R, Lawler AM, Lee SJ (1999) Characterization of GDF-10 expression patterns and null mice. *Dev Biol* 212: 68–79
137 Katagiri T, Boorla S, Frendo JL, Hogan BL, Karsenty G (1998) Skeletal abnormalities in doubly heterozygous Bmp4 and Bmp7 mice. *Dev Genet* 22: 340–348
138 Solloway MJ, Robertson EJ (1999) Early embryonic lethality in Bmp5;Bmp7 double mutant mice suggests functional redundancy within the 60A subgroup. *Development* 126: 1753–1768
139 Brunet LJ, McMahon JA, McMahon AP, Harland RM. (1998) Noggin, cartilage morphogenesis, and joint formation in the mammalian skeleton. *Science* 280: 1455–1457
140 Lemaire P, Yasuo H (1998) Developmental signalling: A careful balancing act. *Curr Biol* 8: R228–R231
141 Drossopoulou G, Lewis KE, Sanz-Ezquerro JJ, Nikbakht N, McMahon AP, Hofmann C, Tickle C (2000) A model for anteroposterior patterning of the vertebrate limb based on sequential long- and short-range Shh signalling and BMP signalling. *Development* 127: 1337–1348
142 Dahn RD, Fallon JF (1999) Limbiting outgrowth: BMPs as negative regulators in limb development. *Bioessays* 21: 721–725
143 Hurle JM (1999) Role of BMPs in digit morphogenesis. First European Conference on BMPs, Zagreb, 1998, A70. *Bone* 24: 426
144 Enomoto-Iwamoto M, Nakamura T, Aikawa T, Higuchi Y, Yuasa T, Yamaguchi A, Nohno T, Noji S, Matsuya T, Kurisu K et al (2000) Hedgehog proteins stimulate chondrogenic cell differentiation and cartilage formation. *J Bone Miner Res* 15: 1659–1668
145 Merino R, Gana Y, Macias D, Economides AN, Sampath TK, Hurle JM (1998) Morphogenesis of digits in the avian limb is controlled by FGFs, TGFbetas, and noggin through BMP signaling. *Dev Biol* 200: 35–45
146 Capdevila J, Tsukui T, Rodriquez Esteban C, Zappavigna V, Izpisua Belmonte JC (1999) Control of vertebrate limb outgrowth by the proximal factor Meis2 and distal antagonism of BMPs by Gremlin. *Mol Cell* 4: 839–849
147 Nifuji A, Kellermann O, Kuboki Y, Wozney JM, Noda M (1997) Perturbation of BMP

signaling in somitogenesis resulted in vertebral and rib malformations in the axial skeletal formation. *J Bone Miner Res* 12: 332–342

148 Nifuji A, Kellermann O, Noda M (1999) Noggin expression in a mesodermal pluripotent cell line C1 and its regulation by BMP. *J Cell Biochem* 73: 437–444

149 Nifuji A, Noda M (1999) Coordinated expression of noggin and bone morphogenetic proteins (BMPs) during early skeletogenesis and induction of noggin expression by BMP-7. *J Bone Miner Res* 14: 2057–2066

150 Pizette S, Niswander L (1999) BMPs negatively regulate structure and function of the limb apical ectodermal ridge. *Development* 126: 883–894

151 Merino R, Rodriguez-Leon J, Macias D, Ganan Y, Economides AN, Hurle JM (1999) The BMP antagonist Gremlin regulates outgrowth, chondrogenesis and programmed cell death in the developing limb. *Development* 126: 5515–5522

152 Grimsrud CD, Romano PR, D'Souza M, Puzas JE, Reynolds PR, Rosier RN, O'Keefe RJ (1999) BMP-6 is an autocrine stimulator of chondrocyte differentiation. *J Bone Miner Res* 14: 475–482

153 Ito H, Akiyama H, Shigeno C, Nakamura T (1999) Bone morphogenetic protein-6 and parathyroid hormone-related protein coordinately regulate the hypertrophic conversion in mouse clonal chondrogenic EC cells, ATDC5. *Biochim Biophys Acta* 1451: 263–270

154 Monsoro-Burq A, Le Douarin N (2000) Left-right asymmetry in BMP4 signalling pathway during chick gastrulation. *Mech Dev* 97: 105–108

155 Monsoro-Burq A, Le Douarin (2001) BMP4 plays a key role in left-right patterning in chick embryos by maintaining sonic hedgehog asymmetry. *Molecular Cell* 7: 789–799

156 Hartmann C, Tabin CJ (2001)Wnt-14 plays a pivotal role in inducing synovial joint formation in the developing appendicular skeleton. *Cell* 104: 341–351

157 Murtaugh LC, Chyung JH, Lassar AB (1999) Sonic hedgehog promotes somitic chondrogenesis by altering the cellular response to BMP signaling. *Genes Dev* 13: 225–237

158 Zhang Y, Zhang Z, Zhao X, Yu X, Hu Y, Geronimo B, Fromm SH, Chen YP (2000) A new function of BMP4: dual role for BMP4 in regulation of Sonic hedgehog expression in the mouse tooth germ. Development 127: 1431–1433

159 Wang YH, Rutherford B, Upholt WB, Mina M (1999) Effects of BMP-7 on mouse tooth mesenchyme and chick mandibular mesenchyme. *Dev Dyn* 216: 320–335

160 Rice DP, Kim HJ, Thesleff I (1999) Apoptosis in murine calvarial bone and suture development. *Eur J Oral Sci* 107: 265–275

161 Thesleff I (1999) The role of BMPs in craniofacial and tooth development. First European Conference on BMPs. Zagreb, 1998, A69. *Bone* 24: 426

162 Thesleff I, Aberg T (1999) Molecular regulation of tooth development. *Bone* 25: 123–125

163 Rosen V, Cox K, Hattersley G (1996) Bone morphogenetic proteins. In: JP Bilezikian, LG Raisz, GA Rodan (eds): *Principles of bone biology*. Academic Press, San Diego, 661–671

164 Knutsen R, Wergedal JE, Sampath TK, Baylink DJ, Mohan S (1993) Osteogenic protein-

1 stimulates proliferation and differentiation of human bone cells *in vitro*. *Biochem Biophys Res Commun* 194: 1352–1358
165 Kim GY, Lee HH, Cho SW (1994) Differential effects of transforming growth factor-beta 1 and bone morphogenetic proteins in cultured rat osteogenic sarcoma and mink lung epithelial cells. *Biochem Mol Biol Int* 33: 253–261
166 Iwasaki M, Nakahara H, Nakase T, Kimura T, Takaoka K, Caplan AI, Ono K (1994) Bone morphogenetic protein 2 stimulates osteogenesis but does not affect chondrogenesis in osteochondrogenic differentiation of periosteum-derived cells. *J Bone Miner Res* 9: 1195–1204
167 Chen P, Vukicevic S, Sampath TK, Luyten FP (1995) Osteogenic protein-1 promotes growth of chick sternal chondrocytes in serum-free cultures. *J Cell Sci* 108: 105–114
168 Harada S, Sampath TK, Aubin JE, Rodan GA (1997) Osteogenic protein-1 up-regulation of the collagen X promoter activity is mediated by a MEF-2-like sequence and requires an adjacent AP-1 sequence. *Mol Endocrinol* 11: 1832–1845
169 Chen P, Vukicevic S, Sampath TK, Luyten FP (1993) Bovine articular chondrocytes do not undergo hypertrophy when cultured in the presence of serum and osteogenic protein-1. *Biochem Biophys Res Commun* 197: 1253–1259
170 Vukicevic S, Luyten FP, Reddi AH (1989) Stimulation of the expression of osteogenic and chondrogenic phenotypes *in vitro* by osteogenin. *Proc Natl Acad Sci USA* 86: 8793–8797
171 Martinovic S, Basic N, Dorai H, Sampath TK, Vukicevic S (2001) The requirement of bone morphogenetic protein for maintenance and stimulation of osteoblastic differentiation in mouse osteoblastic MC3T3-E1 cells (unpublished observations)
172 Dorai H, Sampath TK (2001) Bone morphogenetic protein-7 modulates genes that maintain the vascular smooth muscle cell phenotype in culture. J *Bone Joint Surg Am* 83–A (Suppl 1): S70–S78
173 Vukicevic S, Basic V, Rogic D, Basic N, Shih MS, Shepard A, Jin D, Dattatreyamurty B, Jones W, Dorai H, Ryan S, Griffiths D, Maliakal J, Jelic M, Pastorcic M, Stavljenic A, Sampath TK (1998) Osteogenic protein-1 (bone morphogenetic protein-7) reduces severity of injury after ischemic acute renal failure in rat. *J Clin Invest* 102: 202–214
174 Willette RN, Gu JL, Lysko PG, Anderson KM, Minehart H, Yue T (1999) BMP-2 gene expression and effects on human vascular smooth muscle cells. *J Vasc Res* 36: 120–125
175 Bhatia M, Bonnet D, Wu D, Murdoch B, Wrana J, Gallacher L, Dick JE (1999) Bone morphogenetic proteins regulate the developmental program of human hematopoietic stem cells. *J Exp Med* 189: 1139–1147
176 Detmer K, Steele TA, Shoop MA, Dannawi H (1999) Lineage-restricted expression of bone morphogenetic protein genes in human hematopoietic cell lines. *Blood Cells Mol Dis* 25: 310–323
177 Church VL, Harvey B, Ashton BA (1998) Differential bone morphogenetic expression by pluripotent bone marrow stromal stem cells. *Biochem Soc Transactions* 26: S25
178 Kaplan FS, Glaser DL, Shlomchik W, Emerson SG, Cannon FH, Shore EM (1998) Osteogenic morphogens in hematopoietic cells: rare genetic insights into the origin of

heterotopic bone and marrow. First European Conference on BMPs, Zagreb, 1998, A27. *Bone* 24: 415

179 Ploemacher RE, Engels LJ, Mayer AE, Thies S, Neben S (1999) Bone morphogenetic protein 9 is a potent synergistic factor for murine hemopoietic progenitor cell generation and colony formation in serum-free cultures. *Leukemia* 13: 428–437

180 Friedlaender GE, Perry CR, Cole JD, Cook SD, Cierry G, Muschler GF, Zych GA, Calhoun JH, LaForte AJ, Yin S (2001) Osteogenic protein-1 (bone morphogenetic protein-7) in the treatment of tibial nonunions. *J Bone Joint Surg (Am)* 83: 151–158

181 Vukicevic S, Stavljenic A, Pecina M (1995) Discovery and clinical applications of bone morphogenetic proteins. *Eur J Clin Chem Clin Biochem* 33: 661–671

182 Katic V, Majstorovic L, Maticic D, Pirkic B, Yin S, Kos J, Martinovic S, McCartney JE, Vukicevic S (2000) Biological repair of thyroid cartilage defects by osteogenic protein-1 (bone morphogenetic protein-7) in dog. *Growth Factors* 17: 221–232

# Preclinical models of recombinant BMP induced healing of orthopedic defects

*Stephen D. Cook*[1] *and David C. Rueger*[2]

[1]Tulane University School of Medicine, Department of Orthopaedic Surgery, 1430 Tulane Avenue – SL32, New Orleans, LA 70112, USA; [2]Stryker Biotech, Research Department, 35 South Street, Hopkinton, MA 01748, USA

## Introduction

Segmental bone loss and nonunion, whether after reconstructive surgery, lesion excision, or fracture, can present complex orthopedic problems. An important part of the therapeutic approach to bone defects is the implantation of materials that support new bone formation. Such implants may hasten healing by three mechanisms: osteoconduction, osteogenesis, and osteoinduction [1–4]. In osteoconduction, the implanted material serves as an inert scaffold, or trellis, for the ingrowth of host bone. This includes the differentiation and maturation within the implant of host osteoprogenitor cells, with ingrowth of vascular elements. Ideally, "creeping substitution" then replaces the implant with new bone to form a functional skeletal element. Osteogenesis is the synthesis of new bone by surviving pre-osteoblasts and osteoblasts within a bone autograft. These cells proliferate and mature into centers of new bone formation. Osteoinduction is the formation of new bone by the active recruitment of host pluripotent cells that differentiate into chondroblasts and osteoblasts. This review focuses on the osteoinduction process produced as a result of the biological activity of certain members of the family of proteins called bone morphogenetic proteins (BMPs).

In recent years, the search for an acceptable substitute for autogenous and allograft bone has involved proteins that induce bone formation *in vivo*. It is now well accepted that osteoinduction is controlled, at least in part, by bone matrix proteins referred to as BMPs or OPs (osteogenic proteins) [5–7]. These proteins have been isolated from the bones of a variety of mammalian species, including mouse, rat, bovine, monkey, and man [8–15] as well as from clonal osteogenic sarcoma cell lines [16, 17]. In addtion, the genes for BMPs have been identified and the proteins produced by recombinant DNA methods [18–21].

The BMPs comprise a subgrouping of the TGF-β superfamily of proteins [6, 7, 22, 23] and number about 15 members. Not all BMPs have been shown to be osteoinductive; those that have been demonstrated to have such biological activity are BMP-2, 4, 5, 6, 7 (OP-1) and GDF5 (CDMP-1 or MP52). Other BMP members

Bone Morphogenetic Proteins, edited by Slobodan Vukicevic and Kuber T. Sampath

are either inactive in osteoinductive assays or have not yet been evaluated. Comparison of the amino acid sequences of the osteoinductive BMPs within their highly conserved seven cysteine domain to those of OP-1 (BMP-7) reveals that OP-1 (BMP-7) is most closely related to the BMP-5/6 gene products (88%/87%), to a lesser extent BMP 2/4 (60%/58%), and to a much lesser extent GDF-5 (51%). In regard to the TGF-β's themselves, OP-1 shows 35–78% homology [24].

The biological activity of BMPs was initially evaluated by implantation of the BMPs with a collagen carrier in subcutaneous sites in rats. Osteoinductive BMPs induce a sequence of cellular events which leads to the formation of fully functional new bone [24]. The BMP containing implants recruit nearby mesenchymal stem cells and trigger their differentiation into chondrocytes within 5 to 7 days. Upon capillary invasion, the chondrocytes become calcified, hypertrophied and are subsequently replaced by newly formed bone within 9 to 12 days. The mineralized bone is extensively remodeled, and becomes occupied by ossicles filled with functional bone marrow elements by 14 to 21 days.

Several recombinant BMPs have also been tested in bony defect models to evaluate their ability to induce bone to accomplish repair. OP-1 (BMP-7) and BMP-2 have been tested in a variety of animal species including rats, rabbits, dogs, goats, sheep and non-human primates. These BMPs were observed to induce new bone successfully in each of these species. More recently, a third member of the BMP family, GDF-5, has also been shown to repair defects in bony models. This chapter reviews the highlights from these studies which include repair of large segmental gaps, acceleration of fracture healing, enhancement of bone graft incorporation, improvement of osseointegration of metal prostheses, acceleration of the distraction osteogenesis process, and promotion of spinal fusion.

## Restoration of large diaphyseal segmental bone defects

The evaluation of the inductive properties of recombinant BMPs in bony sites was first done in surgically created large critical size diaphyseal segmental defects. Implantation of BMPs with carrier matrices in these defects led to the regeneration of new bone which is fully functional both biologically and biomechanically. These results have been demonstrated in rats, rabbits, dogs, sheep and nonhuman primates [25–36]. Table 1 describes the large animal studies that have been published; these include studies with OP-1, BMP-2 or GDF-5 and either collagen or polylactic acid/polyglycolic acid polymers as delivery materials.

OP-1 studies primarily used highly purified bone-derived collagen particles as the carrier material (Fig. 1). The large animal models included both ulna and tibia segmental defect. The dog has been the species used for most investigations although the healing in non-human primates has also been evaluated. The study results demonstrated that both the rate and quality of the osseous union were better than

*Table 1 - Critical size segmental gap studies in large animal*

| Citation | Year | Species | BMP | Carrier | Model |
|---|---|---|---|---|---|
| Gerhart et al. [29] | 1993 | Sheep | BMP-2 (1.5 mg) | Demin./GuHCl-extracted sheep bone collagen with autologous blood | 2.5 cm osteotomy in the femur; plate fixation; 12 week evaluation |
| Cook et at. [30] | 1994 | Dog | OP-1/BMP-7 (0.62, 1.2 or 2.0 mg) | Purified bovine bone collagen | 2.5 cm osteotomy in the ulna; no fixation; 16 week evaluation |
| Kirker-Head et al.[31] | 1995 | Sheep | BMP-2 (1.5 mg) | Demin./GuHCl-extracted sheep bone collagen with autologous blood | 2.5 cm osteotomy in the femur; plate fixation; 12 month evaluation |
| Cook et al. [32] | 1995 | Monkey | OP-1/BMP-7 (0.25, 0.5 1.0 or 2.0 mg) | Purified bovine bone collagen | 2 cm osteotomy in the ulna or tibia; no fixation for ulna and intra-medullary rod fixation for the tibia; 20 week evaluation |
| Kirker-Head et al. [33] | 1998 | Sheep | BMP-2 (2 or 4mg) | Polylactic/polyglycolic acid polymer with autologous blood | 2.5 cm osteotomy in the femur; plate fixation; 12 month evaluation |
| Itoh et al. [34] | 1998 | Dog | BMP-2 (0.04, 0.16 or 0.64 mg) | Polylactic/polyglycolic acid/ gelatin sponge | 2 cm osteotomy in the ulna; plate fixation; 16 week evaluation |
| Sciadini et al. [35] | 2000 | Dog | BMP-2 (0.15, 0.6 or 2.4 mg) | Bovine collagen sponge | 2.5 cm osteotomy in the radius; external fixators; 48 week evaluation |
| Spiro et al. [36] | 2000 | Baboon | GDF-5 (0.022, 0.22 or 2.22 mg) | Mineralized bovine collagen matrix (Healos) | 1.5 cm osteotomy in fibula; plate fixation; 21 week evaluation |

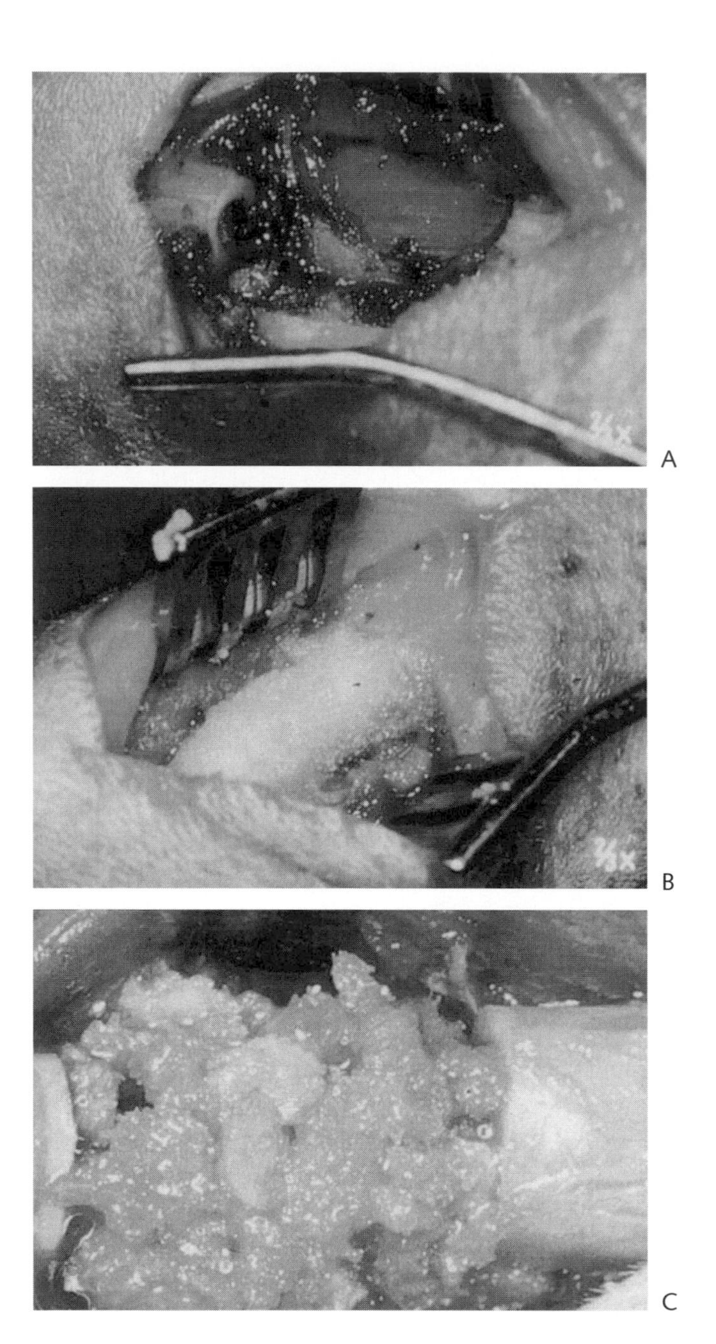

*Figure 1*
*Application of the OP-1/Collagen Implant into a segmental defect in the primate ulna. (A) shows the fresh defect of 2 cm. (B) shows the defect filled with the OP-1/collagen implant. (C) shows the defect filled with morselized autograft.*

that achieved by autogenous bone graft controls. There was a dose dependence relating the amount of OP-1 to the amount of bone formed in the range of 0.25 to 2.0 mg OP-1 per defect. Implantation of the carrier material alone, or no implant material in the defects resulted in fibrous unions in all cases. In the primate ulna defect model, OP-1 was shown to be capable of healing defects which did not heal with autogenous bone [32]. In addition to the bone derived Type 1 collagen matrix carrier, other carriers such as polylactic/polyglycolic polymers and calcium phosphate materials have been evaluated in the segmental defect models although with less acceptable healing rates and characteristics [37]. Finally, OP-1 alone without a carrier material has been implanted in critical size defects and shown to result in healing similar to that obtained with the collagen carrier material [38].

Radiographically, bone formation first appears in segmental defects implanted with OP-1 as calcifications with a diffuse pattern resembling fracture callus at 2 to 3 weeks postoperative. This occurs at similar times in rabbits, dogs and primates (Fig. 2). The island of newly formed calcified tissue then coalesces and remodels to form normal appearing bone which bridges and fills the defect. By 4 to 8 weeks, the new bone is sufficiently remodeled such that the beginning of new cortices have formed. The mass of bridging new bone continues to remodel with the new cortices being fully integrated and continuous with the cortices of the ulna or tibia at later time periods. The quantity and rate of bone formation is dependent upon the amount of OP-1 implanted; although the end result is equivalent above a threshold concentration which is both species and carrier dependent [26, 30, 32].

The explanted ulna and tibia have contours and appearance similar to that of the intact limb. Mechanically, when tested in torsion, the OP-1 treated defects restore a high degree of mechanical strength. In all animal models, close to 100% of the intact limb strength is achieved in OP-1 treated defects which is significantly greater than that achieved in equivalent defects treated with autogenous bone.

Histologically, at 2 to 3 weeks in OP-1 treated sites, cell proliferation is evident and phenotype differentiation is observed. At later time periods, calcifying tissue and plump chondrocytes, as well as osteoblasts are present. By 12 weeks, healed OP-1 treated defects reveal dense lamellar bone with some areas of woven bone present. Bone continuity is observed at the original cortex-new bone interface. From 12 to 20 weeks, well remodeled new cortices with a medullary canal are observed. The medullary canal is filled with fully functional marrow elements (Fig. 3).

Several large animal studies were published using BMP-2 in sheep and dog models [29, 31, 33–35]. These involved critical sized defects in the femur, ulna or radius and used a variety of carrier materials. The results suggested that BMP-2 is similar to OP-1 (BMP-7) in being able to achieve union across the defect, both by radiographic analysis and by mechanical strength testing. In addition, a similar dose of BMP-2 was used to achieve union, that being 1–2 mg per defect. In one study, the healing process was followed for 12 months using the sheep model. The results demonstrated that the bone healing process initiated by BMP-2 resulted in stable

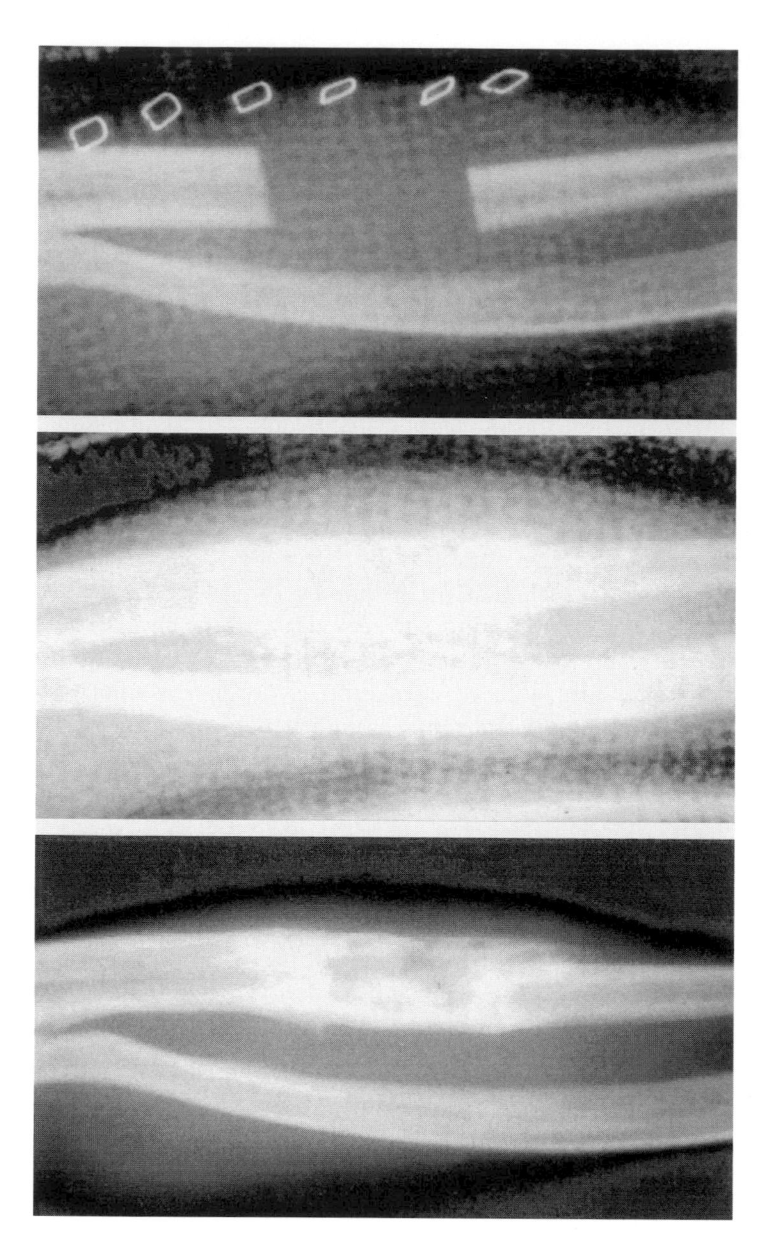

*Figure 2*
*Bone formation is rapid with the OP-1 device. The top radiograph shows a primate ulna critical size defect model immediately postop after implantation of the OP-1/collagen implant. By 6 weeks, new bone completely fills the defect space and the cast is removed (middle radiograph). The new bone continues to remodel until sacrifice at 20 weeks (bottom radiograph).*

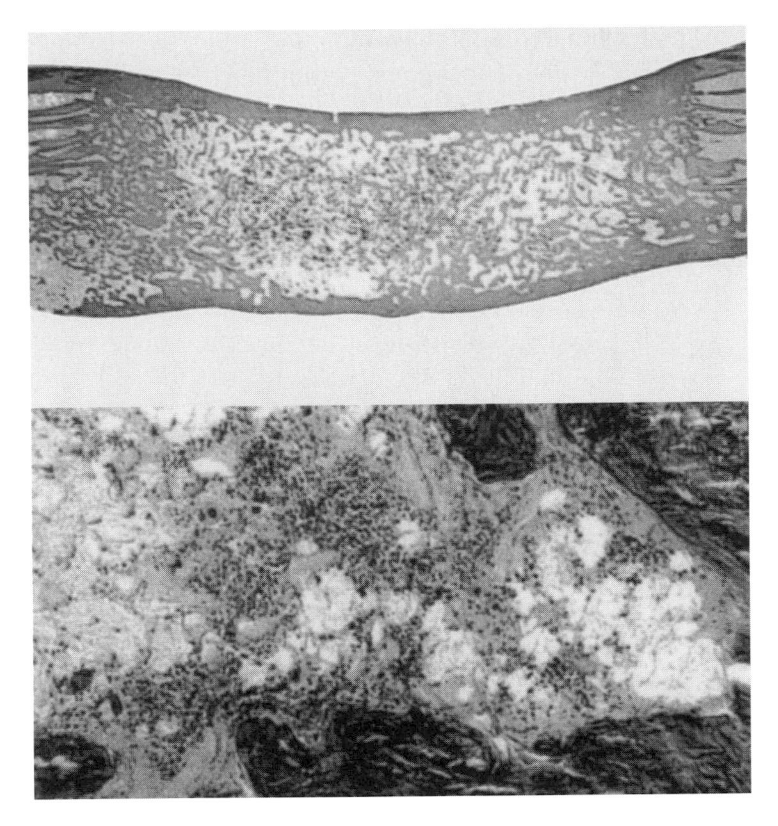

*Figure 3*
*Histological analysis shows that the newly formed bone remodels into new cortices with a well developed medullary cavity. Functional marrow elements are present in the primate ulna critical site defect model at 20 weeks.*

bone that was physiologically normal at 12 months with no adverse responses observed. This study also demonstrated that although the remodeling achieved was extensive, it did not yet appear to be complete. Finally, comparisons of carrier materials such as polylactic/polyglycolic acid (PL/PG) polymers and a bovine collagen sponge demonstrated the ability of these materials to serve as delivery materials for BMP. However, an early inflammatory response was observed for the PL/PG polymers and cyst-like void formation was reported in defects treated with higher doses of BMP-2 using the collagen sponge; these were not observed in the studies using demineralized, guanidine-extracted sheep bone collagen.

GDF-5 has also been reported to be successfully used to achieve union across a critical size long bone defect [36]. Although reported to be less active than other BMPs in subcutaneous or intramuscular sites, GDF-5 in combination with a mineralized bovine collagen matrix induced bone formation and union in a baboon fibula defect using a dose range similar to that employed in the OP-1 and BMP-2 studies.

## Acceleration of healing in a noncritical size defect

The use of BMPs in the repair of noncritical size defects has been shown to accelerate the repair process. Studies have evaluated injectable formulations of BMPs using both diaphyseal segmental defect models and closed fracture models. In both models new bone formed significantly faster and restored strength and stiffness earlier than nontreated controls [38-44]. This data suggests a clinical potential for BMPs to be used for injecting into acute fractures to speed the bone healing process.

OP-1 has been evaluated in both the noncritical size segmental and fracture models [38–40]. Most studies have been done using the segmental defect [38]. Bilateral 3.0 mm noncritical size defects were surgically created in the mid-ulna of adult male dogs. After soft tissue closure, the defects were injected with 0.35 mg of OP-1 in an acetate buffer solution on one side while the contralateral defect received a control acetate buffer solution or received no injection. Radiographically, new bone formation was evident at 2 weeks postoperative in OP-1 treated defects. By 4 weeks, new bone had bridged the defect end and continued to increase in density to 8 weeks. By 12 weeks, new radiodense bone filled and bridged the defects and began to remodel. Nontreated and vehicle controls showed little bone activity at 2 and 4 weeks. At 8 weeks, periosteal new bone formed from the host bone ends although bony bridging was not complete until 12 weeks. Torsional strengths of defects treated with OP-1 were significantly greater than controls and approached the strength of the intact ulnae between 4 and 8 weeks (Fig. 4). Histologic findings correlated with radiographic and mechanical testing results. In OP-1 treated defects, maturing bone was well incorporated with the host bone at early time periods. At later time periods, dense bone filled and bridged the defects. In controls, similar repair was not observed until 12 to 16 weeks.

OP-1 has also been evaluated in closed fractures created in the tibia of goats [40]. Using external fixators for stabilization, a single injection of 1 mg OP-1 was introduced into the fracture gap immediately after the fracture occurred. The results demonstrated that OP-1 accelerated the healing by means of stimulation of the normal fracture healing process observed at 2 weeks.

The use of BMP-2 has also been studied in closed fracture models [41-44]. Data have been reported using rat, rabbit and goat models using either implantable or injectable formulations. In general, the data support the use of BMP-2 to accelerate the rate of fracture repair. However, the data also indicated that the method of application can affect the outcome. The data from a rabbit tibia study suggested that solid carriers inhibit callus formation by acting as a mechanical barrier to the

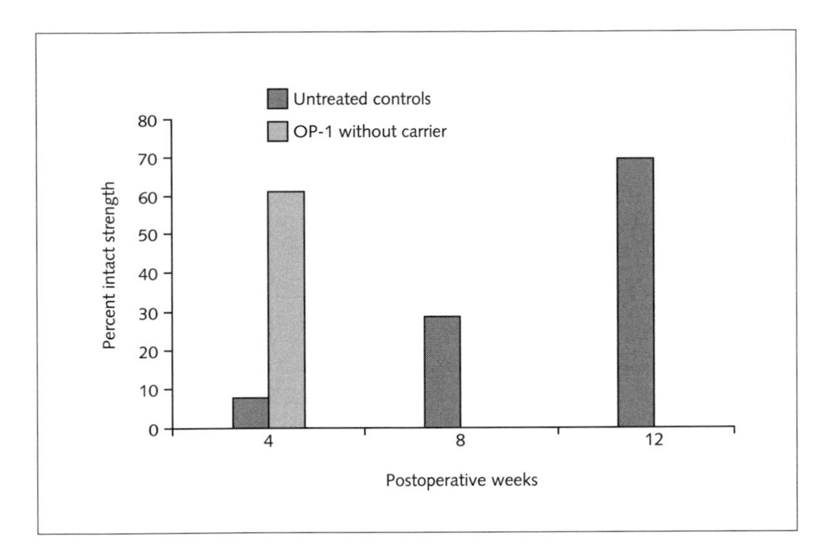


*Figure 4*
*Torsional mechanical strength of OP-1 treated and controls using a canine non-critical size defect model. (Percent of intact ulna strength.)*

migration of cells into the defect site, but when BMP-2 is injected into the fracture site without these carriers the callus develops more rapidly so that the rate of union is accelerated. However, studies using a goat tibia model demonstrated increased callus associated with BMP-2 treatment using an implantable collagen sponge. It is clear that more studies are necessary to fully evaluate the effect of delivery formulations in these models.

## Enhancement of autograft and allograft incorporation

Most of the studies conducted to date utilized collagen or a variety of other carrier materials that provide no initial biomechanical structure or stability. In order to provide such support, the use of BMPs in conjunction with autograft and allograft bone has been investigated [45-48]. This type of application may be especially useful in large defects associated with trauma or in revision total joint procedures. In studies with OP-1 there was observed a dramatic improvement of the biological activity of both autograft and allograft bone resulting in greater new bone formation and earlier graft incorporation [45, 46]. This activity has been observed with both morselized and strut grafts.

Morselized graft studies have been done using a bilateral 2.5 cm critical size osteoperiosteal segmental defect model created in the mid-ulna of dogs. Defects

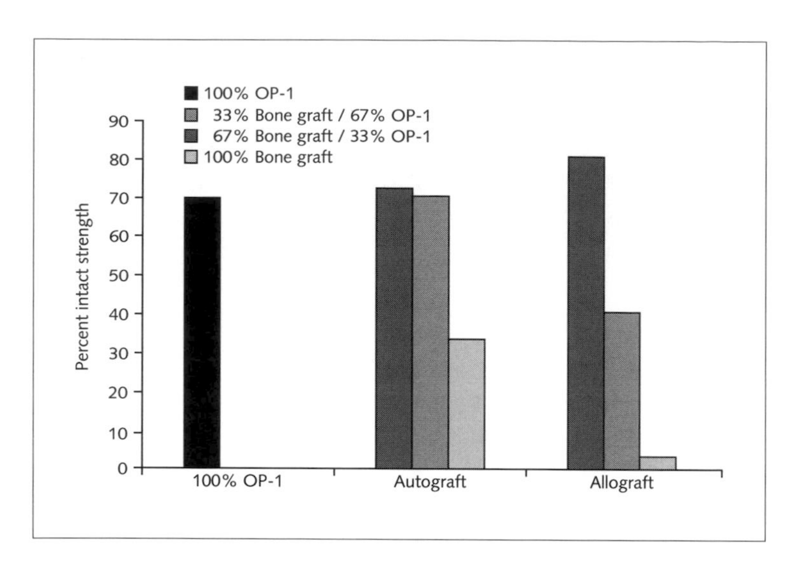


*Figure 5*
*Torsional mechanical strength of canine critical size segmental defects treated with OP-1 and autogenous or allograft bone.*

were treated with the OP-1/collagen implant alone; 1/3 OP-1/collagen implant and 2/3 freeze dried cancellous autograft or allograft; 2/3 OP-1/collagen implant and 1/3 autograft or allograft bone; 100% allograft or 100% autograft bone. The healing was studied radiographically until sacrifice at 12 weeks. After sacrifice, mechanical testing and a histological assessment was conducted. Radiographically, as early as 2–4 weeks significant bone formation was observed in all sites containing OP-1, whereas defects filled with 100% graft material showed no new bone formation until 6 weeks. Defects treated with any amount of OP-1 combined with allograft or autograft demonstrated earlier and greater volume of new bone formation compared to bone graft alone. The amount of new bone formed was proportional to the amount of OP-1 implanted. OP-1 enhanced graft incorporation with the new bone and remodeling of the graft. Only 22% of allograft alone and 67% of autograft alone defects were completely healed at 12 weeks. Defects treated with bone graft and OP-1 or the OP-1 device alone healed in 83% to 100% of cases. Defects treated with any amount of OP-1 were stronger in torsion than the 100% bone graft defects (Fig. 5). Histologically, the amount of new bone, degree of remodeling, and graft incorporation were proportional to the amount of OP-1 implanted. Segmental defects treated with 1/3 bone graft and 2/3 OP-1 demonstrated the most advanced graft incorporation, remodeling and greatest strength compared to the other treatment groups.

Strut graft studies have been done using an allograft strut onlayed to the mid-femur in adult dogs. Each defect received the OP-1/Collagen Implant interposed between the graft and host bone and the graft secured using stainless steel cables. The results demonstrated that the healing of the struts to the femur was dramatically enhanced by the addition of the OP-1. The OP-1 treated sites had significantly greater radiographic and histologic scores at all time period (4, 6 or 8 weeks). Strut healing with the OP-1/Collagen Implant at 4 weeks was superior to control sites at 8 weeks.

Studies have also been reported evaluating the use of BMPs in conjunction with impacted allograft in unloaded bone chamber models in rat and goat tibia [47, 48]. Solutions of BMP-2 or OP-1 were added to freeze dried allograft morsels prior to implantation. The results suggested that the addition of either BMP produced a strong stimulatory effect on bone graft incorporation by increasing bone ingrowth.

## Improvement of osseointegration of prosthetic devices

The use of BMPs may provide a means to obtain both early and long-term prosthesis stabilization due to increased amounts of bone apposition and/or ingrowth to the implant. BMPs have been investigated both as a coating on implants and in conjunction with a carrier material using porous and smooth surfaced metal implants [49–54]. The results of these studies have indicated that BMPs can promote enhanced osseointegration of metal implants by inducing significant new bone formation in implant-bone interface gap spaces.

Initial OP-1 studies evaluated treated and untreated porous 6.0 mm cobalt-chromium alloy implants after placement transcortically through the femoral diaphysis of adult dogs bilaterally [49]. The OP-1 was deposited on the surface and internal pores of the metal implant. At 3 and 6 weeks post-sacrifice, the implants were subjected to axial push out testing and quantitative histologic analysis of bone ingrowth. The porous metal implants treated with OP-1 demonstrated greater surface bone ingrowth and apposition compared to control implants although little difference was observed in mechanical attachment strength. Bone ingrowth was found to be present throughout the porous structure in implants treated with OP-1 rather than only at or near the surface as observed in nontreated specimens.

In several studies OP-1 formulated with collagen carrier was evaluated [50, 51]. In one study the right and left mandibular premolars of adult dogs were extracted and HA coated and uncoated dental implants were placed into the fresh extraction sites [50]. This model creates 1 to 3 mm gaps around the top of the implant. The left side implants were placed with the OP-1 collagen implant packed into the interface gap spaces. All animals were sacrificed at 12 weeks post-operative and evaluated histologically. The use of the OP-1 collagen implant resulted in increased new bone formation in close apposition to the implant surface. In both smooth and grooved surfaced implants greater bone apposition and filling of interface gap spaces was observed in sites treated with OP-1 compared to sites in which the device was not present. A similar study has been done using unloaded cylindrical titanium alloy implants surrounded by a 3-mm gap in the femoral condyles of dogs [51]. This study also demonstrated that the OP-1/collagen implant is capable of enhancing mechanical fixation and peri-implant bone formation. More recently a similar 3-mm gap model in the dog humerus was used to evaluate a mixture of the OP-1/collagen implant and impacted allograft [52]. The data demonstrated that the composite containing the low dose of the OP-1/ collagen had some effect on peri-implant bone formation, but no effect on implant fixation, and suggested that more model development is necessary since access to a blood supply and stem cells was limited in this model.

OP-1 has also been evaluated using a natural mineral material (BioOss) as carrier. In this study in the miniature pig, the OP-1/BioOss implant was evaluated in a sinus floor augmentation model that included simultaneous placement of titanium dental implants [53]. It was concluded that the application of OP-1 produced a more rapid and enhanced osseointegration of the implants when compared to the BioOss alone.

Studies have also been reported evaluating the use of BMP-2 to enhance osseointegration of metal implants [54, 55]. In monkey or dog models the BMP-2/Collagen sponge material was used in conjunction with titanium dental implants in mandibular defects. The data from these studies showed that the BMP-2 stimulated bone formation and osseointegration of the implant.

## Acceleration of distraction osteogenesis

A new area of investigation is the use of BMPs in conjunction with distraction osteogenesis. The process of limb lengthening is an extremely long and painful procedure and thus a procedure that could accelerate the bone formation process would be of tremendous therapeutic value. Preliminary reports have been described using both OP-1 and BMP-2 [56–58]. Formulations have been injected at various times during the process, including prior to distraction and during the bone consolidation phase. Models are being evaluated in the tibia or femur of rat, rabbit or sheep. The preliminary results in each of these models support the use of BMPs for increasing the rate of bone formation and shortening the treatment period.

## Promotion of spinal fusions

Although the evaluation of the inductive properties of recombinant BMPs has been done in a variety of bony sites, no one area has received as much attention as the spinal column. Implantation of BMPs with carrier matrices in a variety of models leads to the generation of new bone which has been observed to effectively promote both intertransverse process and interbody fusions [59, 60]. These results have been demonstrated in many different animal species, including rabbits, dogs, goats, sheep, and non-human primates [61–87]. Table 2 describes the large animal studies that have been published; these include investigations with OP-1, BMP-2 or GDF-5 using primarily collagen as the delivery vehicle. However, some studies also describe the use of synthetic polymers, ceramics or autograft bone itself, as alternative materials to deliver BMPs.

Spinal fusions are one of the most common clinical indications where bone grafting is utilized. Thus the spinal column was an appropriate site to evaluate the use of BMPs to replace autolograft bone. Numerous studies have been done in animals to define the dose and delivery material, to determine the long-term outcome of the fusion site and to evaluate the safety in sites that can be exposed to the spinal cord. Most of these studies have been done using intertransverse process fusion models, although, more recently the use of BMPs to achieve interbody fusions has also been evaluated.

The first studies with recombinant BMPs were done using posterolateral intertransverse process fusions in dog models with OP-1 or BMP-2 [61–66]. This type of spinal fusion is a commonly performed procedure and generally utilizes onlay grafting of autogenous corticocancellous bone after decortication of the bony surfaces of the vertebral elements. In the earliest OP-1 study, OP-1 delivered with bone-derived collagen particles was evaluated in a canine posterior spinal fusion model and the results compared to those obtained with autograft bone as well as carrier alone and no implant controls at different levels on the same spine [61]. No instrumentation was used in this model. Radiographic analysis, including computed tomography (CT) and magnetic resonance imaging (MRI), demonstrated a marked difference in the rate in which spinal fusion was obtained. The OP-1 treated fusion segments attained a stable fusion by 6 weeks and were completely fused by 12 weeks post-implantation. The autograft treated sites did not demonstrate complete fusion until 26 weeks post-implantation. The carrier alone and no implant control displayed minimal evidence of new bone formation and did not promote fusion. Mechanically, the OP-1 fusion sites demonstrated excellent torsional stability as early as 6 weeks, which continued to increase with time *in situ*. Autograft sites demonstrated less mechanical stability compared to OP-1 at all time periods. The carrier alone and no implant controls exhibited minimal mechanical stability at all time periods. Histologically, extensive new bone formation was present at 6 weeks in OP-1 sites. A well organized network and complete trabecular incorporation of the spinous process and facets were observed at the 12- and 26-week time periods. In contrast, the autograft bone treated sites did not demonstrate complete graft incorporation or fusion until 26 weeks postoperative. At 6 weeks, some new bone formation was evident with increased amounts present at 12 weeks. The autogenous bone graft treat-

*Table 2 - Spinal fusion studies in large animals*

| Citation | Year | Species | BMP | Carrier | Model |
|---|---|---|---|---|---|
| Cook et al. [61] | 1994 | Dog | OP-1/BMP-7 (2.0 mg) | Purified bovine bone collagen | Posterolateral intertransverse process fusion 4 levels (between T13 and L7); no instrumentation; 26 week evaluation |
| Muschler et al. [66] | 1994 | Dog | BMP-2 (0.4 mg) | Polylactic/polyglycolic acid polymer | Posterolateral lumbar intertransverse process fusion; three levels (L1-L2, L3-L4 and L5-L6); internal plate fixation; 12 week evaluation |
| Sandhu et al. [68] | 1995 | Dog | BMP-2 (2.3 mg) | Porous polylactic acid polymer | Posterolateral lumbar intertransverse process fusion; one level (L4-L5); non-instrumented; 12 week evaluation |
| Sheehan et al. [69] | 1996 | Dog | BMP-2 (1.6 mg) | Tendon collagen and autogenous bone | Posterolateral lumbar intertransverse process process fusion; two levels (T13-L1 and L4-L5 or L2-L3 and L6-L7); non-instrumented; 12 week evaluation |
| Sandhu et al. [71] | 1996 | Dog | BMP-2 (0.058, 0.115, 0.23, 0.46 or 0.92 mg) | Porous polylactic acid polymer | Posterolateral lumbar intertransverse process fusion; single level (L4-L5); non-instrumented; 12 week evaluation |
| Helm et al [93]. | 1997 | Dog | BMP-2 (2 mg) | Tendon collagen or de mineralized bone matrix | Lumbar decompression with contralateral posterior fusion; L3-L7; non-instrumented; 24 week evaluation |
| Sandhu et al. [75] | 1997 | Dog | BMP-2 (0.058, 0.23 or 0.92 mg) | Porous polylactic acid polymer | Posterolateral lumbar intertransverse process process fusion; single level (L4-L5); non-instrumented; 12 week evaluation; with and without decortication |
| Fischgrund et al. [74] | 1997 | Dog | BMP-2 | Combinations of autogenous bone graft, bovine collagen sponge, polyactic/polyglycolic acid sponge and porous polylactic acid polymer | Posterolateral lumbar intertransverse process fusion; three levels (L1-L2, L3-L4, L5-L6); non-instrumented; 8 week evaluation |
| Zdeblick et al. [77] | 1998 | Goat | BMP-2 (0.2 mg) | Bovine collagen sponge | Anterior cervical interbody fusion; three levels (C2-C3, C4 and C4-C5); BAK device; 12 week evaluation |

*Table 2 - continued*

| Citation | Year | Species | BMP | Carrier | Model |
|---|---|---|---|---|---|
| Boden et al. [76] | 1998 | Monkey | BMP-2 (multiple doses) | Bovine collagen sponge | Laparoscopic anterior lumbar interbody fusion; single level (L6-S1); titanium threaded cylindrical cage; 24 week evaluation |
| Paramore et al. [62] | 1999 | Dog | OP-1/BMP-7 (2.0 mg) | Purified bovine bone collagen with or without carboxymethyl cellulose | Dorsolateral lumbar intertransverse process fusion; single level (L2-L3); non-instrumented; 16 week evaluation; safety study |
| Cunningham et al. [63] | 1999 | Sheep | OP-1/BMP-7 (2.5 mg) | Purified bovine bone collagen | Thoracic interbody fusion; three levels (T5-T6, T7-T8, T9-T10); BAK device; 4 month evaluation |
| David et al. [85] | 1999 | Dog | BMP-2 (0.054, 0.215 or 0.86 mg) | Bovine collagen sponge or O-polylactic acid sponge | Posterior lumbar intertransverse process fusion; single level (L4-L5) non-instrumented; 12 week evaluation |
| Takahashi et al. [84] | 1999 | Goat | BMP-2 (0.005 or 0.05 mg) | Porous hydroxyapatite block | Anterior cervical interbody fusion; three levels (L3-L4, C4-C5, or C5-C6) non-instrumented; 12 week evaluation |
| Martin et al. [82] | 1999 | Monkey | BMP-2 (various doses 2 to 32 mg) | Bovine collagen sponge or porous polylactic acid polymer | Posterolateral intertransverse process fusion; single level (L4-L5); non-instrumented; 24 week evaluation |
| Hecht et al. [81] | 1999 | Monkey | BMP-2 (0.4 mg) | Bovine collagen sponge and freeze dried cortical dowel allograft cylinder | Anterior lumbar interbody fusion; single level (L7-S1); non-instrumented; 6 month evaluation |
| Boden et al. [80] | 1999 | Monkey | BMP-2 (6, 9 or 12 mg) | Hydroxyapatite/Tricalcium phosphate blocks | Posterolateral lumbar intertransverse process fusion; single level (L4-L5); 24 week evaluation |
| Spiro et al. [36] | 2000 | Baboon | GDF-5 (5 or 15 mg) | Mineralized bovine collagen matrix (Healos) | Posterolateral lumbar intertransverse process fusion; single level (L4-L5); non-instrumented; 20 week evaluation |
| Magin et al. [65] | 2001 | Sheep | OP-1/BMP-7 (2.5 mg) | Purified bovine bone collagen | Lumbar interbody L4-L5 fusion; single level (L4-L5); internal fixation; 6 month evaluation |

ed sites did not attain the degree of remodeling observed in OP-1 sites at 26 weeks. The study results demonstrated that OP-1 is an effective bone graft substitute for achieving stable spine fusion in a significantly more rapid fashion than could be achieved with autogenous bone graft.

In the earliest study with BMP-2, efficacy was also demonstrated in a dog posterolateral intertransverse process fusion model [66]. However, the difference in this model with that described above with OP-1 was that each fusion site was internally fixed with plates and the carrier material was a polylactic/polyglycolic acid (PLGA) polymer. The BMP-2/PLGA polymer was compared to autogenous cancellous bone and carrier alone. The results showed equivalency between the autogenous bone and the BMP-2 implant, while the carrier alone was clearly inferior. In this model, a site effect was observed with the BMP-2 that was not evident with autograft bone; it was suggested that the L1-L2 site produced lower union sites than the L3-L4 and L5-L6 sites.

Since the original two studies numerous laboratories have confirmed the ability of recombinant BMPs to promote successful intertransverse process fusions [62, 64, 67–75, 78–80, 82, 83, 85–87]. OP-1, BMP-2 and GDF-5 have been investigated with BMP-2 the most extensively studied of the group. BMP-2 has been evaluated in a variety of delivery materials, including collagen sponges, biodegradable polymers, calcium phosphate materials and autograft bone. Although differences are observed with the carriers, the data demonstrate that BMP-2 is effective at achieving fusions at different intertransverse process sites and, for the most part, results in more rapid and reliable healing than seen using autogenous bone. However, the long-term outcome of the fusion masses with different carriers has not been sufficiently evaluated and it needs to be determined whether slow resorbing ceramic materials and voids remaining in polylactic/polyglycolic acid polymer implants are significant. In regard to the dose effects of BMP-2, such studies are highly dependent upon the experimental model and additional studies need to be done to further evaluate whether milligram doses are required.

More recently, the use of OP-1 or BMP-2 to promote interbody fusions has been investigated [63, 65, 76, 77, 81, 84]. One such study assessed OP-1 delivered with bone-derived collagen particles as an autograft substitute for thoracic interbody spinal fusion in a sheep model [63]. Twelve sheep underwent a multi-level thoracic spinal decompression *via* a video-assisted thoracoscopic approach. Three noncontiguous destabilization sites (T5-6, T7-8, T9-10) were prepared and randomly assigned to either a control or treatment group. Control groups were either disk destabilization alone, an empty BAK cage or no surgical intervention at all. The treatment groups were either autograft alone, BAK cage packed with autograft or a BAK cage packed with OP-1 device. Four months postoperatively, the animals were euthanized, and the interbody fusion sites were analyzed using biomechanical testing, computed tomography, microradiography and histomorphometry. Biomechanical testing demonstrated higher segmental stiffness levels when comparing the experimental groups to the control groups ($p < 0.05$). There were no quantifiable differences when comparing functional unit stability within the three experimental techniques or the three control groups ($p > 0.05$). Fusion was assessed by computed tomography and microradiography. In the control groups, destabilization alone had a 16% fusion rate and the empty BAK cage had a 33% fusion rate. In the treatment groups, the autograft treated group had a fusion rate of 50%, the BAK cage with autograft had a fusion rate of 63% and the BAK cage with the OP-1 Collagen Implant had a fusion rate of 75%. In all the treatment groups, the histological characterization of the fusion sites was in agreement with the radiographic findings. In the fused sections treated with the OP-1 collagen implant in the BAK cage, the bone present in the cage was a dense, well organized, woven trabecular bone. None of the original collagen matrix was present. Overall, the autograft sites did not demonstrate the same degree of bone remodeling and incorporation that was observed in the OP-1 treated group. Histomorphometric analysis showed significantly more trabecular bone formation at the fusion site for the experimental groups when compared to the controls ($p < 0.05$). The results of this study demonstrated that the use of the OP-1 collagen implant with an interbody fusion cage could promote vertebral interbody fusion. The OP-1 Collagen performed as well as the conventional autologous iliac crest bone.

Cervical and lumbar interbody fusion models have also been investigated with BMPs [65, 76, 77, 81, 84]. For the most part, these studies have utilized BMP-collagen materials placed inside titanium interbody cages. In one study, the BMP material was placed inside a freeze-dried cortical allograft cylinder. With each of the models the results have demonstrated that the BMP materials can be effective in promoting fusion. However, these models are more complex than the intertransverse process models and it is apparent that some fusion sites may be more difficult to fuse than others. Additional evaluation need to be done using different cages and instrumentation and different BMP delivery materials.

## Conclusion

Recombinantly produced osteoinductive BMPs, when implanted locally at subcutaneous or bony sites, initiate the recruitment, attachment, proliferation and differentiation of mesenchymal cells leading to new bone formation containing fully functional bone marrow components. Implantable formulations containing these BMPs have demonstrated an exciting therapeutic potential to replace conventionally employed autogenous bone grafts in the repair of a variety of defects, including large gaps, nonunions and bone fractures, and to promote spinal fusions and the osseointegration of metallic implant devices. Most data in the field have resulted from research on two members of the BMP family, OP-1 (BMP-7) and BMP-2.

Numerous studies have now been published demonstrating that BMPs can reproducibly repair large critical size defects in long bones in many different animal species. These preclinical studies suggest that BMPs not only can repair bone similar to autograft, but in fact can speed the process significantly. Studies have also demonstrated that BMPs are also very effective when combined with autograft or allograft bone to increase the rate and extent of graft incorporation. However, it is clear that important areas for future investigation involve delivery materials that provide containment for the BMP and structural support for the defect site.

The majority of preclinical studies that have been published involve the use of BMPs to replace autogenous bone graft for spinal fusions. Important variables, such as dose, delivery material and site effects have been examined in both intertransverse process and interbody fusion models. These studies have demonstrated that BMP-containing materials can be very effective in promoting spinal fusions although interbody fusion models are clearly more complex and need additional studies.

The results of preclinical studies with injectable formulations of BMPs demonstrate a potential application in accelerating fracture repair. The ability to speed repair with an earlier return to function is an attractive benefit. Similar formulations of BMPs are also being evaluated in conjunction with accelerating the process of distraction osteogenesis (limb lengthening). However, many variables including the optimal time for injection and the delivery formulations need to be evaluated.

The use of BMPs in conjunction with metal prostheses has been an area of investigation that suggests a potential application for improving and speeding up osseointegration. However, development of preclinical models involving impacted graft materials as well as metal implants is challenging, and much work is needed to determine the clinical relevance of this indication.

Taken together, preclinical testing has demonstrated that BMPs can induce bone in a variety of orthopedic defects. None of these studies has reported any adverse effects from the BMPs that would diminish the clinical potential. The efficacy of the BMPs is related to a number of factors, including bony location, BMPs dose and carrier material. Based upon preclinical studies to date, the potential therapeutic applications of BMPs appear to be large and diverse and most importantly, the initial experience in human studies has confirmed the usefulness of the animal experience. Additional animal studies will need to be done to optimize delivery to specific defect sites and to investigate new applications.

## References


1 Reddi A, Weintroub S, Muthukumaran N (1987) Biological principles of bone induction. *Orthop Clin North Am* 18: 207–212

2 Urist M (1965) Bone: formation by autoinduction. *Science* 150: 893–899

3 Reddi A (1981) Cell biology and biochemistry of endochondral bone development. *Collagen Rel Res* 1: 209–167
4 Urist M, Iwata H (1973) Preservation and biodegradation of morphogenetic property of bone matrix. *J Theor Biol* 38: 155–167
5 Reddi A (1998) Role of morphogenetic proteins in skeletal tissue engineering and regeneration. *Nature Bio* 16: 247–252
6 Wozney J (1998) The bone morphogenetic protein family: multifunctional cellular regulators in the embryo and adult. *Eur J Oral Sci* 106: 160–166
7 Sakou T (1998) Bone morphogenetic proteins: from basic studies to clinical approaches. *Bone* 22: 591–603
8 Wang E, Rosen V, Cordes P, Hewick R, Kriz M, Luxenberg D, Sibley B, Wozney J, (1988) Purification and characterization of other distinct bone-inducing proteins. *Proc Nat Acad Sci* 85: 9484–9488
9 Celeste A, Iannazzi J, Taylor R, Hewick R, Rosen V, Wang E, Wozney J (1990) Identification of transforming growth factor-β superfamily members present in bone-inductive protein purified from bovine bone. *Proc Natl Acad Sci USA* 87: 9843–9847
10 Sampath T, Reddi A (1981) Dissociative extraction and reconstitution of extracellular matrix components involved in local bone differentiation. *Proc Natl Acad Sci USA* 78: 7599–7602
11 Urist M, DeLange R, Finerman G (1983) Bone cell differentiation and growth factors. *Science* 220: 680–686
12 Sampath T, Muthukumaran N, Reddi A (1987) Isolation of osteogenin, an extracellular matrix-associated bone inductive protein by heparin affinity chromatography. *Proc Natl Acad Sci USA* 84: 7109–7113
13 Sampath T, Rashka K, Doctor J, Tucker R, Hoffmann F (1993) *Drosophila* TGF-β superfamily proteins induce endochondral bone formation in mammals. *Proc Natl Acad Sci USA* 90: 6004–6008
14 Sampath T, Reddi A (1983) Homology of bone-inductive proteins from human, monkey, bovine, and rat extracellular matrix. *Proc Natl Acad Sci USA* 80: 6591–6595
15 Sampath T, Coughlin J, Whetstone R, Banach D, Corbet C, Ridge R, Özkaynak E, Oppermann H, Rueger D (1990) Bovine osteogenic protein is composed of dimers of OP-1 and BMP-2A, two members of the transforming growth factor-β superfamily. *J Biol Chem* 265: 13198–13205
16 Takaoka K, Yoshikawa H, Masuhara K, Sugamoto K, Tsuda T, Aoki Y, Ono Y, Sakamoto Y (1989) Establishment of a cell line producing bone morphogenetic protein from a human osteosarcoma. *Clin Orthop* 244: 258–264
17 Tsuda T, Masuhara K, Yoshikawa H, Shimuzu N, Takaoka K (1989) Establishment of an osteoinductive murine osteosarcoma clonal cell line showing osteoblast phenotypic traits. *Bone* 10: 195–200
18 Wozney J, Rosen V, Celeste A, Mitsock L, Whiters M, Kriz R, Hewick R, Wang E, (1988) Novel regulators of bone formation: Molecular clones and activities. *Science* 242: 1526–1534

19 Wang E, Rosen V, D'Alessandro J, Bauduy M, Cordes P, Harada T, Israel D, Hewick R, Kerns K, LaPan P et al (1990) Recombinant human bone morphogenetic protein induces bone formation. *Proc Natl Acad Sci USA* 87: 2220–2224
20 Sampath T, Özkaynak E, Jones W, Sasa H, Tucker R, Tucker M, Kusmik W, Lightholder J, Pang R, Corbett C, Oppermann H, Rueger D (1992) Recombinant human osteogenic protein-1 (hOP-1) induces new bone formation *in vivo* with a specific activity comparable with natural bovine osteogenic protein and stimulates osteoblast proliferation and differentiation *in vitro*. *J Biol Chem* 267: 20352–20362
21 Özkaynak E, Rueger D, Drier E, Corbett C, Ridge R, Sampath T, Oppermann H (1990) OP-1 cDNA encodes an osteogenic protein in TGF-β family. *EMBO J* 9: 2085–2093
22 Luyten F (1997) Cartilage-derived morphogenetic protein-1. *Int J Cell Biol* 29: 1241–1244
23 Storm E, Kingsley, D (1999) GDF5 coordinates bone and joint formation during digit development. *Dev Biol* 209: 11–27
24 Sampath T, Rueger D (1994) Structure, function, and orthopaedic applications of osteogenic protein (OP-1). *Complications in Orthopaedics* 9 (Winter): 101–107
25 Yasko A, Lane J, Fellinger E, Rosen V, Wozney J, Wang E (1992) The healing of segmental defects induced by recombinant human bone morphogenetic protein (rhBMP-2). *J Bone Joint Surg* 74A: 59–671
26 Cook S, Baffes G, Wolfe M, Sampath T, Rueger D,Whitecloud T (1994) The effect of recombinant human osteogenic protein-1 on healing of large segmental bone defects. *J Bone Joint Surg* 76A: 27–838
27 Bostrom M, Lane J, Tomin E, Browne M, Berberian W, Turek T, Smith J, Wozney J, Schildhauer T (1996) Use of bone morphogenetic protein-2 in the rabbit ulnar nonunion model. *Clin Orthop Rel Res* 327: 272–282
28 Zegzula H, buck D, Brekke J, Wozney J, Hollinger J (1997) Bone formation with use of rhBMP-2 (Recombinant human bone morphogenetic protein-2). *J Bone Joint Surg* 72: 1778–1790
29 Gerhart T, Kirker-Head C, Kriz M (1993) Healing segmental femoral defects in sheep using recombinant human bone morphogenetic protein. *Clin Orthop* 293: 317–326
30 Cook S, Baffes G, Wolfe M, Sampath T, Rueger D (1994) Recombinant human bone morphogenetic protein-7 induces healing in a canine long bone segmental defect model. *Clin Orthop* 301: 302–312
31 Kirker-Head C. Gerhart T. Schelling T, Hennig G, Wang E Holtrop M (1995) Long term healing of bone using recombinant human bone morphogenetic protein 2. *Clin Orthop* 318: 222–230
32 Cook S, Wolfe M, Salkeld S, Rueger D (1995) Recombinant human osteogenic protein-1 (rhOP-1) heals segmental defects in nonhuman primates. *J Bone Joint Surg* 77A: 734–750
33 Kirker-Head C, Gerhart T, Arstrong R, Schelling T, Carmel L (1998) Healing bone using recombinant human bone morphogenetic protein 2 and copolymer. *Clin Orthop* 349: 205–217

34 Itoh T, Mochizuki M, Nishimura R (1998) Repair of ulnar segmental defect by recombinant human bone morphogenetic protein-2 in dogs. *J Vet Med Sci* 60: 451–458
35 Sciadini M, Johnson K (2000) Evaluation of recombinant human bone morphogenetic protein-2 as a bone-graft substitute in a canine segmental defect model. *J Orthop Res* 18: 289–302
36 Spiro R, Liu L., Heidaran M, Thompson A, Ng C, Pohl J, Poser J (2000) Inductive activity of recombinant human growth and differentiation factor-5. *Biochem Soc Trans* 28: 362–368
37 Salkeld S, Cook S, Rueger D (1995) Synthetic polymers as carriers for osteogenic proteins. Abstract in *Transactions of the 21st Annual Meeting of the Society for Biomaterials and 27th International Biomaterials Symposium*, San Francisco, CA
38 Popich L, Salkeld S, Rueger D, Tucker M, Cook S (1997) Critical and noncritical size defect healing with osteogenic protein-1. Abstract in *Transactions of the 23rd Annual Meeting of the Society for Biomaterials and 29th International Biomaterials Symposium*, New Orleans, LA, 20: 244
39 Reddi A (1998) Fracture repair process: Initiation of fracture repair by bone morphogenetic proteins. *Clin Orthop Rel Res* 355S: S66–S72
40 Blokhuis T, den Boer F, Bramer J, Jenner J, Bakker F, Patka P, Haarman H (2001) Biomechanical and histological aspects of fracture healing, stimulated with osteogenic protein-1. *Biomaterials* 22: 725–730
41 Einhorn T, Majeska R, Oloumi G (1997) Enhancement of experimental fracture healing with a local percutaneous injection of rhBMP-2. Abstract in *American Academy of Orthopaedic Surgeons Annual Meeting*, San Francisco, CA, 64: 216
42 Bax B, Wozney J, Ashhurst D (1998) Bone morphogenetic protein-2 increases the rate of callus formation after fracture of the rabbit tibia. *Calcif Tissue Int* 65: 83–89
43 Bostrom M, Camacho N (1998) Potential role of bone morphogenetic proteins in fracture healing. *Clin Orthop Rel Res* 355S: S274–S282
44 Welch R, Jones A, Buchloz R, Reinert C, Tjia J, Pierce W, Wozney J, Li X (1998) Effect of recombinant human bone morphogenetic protein-2 on fracture healing in a goat tibial fracture model. *J Bone Min Res* 13: 1483–1490
45 Salkeld S, Patron L, Barrack R, Cook S (2001) The effect of osteogentic protein-1 (OP-1:BMP-7) on the healing of segmental bone defects treated with autograft or allograft bone. *J Bone Joint Sur* 83 A: 803–816
46 Cook S, Barrack R, Santman M, Popich-Patron L, Salkeld S, Whitecloud T (2000) Strut allograft healing to the femur with recombinant human osteogenic protein-1. *Clin Orthop Rel Res* 381: 47–57
47 Lamerigts N, Buma P, Aspenberg P, Schreurs B, Slooff T (1999) Role of growth factors in the incorporation of unloaded bone allografts in the goat. *Clin Orthop Rel Res* 368: 260–270
48 Tagil M, Jeppsson C, Aspenberg P (2000) Bone graft incorporation: effects of osteogenic protein-1 and impaction. *Clin Orthop Rel Res* 371: 240–245

49 Cook S, Rueger D (1996) Osteogenic Protein-1: Biology and applications. *Clin Orthop* 324: 29–38

50 Cook S, Salkeld S, Rueger D (1996) Evaluation of osteogenic protein-1 to aid in the initial fixation of implants. *J Oral Implantology* 21: 1–6

51 Lind M, Overgaard S, Song Y, Goodman S, Bunger C, Soballe K (2000) Osteogenic Protein 1 device stimulates bone healing to hydroxyapaptite-coated and titanium implants. *J Arthroplasty* 15: 339–346

52 Lind M, Overgaard S, Jensen T, Song Y, Goodman S, Bunger C, Soballe K (2001) Effect of osteogenic protein 1/collagen composite combined with impacted allograft around hydroxyapatite-coated titanium alloy implants is moderate. *J Biomedical Mat Res* 55: 89–95

53 Terheyden H, Jepsen S, Moller B, Tucker M, Rueger D (1999) Sinus floor augmentation with simultaneous placement of dental implants using a combination of deproteinized bone xenografts and recombinant human osteogenic protein-1. *Clin Oral Impl Res* 10: 510–521

54 Sigurdsson T, Fu E, Tatakis D, Rohrer M, Wikesjo R (1997) Bone morphogenetic protein-2 for peri-implant bone regeneration and osseointegration. *Clin Oral Impl Res* 8: 367–374

55 Hanisch O, Tatakis D, Rohrer M, Wohrle P, Wozney J, Wikesjo U (1997) Bone formation and osseointegration stimulated by rhBMP-2 following subantral augmentation procedures in nonhuman primates. *Int J Oral Maxillofac Imp* 12: 785–792

56 Mizumoto Y, Moseley T, Cooper V, Drew M, Reddi H (2001) A single injection of rhBMP-7 accelerates callus formation in a rat model of distraction osteogenesis. Abstract in *47th Annual Meeting, Orthopaedic Research Society*, San Francisco, CA, 26: 145

57 Amako M, Hamdy R, Steffen T (2001) The effects of a single injection of OP-1 on stimulating of new bone formation in distraction osteogenesis in the rabbit. Abstract in *47th Annual Meeting, Orthopaedic Research Society*, San Francisco, CA, 26: 582

58 Windhagen H, Wagner L, Brunger J, Nolle O, Thorey F, Luppen C, Seeherman H, Bouxsein M (2001) RhBMP-2 in an injectable gelfoam carrier enhances consolidation of the distracted callus in a sheep model of distraction osteogenesis. Abstract in *47th Annual Meeting, Orthopaedic Research Society*, San Francisco, CA, 26: 146

59 Zlotolow D, Vaccaro A, Salamon M, Albert T (2000) The role of human bone morphogenetic proteins in spinal fusion. *J Am Acad Orthop Surg* 8: 3–9

60 Lewandrowski K-U, Ozuna R, Pedlow F, Hecht A (2000) Advances in the biology of spinal fusion: growth factors and gene therapy. *Curr Opin Orthop* 11: 167–175

61 Cook S, Dalton J, Tan E, Whitecloud T, Rueger D (1994) *In vivo* evaluation of recombinant human osteogenic protein (rhop-1) implants as a bone graft substitute for spinal fusions. *Spine* 19: 1655–1663

62 Paramore C, Lauryssen C, Rauzzino M, Wadlington V, Palmer C, Brix A, Cartner S, Hadley M (1999) The safety of OP-1 for lumbar fusion with decompression – a canine study. *Neurosurgery* 44: 1151–1156

63 Cunningham B, Kanayama M, Parker L, Weis J, Sefter J, Fedder I, McAee P (1999) Osteogenic protein *versus* autologous interbody arthrodesis in the sheep thoracic spine. A comparative endoscopic study using the bagby and kuslich interbody fusion device. *Spine* 24: 509–518

64 Grauer J, Patel T, Erulkar J, Troiano N, Panjabi M, Friedlaender G (2001) Evaluation of OP-1 as a graft substitute for intertransverse process lumbar fusion. *Spine* 26: 127–133

65 Magin M, Delling G (2001) Improved lumbar vertebral interbody fusion using rhOP-1. *Spine* 26: 469–478

66 Muschler G, Hyodo A, Manning T, Kambic H, Easley K (1994) Evaluation of human bone morphogenetic protein 2 in a canine spinal fusion model. *Clin Orthop* 308: 229–240

67 Schimandle J, Boden S, Hutton W (1995) Experimental spinal fusion with recombinant human bone morphogenetic protein-2 (rhBMP-2). *Spine* 20: 1326–1337

68 Sandhu H, Kanim L, Kabo M, Toth J, Zeegen E, Liu D, Seeger L, Dawson E (1995) Evaluation of rhBMP-2 with an OPLA carrier in a canine posterolateral (transverse process) spinal fusion model. *Spine* 20: 2669–2682

69 Sheehan J, Kallmes D, Sheehan J (1996) Molecular methods of enhancing lumbar spinal fusion. *Neurosurgery* 39: 548–554

70 Boden S, Moskovitz P, Morone M, Toribitake Y (1996) Video-assisted lateral intertransverse process arthrodesis: validation of a new minimally invasive lumbar spinal fusion technique in the rabbit and nonhuman primate (rhesus) models. *Spine* 21: 2689–2697

71 Sandhu H, Kanim L, Kabo M, Toth J, Zeegen E, Liu D, Delamarter R, Dawson E (1996) Effective doses of recombinant human bone morphogenetic protein-2 in experimental spinal fusion. *Spine* 21: 2115–2122

72 Holliger E, Trawick R, Boden S, Hutton W (1996) Morphology of the lumbar intertransverse process fusion mass in the rabbit model: a comparison between two bone graft materials-rhBMP-2 and autograft. *J Spinal Disord* 9: 125–128

73 Helm G, Sheehan J, Sheehan J, Sheehan J, Jane J, DiPierro C, Simmons N, Gillies G, Kallmes D, Sweeney T (1997) Utilization of type I collagen gel demineralized bone matrix, and bone morphogenetic protein-2 to enhance autologous bone lumbar spinal fusion. *J Neurosurg* 86: 93–100

74 Fischgrund J, James S, Chabot M (1997) Augmentation of autograft using rhBMP-2 and different carrier media in the canine spine fusion model. *J Spinal Disord* 10: 467–472 Erratum in *Spine* 22: 2463 (1997)

75 Sandhu H, Kanim L, Toth J, Kabo J, Liu D, Delamarter R, Dawson E (1997) Experimental spinal fusion with recombinant human bone morphogenetic proein-2 without decortication of osseous elements. *Spine* 22: 1171–1180

76 Boden S, Martin Jr. G, Horton W, Truss T, Sandhu H (1998) Laparoscopic anterior spinal arthrodesis with rhBMP-2 in a titanium interbody threaded cage. *J Spinal Disord* 11: 95–101

77 Zdeblick T, Ghanayem A, Rapoff A (1998) Cervical interbody fusion cages: an animal model with and without bone morphogenetic protein. *Spine* 23: 758–766
78 Minimide A, Tamaki T, Kawakami M, Hashizume H, Yoshida M, Sakata R, Friedlaender G (1999) Experimental spinal fusion using sintered bovine bone coated with type I collagen and recombinant human bone morphogenetic protein-2. *Spine* 24: 1863–1870
79 Itoh H, Ebara S, Kamimura M, Tateiwa Y, Kinoshita T, Yuzawa Y, Takaoka K (1999) Experimental spinal fusion with use of recombinant human bone morphogenetic protein 2. *Spine* 24: 1402–1409
80 Boden S, Martin G, Morone M, Ugbo J, Moskovitz P (1999) Posterolateral lumbar intertransverse process spine arthrodesis with recombinant human bone morphogenetic protein 2/hydroxyapatite-tricalcium phosphate after laminectomy in the nonhuman primate. *Spine* 24: 1179–1185
81 Hecht B, Fishgrund J, Herkowitz H, Penman L, Toth J, Shirkhoda A (1999) The use of recombinant human bone morphogenetic protein 2 (rhBMP-2) to promote spinal fusion in a nonhuman primate anterior interbody fusion model. *Spine* 24: 629–636
82 Martin G, Boden S, Morone M, Moskovitz (1999) Posterolateral intertransverse process spinal arthrodesis with rhBMP-2 in a nonhuman primate: important lessons learned regarding dose, carrier, and safety. *J Spinal Disorders* 12: 179–186
83 Martin G, Boden S, Titus L (1999) Recombinant human bone morphogenetic protein-2 overcomes the inhibitory effect of ketorolac, a nonsteroidal anti-inflammatory drug (NSAID), on posterolateral lumbar intertransverse process spine fusion. *Spine* 24: 2188–2194
84 Takahashi T, Tominaga T, Watabe N, Yokobori Jr A, Sasada H, Yoshimoto T (1999) Use of porous hydroxyapatite graft containing recombinant human bone morphogenetic proein-2 for cervical fusion in a caprine model. *J Neurosurg* 90: 224–230
85 David S, Gruber H, Meyer R, Murakami T, Tabor W, Howard B, Wozney J, Hanley E (1999) lumbar spinal fusion using recombinant human bone morphogenetic protein in the canine. A comparision of three dosages and two carriers. *Spine* 24: 1973–1979
86 Meyer R, Gruber H, Howard B, Tabor W, Murakami T, Kwiatkowski T, Wozney J, Hanley E (1999) Safety of recombinant human bone morphogenetic proein-2 after spinal laminectomy in the dog. *Spine* 24: 747–754

# Osteogenic protein-1 (OP-1, BMP-7) for stimulation of healing of closed fractures: evidence based medicine and pre-clinical experience

*Taco J. Blokhuis[1], Peter Patka[1], Henk J.Th.M. Haarman[1] and Lex R. Giltaij[2]*

[1]Vrije Universiteit Medical Centre (VUmc), Department of Trauma Surgery, P.O. Box 7057, 1007 MB Amsterdam, The Netherlands; [2]Stryker Biotech, Notengaard 13, 3941 LV Doorn, The Netherlands

## Introduction

Fracture healing is a time-consuming process, especially in fractures of the lower extremity in humans. These fractures are known to heal twice as slowly as fractures located in other places in the human body. Fractures of the lower extremity often occur due to high-energy trauma such as motor vehicle accidents. Consequently, the average age of the patients is generally low, with a high incidence at the age of 30 years. Therefore, most patients are employed and involved in social activities. This means that the long period of healing and recuperation that follows a fracture of the lower extremity is expensive, for both patient and society [1]. Apart from this fact, 5 to 10 per cent of all fracture patients encounter disturbances in the healing process [2], resulting in delayed unions or even nonunion. A stimulation of the fracture healing process would be beneficial for this group of patients as well.

Since the first description of bone morphogenetic proteins (BMP's) in 1965 by Marshal Urist [3], extensive research on the effectiveness of these proteins in the stimulation of bone healing has been performed. Most mechanisms through which these proteins exert their osteoinductive activity have been elucidated [4–7]. Also, the stimulating effect on bone healing of most of these proteins has been well established in animal experiments [8]. The efficacy of BMP's has been demonstrated in the healing of various large bone defects and in spinal fusion [9–17]. Osteogenic protein-1 (OP-1) is a powerful bone morphogenetic protein with a strong osteoinductive capacity [4]. The effectiveness of recombinant human (rh)OP-1 in the treatment of bone defects has been well documented in several animal experiments [4, 10, 18, 19]. The first clinical randomized study describing its effectiveness in a human fibular defect was described by Geesink et al. [13], and its effectiveness in treatment of tibial nonunions was published recently [20].

Bone morphogenetic proteins play a fundamental role during embryogenesis. As the fracture repair process resembles embryogenetic bone formation, BMP's are expected to play an important role during fracture repair. In fact, several BMP's and

Bone Morphogenetic Proteins, edited by Slobodan Vukicevic and Kuber T. Sampath

receptors for BMP's have already been demonstrated to show an elevated expression during fracture repair [21–24]. The stimulation and acceleration of fresh fracture healing by local application of exogenous OP-1 seems to be a promising new development in treatment of fractures [25, 26].

Direct, single injection of BMP's into the fracture gap in closed fractures would be an ideal application manner for BMP's in fracture repair, as the fracture hematoma is left intact and the risk of infection is minimized. Stimulation of fracture repair with BMP's and the influence of specific carrier materials, such as the frequently used bovine type I collagen, were recently investigated in a closed fracture model in goats, using a single injection of OP-1 [27].

## Experimental studies

All procedures were performed after approval of the animal ethics committee was obtained. With a custom made three-point bending device, a closed midshaft fracture was created under general anesthesia in the left tibia of 40 adult female goats, weighing between 50 and 70 kg. An increasing force was applied by means of a pneumatic device in a mediolateral direction perpendicular to the bone axis until breakage occurred. As the fracture type could not be standardized, oblique and transverse fractures occurred, but comparison of the mean angle of the fractures showed no differences between either of the four treatment groups (data not shown).

The fractures were stabilized with an AO external fixation device (West Meditec, Bilthoven, the Netherlands) with radiolucent bars, which was placed at the lateral side of the tibia. The animals were randomly assigned to four different treatment groups; group I, $n = 10$: no injection; group II, $n = 10$: injection of 1 mg recombinant human osteogenic protein-1 (rhOP-1, Stryker Biotech, Hopkinton, MA, further referred to as OP-1) dissolved in 0.63 ml sodium acetate buffer to create a liquid solution; group III, $n = 11$: injection of 1 mg OP-1 with 400 mg bovine type I collagen matrix in combination with carboxymethylcellulose (CMC) to give a viscous putty consistency; group IV, $n = 9$: injection of 400 mg bovine type I collagen matrix with CMC alone. All injections were given under aseptic conditions and under fluoroscopic control.

Animals were sacrificed either 2 weeks ($n = 21$) or 4 weeks ($n = 19$) post injection. After sacrifice both tibiae were explanted, all soft tissues of the right and left tibia were removed, and all tibiae were kept in alcohol 70% for a standardized period of 14 days until mechanical testing was performed. Fracture healing was evaluated using computed tomography (CT), mechanical testing, and histology.

Axial spiral CT scans were performed using a Somatom Plus CT Scanner (Siemens, Erlangen, Germany), with a slide thickness of 1 mm. With image analysis software (Voxel Q, Picker International, Cleveland, OH), the images were analyzed after three-dimensional reconstructions were made. Using a manual marking approach, the new bone callus was outlined on every CT and the callus volume was calculated. To ensure that the entire callus was included in the analysis, 1 cm adjacent normal bone proximal and distal to the callus was scanned in all animals.

Nondestructive biomechanical evaluation was performed, utilizing a standardized four-point nondestructive bending test [28]. The stiffness of each tibia was measured in twenty-four directions. The twenty-four stiffness values of both tibiae of each animal were then plotted in polar coordinates, and by regression two ellipses were obtained. From these ellipses, area ratio and stiffness index were calculated. The area ratio is the ratio of the ellipses of the left and right tibia, providing a parameter for the total stiffness of the operated bone in comparison with the intact tibia. The stiffness index is the ratio of the stiffness of the operated and the intact tibia in the direction where this ratio is minimal, thereby providing a comparison at the weakest point.

After the nondestructive bending test, a torsional test to failure was performed to determine torsional stiffness and torsional strength. The outcome values of the torsional test were expressed as a percentage of the intact, contralateral bone to account for variability between animals. All specimens were kept moist during testing, since drying could influence the outcome of mechanical tests [29, 30].

After the CT scans and mechanical tests were performed, four longitudinal 2 mm thick slices of the fracture area were prepared from the anterior, posterior, lateral, and medial side. After dehydration in ascending grades of ethanol, they were embedded in polymethylmetacrylate (PMMA). Using a motor-driven microtome (Jung K, Heidelberg, Germany), 5 µm sections were cut and stained with Goldner's trichrome and toluidin blue 0.2%. All sections were examined by two reviewers who were blinded to treatment and survival period of the animals (2 or 4 weeks). In case of any disagreement between the two investigators, the final score for the histology was obtained by discussion with a third investigator.

Several histological aspects were scored: bony bridging of the fracture gap, amount of woven bone in the callus, presence and amount of cartilage in the callus, and inflammatory reactions. If remnants of collagen particles were seen, the incorporation of these particles in a newly formed bone, the interface between these remnants and the newly formed bone, especially any encapsulation by fibrous tissue, and presence and severity of inflammatory reactions aimed at these particles were scored as well. The definitions of all histological parameters are given in Table 1.

## Statistics

For comparing the outcome of both CT and biomechanical testing, comparisons between the groups were made with the Mann-Whitney test. Since the Mann-Whitney test does not correct for multiple comparisons, a restricted number of compar-

*Table 1 - Definition of the histological parameters*

| Parameter | Definition | Value* |
|---|---|---|
| Bony bridging of the fracture gap | A continuous field of woven bone between the old cortices, bridging the fracture gap | 0 (no bridging) – 4 (four sides of the fracture) |
| Amount of woven bone in the callus | No woven bone, small, moderate, or large amount of woven bone | 0, 1, 2, or 3 |
| Presence of cartilage | No cartilage, cartilage in the center of the fracture gap, of cartilage throughout the fracture gap | 0, 1, or 2 |
| General inflammatory reactions | No reaction, or a general inflammatory reaction consisting of granulocytes and/or lymphocytes | 0 or 1 |
| Presence of particles | No remnants found, small amount, or large amount of remnants found | 0, 1, or 2 |
| Inflammatory reactions aimed at the particles | No reaction, mild inflammatory reaction, or abundant inflammatory reaction | 0, 1, or 2 |
| Incorporation of the particles | No incorporation or incorporation in woven bone | 0 or 1 |
| Resorption of the particles | No resorption, some resorption, or active resorption by a large amount of resorbing cells | 0, 1, or 2 |

**The given order of values corresponds with the given order of definitions, e.g. for amount of woven bone: 0 = no woven bone, 1 = small amount, 2 = moderate amount, 3 = large amount of woven bone*

isons considered to be clinically relevant were chosen: group I vs. group II, III, and IV, and group III vs. IV. All histological parameters were examined using the same comparisons. Significance was set at $p < 0.05$. All calculations were performed using the statistical package SPSS version 9.0 (SPSS Inc., Chicago).

## Results of experimental studies

### Computed tomography

After 2 weeks, a larger callus volume was seen in both groups treated with OP-1 compared to no injection ($p = 0.009$ for OP-1 and $p = 0.002$ for OP-1 + collagen

*Table 2 - Results of the computed tomography, mechanical tests, number of sides with bony bridging, and amount of woven bone at two weeks*

| Parameter | No injection (group I) | OP-1 alone (group II) | OP-1 + Matrix (group III) | Matrix alone (group IV) |
|---|---|---|---|---|
| Callus volume | 5.54 ± 1.0 | 12.65 ± 1.5 | 13.57 ± 2.9 | 9.70 ± 1.4 |
| Stiffness index | 0.03 ± 0.01 | 0.17 ± 0.04 | 0.06 ± 0.01 | 0.07 ± 0.01 |
| Area ratio | 0.002 ± 0.001 | 0.05 ± 0.01 | 0.007 ± 0.002 | 0.009 ± 0.002 |
| Torsional strength (%) | 9.2 ± 4.5 | 9.6 ± 1.4 | 10.8 ± 3.5 | 7.5 ± 1.1 |
| Torsional stiffness (%) | 7.6 ± 2.5 | 21.8 ± 3.3 | 11.8 ± 1.6 | 8.7 ± 1.6 |
| Total number of sides with bony bridging | 1 | 15 | 5 | 1 |
| Amount of woven bone | 1.20 ± 0.15 | 2.18 ± 0.24 | 1.68 ± 0.16 | 1.50 ± 0.05 |

carrier). The difference between collagen alone and no injection was also significant ($p$ = 0.03). The values of callus volume after 2 weeks, expressed as means ± standard error of the mean (SEM) are shown in Table 2. After 4 weeks, the callus volume in the group treated with OP-1 + collagen carrier (35.7 ml ± 4.6) was significantly higher than the volume in the group treated with collagen alone (12.9 ml ± 2.7, $p$ = 0.01) and no injection (17.7 ml ± 2.6, $p$ = 0.02).

## Mechanical testing

The results of the mechanical tests at 2 weeks are also summarized in Table 2. The data are given as means ± SEM. At 2 weeks, group II (OP-1 alone) showed the highest mean stiffness index and area ratio (0.17 ± 0.04 and 0.05 ± 0.01, respectively). These values were significantly higher in comparison with no injection (group I, 0.03 ± 0.01 and 0.002 ± 0.001, respectively, $p$ = 0.009 for both). Other comparisons for stiffness index and area ratio were not significant. The highest mean value for torsional stiffness, 21.8 ± 4.0, was observed in group II. The difference of group I (torsional stiffness 7.6 ± 2.5) was significant ($p$ = 0.03). The outcome of the torsional strength showed no differences between groups. At 4 weeks, there were no differences for any mechanical parameter (data not shown).

## Histology

### *Bridging of the fracture gap and bone formation*

The results of the two parameters after 2 weeks associated with stimulation of bone healing, bony bridging of the fracture gap and amount of woven bone, are summarized in Table 2. Also, the frequency of bridging of the fracture gap after 2 and 4 weeks is shown in Table 3. Group II showed bony bridging of the fracture gap significantly more frequently (15 sides with bony bridging) in comparison with group I (1 side, $p = 0.007$). The mean score (± SEM) for the amount of woven bone in the callus was 2.18 ± 0.24 in group II, which was higher than the score in group I (1.20 ± 0.15, $p = 0.008$). In the matrix + OP-1 group (group III), the callus contained more woven bone (1.68 ± 0.16) compared to group I ($p = 0.01$). Bridging of the fracture gap was seen more often in group III (five sides with bony bridging) compared to group I, but this difference was not significant. After four weeks, no differences between groups were observed. In general, all animals showed normally healed fractures.

### *Mechanisms of bone healing*

At 2 weeks newly formed woven bone was observed in all animals. Also, lamellar bone was seen in most animals. Lamellar bone appeared to be present more often in the animals stimulated with OP-1. There were no specific patterns in the sites of the fracture where woven bone was being formed, as suggested by others in bone defects [13]. Cartilage was observed in all groups, indicating the occurrence of enchondral ossification in all treatment groups. Cartilage was usually confined to small fields in the middle of the fracture gap. The only significant difference in cartilage formation was found between group III (OP-1 + matrix) and group IV (matrix alone), the latter having formed less cartilage ($p = 0.02$). At 4 weeks, remodeling was seen frequently. Occasionally, small fields of cartilage were present. No differences were seen between the groups.

### *Behavior of collagen particles*

After 2 weeks remnants of particles were present in a majority of the animals treated with OP-1 + matrix (group III) and matrix alone (group IV). In both groups, mild inflammatory reactions, consisting of a combination of granulocytes and lymphocytes, were seen around these remnants. These reactions, if any, were confined to the areas where the remnants of collagenous matrix were seen. Incorporation of the particles in woven bone was seen in both groups, although it was more often observed in group III ($p = 0.02$). Direct contact between the collagenous material and the newly formed bone could be seen in all animals that showed incorporation of the particles in woven bone, without any fibrous encapsulation of the particles. Also,

*Table 3 - Frequencies of bony bridging of the fracture gap per treatment group at two weeks*

| Sides | No injection (group I) | OP-1 alone (group II) | OP-1 + Matrix (group III) | Matrix alone (group IV) |
|---|---|---|---|---|
| 0 | 4 | 0 | 2 | 4 |
| 1 | 1 | 0 | 3 | 1 |
| 2 | 0 | 2 | 1 | 0 |
| 3 | 0 | 1 | 0 | 0 |
| 4 | 0 | 2 | 0 | 0 |

cell-mediated resorption of the particles was frequently observed, as indicated by the presence of large multinuclear cells adjacent to the remnants. After four weeks, the collagenous particles were mostly phagocytized. Occasionally, small remnants, usually incorporated in the newly formed bone were observed.

## Discussion

Since fractures of the lower extremity are known to heal slowly, and impaired healing occurs in 5 to 10 per cent of all fractures [2], an agent that would assist in fracture healing could result in obvious benefits for patients. Acceleration of the fracture healing process could result in earlier resumption of weightbearing, which has been demonstrated to reduce post-injury bone loss [31]. Theoretically, the rate of impaired healing could also be decreased by stimulation of fracture healing, though this was not specifically explored in this study. Acceleration of fracture healing was observed by administering a single minimally invasive percutaneous injection of OP-1 in the fracture gap immediately after the fracture occurred, as measured by callus volume, biomechanical evaluation, and histology.

After 2 weeks, the amount of woven bone in the callus increased after an injection of OP-1 in the fracture gap. Also, bony bridging of the fracture gap was observed more frequently. The mechanisms of fracture healing appear physiological and undisturbed in all treatment groups, since a combination of direct ossification and enchondral ossification was observed at both timepoints. This indicates that fracture healing can be stimulated with a single injection of OP-1 and that the resulting fracture healing process is normal. As fracture healing was stimulated at such an early timepoint, the question arises what mechanism leads to the acceleration of the healing process. The first step in the healing process is an increase in angiogenesis [32, 33], and evidence is accumulating for a potential role of OP-1 in angiogenesis through a stimulation of vascular endothelial growth factor, a potent angiogenic protein [34–36]. Thus, in conjunction with the differentiation of mesenchymal stem cells, acceleration of the fracture healing cascade may be due to a stimulation of angiogenesis. Further research in this direction is needed before any conclusions can be drawn.

No clear pattern in the localization of bone formation could be distinguished, since fields of woven bone were found throughout the callus in all groups, as well as small fields of cartilage. It therefore seems that bone formation in healing fractures takes place throughout the callus, and no indications were found for a pattern that was described in bone defects [13]. In their study, Geesink et al. described a pattern of bone formation at the outer edges of bone defects, when OP-1 was implanted. This pattern was observed radiographically at an early stage of bone healing. In some cases, abundant bone formation was seen, resulting in irritation of surrounding soft tissues. In our study, no such pattern of bone formation could be observed histologically.

Direct contact between newly formed bone and remnants of the collagenous matrix particles was seen frequently. No fibrous interposition between the bone and the collagenous particles or fibrous encapsulation as the result of a foreign body reaction against the particles was observed in either group treated with the matrix. Therefore, the collagenous carrier material neither led to any adverse effects nor appeared to inhibit the effect of OP-1, even though the callus volume of the group treated with OP-1 + matrix was significantly elevated after 4 weeks, compared to no injection and OP-1 alone. The increase in volume, measured with CT, may have been caused by an inflammatory response, but neither clinical nor histological signs were present, and therefore the presence of an inflammatory reaction is merely speculative.

In the group treated with OP-1 + matrix, incorporation of the particles was seen more often, indicating an acceleration of the incorporation by a rapid, stimulated, bone formation. Mild inflammatory reactions were seen in the immediate vicinity of the remnants of the particles, indicating a mild reaction aimed at the particles. As described above, this reaction was not present at 4 weeks, and therefore any reaction will have been transient and mild, but as the immunological behavior of goats may differ from that in humans, this has to be taken into account and any reaction should be monitored carefully.

## Conclusions

Osteogenic protein 1 (OP-1, BMP-7) accelerates the healing of closed fractures in animals by stimulation of the natural fracture healing processes. At this point, it still remains unclear what mechanism is exactly responsible for this acceleration, but angiogenesis contributes significantly to this acceleration of the fracture healing process. Except for a mild inflammatory response, no adverse effects of either OP-

1 or the collagenous carrier material have been reported. This support the conclusion that a minimally invasive procedure to obtain a fracture healing by a single injection of OP-1 in the fracture gap is an appropriate method to accelerate fracture healing.

## References


1 MacKenzie EJ, Morris JA Jr, Jurkovich GJ, Yasui Y, Cushing BM, Burgess AR, De Lateur BJ, McAndrew MP, Swiontkowski MF (1998) Return to work following injury: the role of economic, social, and job-related factors. *Am J Public Health* 88: 1630–1637
2 Praemer A, Furner S, Rice DP (1992) *Musculoskeletal conditions in the United States.* The American Academy of Orthopaedic Surgeons, Park Ridge, Illinois, 85–124
3 Urist MR (1965) Bone: Formation by autoinduction. *Science* 150: 893–899
4 Cook SD, Rueger DC (1996) Osteogenic protein-1: biology and applications. *Clin Orthop* 324: 29–38
5 Riley EH, Lane JM, Urist MR, Lyons KM, Lieberman JR (1996) Bone morphogenetic protein-2: biology and applications. *Clin Orthop* 324: 39–46
6 Vukicevic S, Stavljenic A, Pecina M (1995) Discovery and clinical applications of bone morphogenetic proteins. *Eur J Clin Chem Clin Biochem* 33: 661–671
7 Wozney JM, Rosen V (1998) Bone morphogenetic protein and bone morphogenetic protein gene family in bone formation and repair. *Clin Orthop* 346: 26–37
8 Blokhuis TJ (2001) New developments in bone healing, imaging and treatment methods. Thesis Vrije Universiteit Medical Centre
9 Cook SD, Dalton JE, Tan EH, Whitecloud TS, Rueger DC (1994) *In vivo* evaluation of recombinant human osteogenic protein (rhOP-1) implants as a bone graft substitute for spinal fusions. *Spine* 19: 1655–1663
10 Cook SD, Wolfe MW, Salkeld SL, Rueger DC (1995) Effect of recombinant human osteogenic protein-1 on healing of segmental defects in non-human primates. *J Bone Joint Surg Am* 77: 734–750
11 Cunningham BW, Kanayama M, Parker LM, Weis JC, Sefter JC, Fedder IL, McAfee PC (1999) Osteogenic protein *versus* autologous interbody arthrodesis in the sheep thoracic spine. A comparative endoscopic study using the Bagby and Kuslich interbody fusion device. *Spine* 24: 509–518
12 Gao TJ, Lindholm TS, Kommonen B, Ragni P, Paronzini A, Lindholm TC, Jamsa T, Jalovaara P (1996) Enhanced healing of segmental tibial defects in sheep by a composite bone substitute composed of tricalcium phosphate cylinder, bone morphogenetic protein, and type IV collagen. *J Biomed Mater Res* 32: 505–512
13 Geesink RG, Hoefnagels NH, Bulstra SK (1999) Osteogenic activity of OP-1 bone morphogenetic protein (BMP-7) in a human fibular defect. *J Bone Joint Surg Br* 81: 710–718
14 Johnson EE, Urist MR (1998) One-stage lengthening of femoral nonunion augmented with human bone morphogenetic protein. *Clin Orthop* 347: 105–116

15 Kirker-Head CA, Gerhart TN, Armstrong R, Schelling SH, Carmel LA (1998) Healing bone using recombinant human bone morphogenetic protein 2 and copolymer. *Clin Orthop* 349: 205–217

16 Kirker-Head CA, Gerhart TN, Schelling SH, Hennig GE, Wang E, Holtrop ME (1995) Long-term healing of bone using recombinant human bone morphogenetic protein 2. *Clin Orthop* 318: 222–230

17 Ripamonti U, Ma SS, Van Dijke H, Reddi AH (1992) Osteogenin, a bone morphogenetic protein, adsorbed on porous hydroxyapatite substrata, induces rapid bone differentiation in calvarial defects of adult primates. *Plast Reconstr Surg* 90: 382–393

18 Cook SD, Baffes GC, Wolfe MW, Sampath TK, Rueger DC (1994) Recombinant human bone morphogenetic protein-7 induces healing in a canine long-bone segmental defect model. *Clin Orthop* 301: 302–312

19 Ripamonti U, van den Heever B, Sampath TK, Tucker MM, Rueger DC, Reddi AH (1996) Complete regeneration of bone in the baboon by recombinant human osteogenic protein-1 (hOP-1, bone morphogenetic protein-7). *Growth Factors* 13: 273–289

20 Friedlaender GE, Perry CR, Cole JD, Cook SD, Cierny G, Muschler GF, Zych GA, Calhoun JH, LaForte AJ, Yin S (2001) Osteogenic protein-1 (bone morphogenetic protein-7) in the treatment of tibial nonunions. *J Bone Joint Surg Am* 83 (Suppl): S151–S158

21 Bostrom MP, Lane JM, Berberian WS, Missri AA, Tomin E, Weiland A, Doty SB, Glaser D, Rosen VM (1995) Immunolocalization and expression of bone morphogenetic proteins 2 and 4 in fracture healing. *J Orthop Res* 13: 357–367

22 Ishidou Y, Kitajima I, Obama H, Maruyama I, Murata F, Imamura T, Yamada N, ten Dijke P, Miyazono K, Sakou T (1995) Enhanced expression of type I receptors for bone morphogenetic proteins during bone formation. *J Bone Miner Res* 10: 1651–1659

23 Nakase T, Nomura S, Yoshikawa H, Hashimoto J, Hirota S, Kitamura Y, Oikawa S, Ono K, Takaoka K (1994) Transient and localized expression of bone morphogenetic protein 4 messenger RNA during fracture healing. *J Bone Miner Res* 9: 651–659

24 Onishi T, Ishidou Y, Nagamine T, Yone K, Imamura T, Kato M, Sampath TK, ten Dijke P, Sakou T (1998) Distinct and overlapping patterns of localization of bone morphogenetic protein (BMP) family members and a BMP type II receptor during fracture healing in rats. *Bone* 22: 605–612

25 Bostrom MP, Camacho NP (1998) Potential role of bone morphogenetic proteins in fracture healing. *Clin Orthop* 355 (Suppl): S274–S282

26 Reddi AH (1998) Initiation of fracture repair by bone morphogenetic proteins. *Clin Orthop* 355 (Suppl): S66–S72

27 Blokhuis TJ, den Boer FC, Bramer JAM, Jenner JMGTh, Bakker FC, Patka P, Haarman HJThM (2001) Biomechanical and histological aspects of fracture healing, stimulated with Osteogenic Protein-1. *Biomaterials* 22: 725–730

28 Bramer JAM, Barentsen RH, vd Elst M, de Lange ESM, Patka P, Haarman HJThM (1998) Representative assessment of long bone shaft biomechanical properties: an optimized testing method. *J Biomech* 31: 741–745

29 Currey JD (1988) The effects of drying and re-wetting on some mechanical properties of cortical bone. *J Biomech* 21: 439–441
30 Linde F, Sorensen HC (1993) The effect of different storage methods on the mechanical properties of trabecular bone. *J Biomech* 26: 1249–1252
31 Van der Wiel HE, Lips P, Nauta J, Patka P, Haarman HJ, Teule GJ (1994) Loss of bone in the proximal part of the femur following unstable fractures of the leg. *J Bone Joint Surg Am* 76: 230–236
32 Oni OO (1997) The early stages of the repair of adult human diaphyseal fractures. *Injury* 28: 521–525
33 Probst A, Spiegel HU (1997) Cellular mechanisms of bone repair. *J Invest Surg* 10: 77–86
34 Yeh LC, Lee JC (1999) Osteogenic protein-1 increases gene expression of vascular endothelial growth factor in primary cultures of fetal rat calvaria cells. *Mol Cell Endocrinol* 153: 113–124
35 Ramoshebi LN, Ripamonti U (2000) Osteogenic protein-1, a bone morphogenetic protein, induces angiogenesis in the chick chorioallantoic membrane and synergizes with basic fibroblast growth factor and transforming growth factor-beta1. *Anat Rec* 259: 97–107
36 Blokhuis TJ, Bronckers ALJJ, Bakker FC, Patka P, Burger EH, Haarman HJThM (2001) Immunolocalisation of OP-1 in human early fracture repair. *J Clin Immun*; *submitted*

# Maxillofacial reconstruction

*Hendrik Terheyden*[1] *and Søren Jepsen*[2]

[1]Department of Oral and Maxillofacial Surgery, University of Kiel, Arnold Heller Str. 16, 24105 Kiel, Germany; [2]Department of Restorative Dentistry and Periodontology, University of Kiel, Arnold Heller Str. 16, 24105 Kiel, Germany

## Introduction

Reconstructive surgery in of the maxillofacial skeleton comprises a large variety of indications which range from dental alveolar surgery to interdisciplinary cranial base interventions, from congenital malformations to acquired traumatic or tumor related defects. In most applications the autogenous bone graft is the clinical gold standard. These grafts range from small intraorally harvested bone particles to large composite vascularized bone flaps. In the face always reconstruction has functional and esthetic aspects. In both aspects the shape of the reconstructed bone segment is very important. Regarding dental occlusion a correct intermaxillary relation has to be achieved, especially if prosthetic rehabilitation with dental implants is intended. Due to the thin skin coverage shape irregularities will end with a bad esthetic result. Furthermore the regenerated mandibular segment has to resist an occlusal load up to 600 N on a single molar tooth.

From a biomechanical point of view it is useful to distinguish between filling of bone gaps and augmentations above the existing anatomical bone level. As long as the osteoinductive components are used to fill preexisting defects like bone cysts or some kinds of small mandibular continuity defects the stability and space keeping effect of the carrier material is not so important. This changes in all kinds of augmentations or in large defects where the bone inducing implant has to resist soft tissue pressure which occurs during mastication, during movements of the tongue or the mimic muscles. These facts are especially important in alveolar ridge augmentation.

This review will focus on mandibular reconstruction and augmentations in dental implant surgery. These indications are standard situations of maxillofacial reconstruction, which frequently occur in clinical routine.

Bone Morphogenetic Proteins, edited by Slobodan Vukicevic and Kuber T. Sampath

## Mandibular reconstruction

### Direct application of BMP in mandibular continuity defects

The key study on mandibular reconstruction with bone morphogenetic proteins was performed by Toriumi and coworkers [1]. In dogs a predictable and load bearing bridging of the defect occurred using rhBMP-2 and a collagen sponge carrier. However some narrowing an reduction of height of the regenerated bone was noticed due to soft tissue pressure on the soft carrier material. A subsequent study with similar long-term results was later reported by the same group using rhBMP-2 and a biodegradable particular polylactide carrier [2]. Complete bridging as well and osseointegration of dental implants was observed in a monkey study by Boyne and coworkers using rhBMP-2 on a collagen sponge carrier [3]. In the latter study a wound dehiscence problem and impairment of bone formation occurred with the intraoral approach which is also typical for clinical work. Although the authors of these studies tried, it is practically impossible to strip all periosteum especially in the alveolar parts in this kind of defect. Thus, the studies resemble clinically more a subperiosteal resection of a benign tumor. The prerequisites of bone healing are good because of a very good receptor bed for bone inducing substances including differentiated cells.

In conclusion, treatment of such defects with rhBMP leads to a predictable bone bridging. Improvements are required regarding the shape of the reconstructed segment, which showed some irregularities in all studies. Clinically such irregularities are not desired since the tolerance of intermaxillary relation of the ridges for prosthetic treatment is not more than a few millimeter and for esthetic reasons. The intraoral approach is a clinical standard for autologous bone grafting. In this case bacterial contamination through the saliva is an additional factor. For application of the recombinant osteoinductive technology supplementation of the carrier material with antibiotic drugs may be a future field of research and development.

### Prefabrication of vascularized bone grafts

Clinically most mandibular defects occur after ablative surgery on malignant tumors. In this case the usually combined intraoral/extraoral approaches are used. Microbiological contamination, extended operative time and extensive scar formation may decrease the success of primary reconstructive procedures. In most cases additional radiotherapy will result in a poor recipient bed for bone grafts and BMP [4]. Clinically, in these cases a revascularized autogenous bone graft is applied. Usually a fibular or iliac bone graft is harvested with a vascular pedicle (and sometimes with additional soft tissue flaps). These vessels are microsurgically connected with facial vessels and blood perfusion of the graft is restored. The disadvantage of such

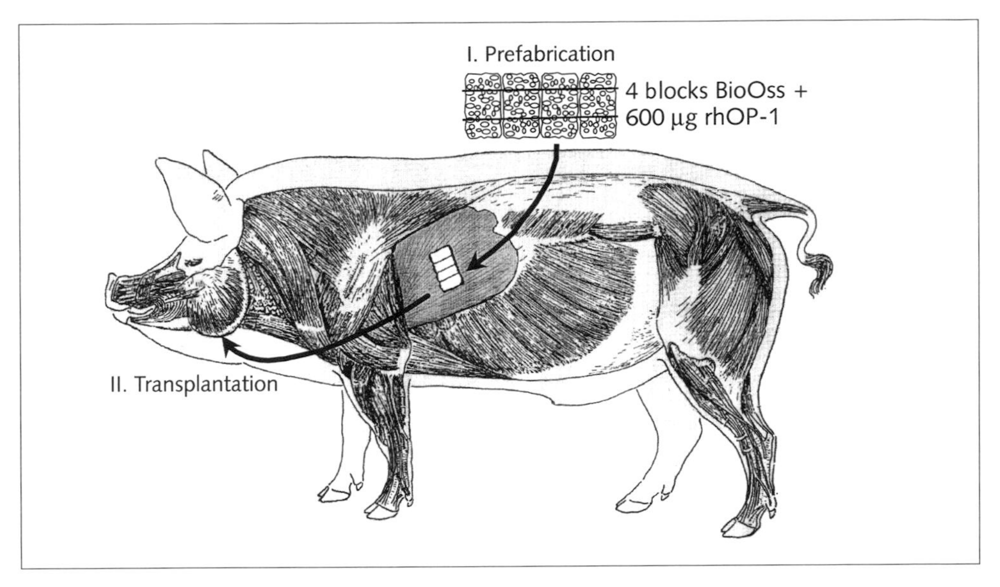


*Figure 1*
*Prefabrication of a bone graft: an osteoconductive scaffold (xenogenic bone blocks) was loaded with 600 µg rhOP-1 and placed in the Latissimus dorsi muscle. After 6 weeks the bone was harvested with a vascular pedicle and grafted to the mandibular defect. Perfusion was restored by microsurgical anastomosis with facial blood vessels.*

technique is that harvesting of a vascularized bone graft is an operative burden and therefore not suitable for every patient. Secondly, problems can occur with donor site morbidity, anatomical limitations of the donor sites and the shape of naturally occurring grafts.

The prefabrication technique allows to create a bone graft in an easily accessible soft tissue area which can be custom shaped according to the requirements of the individual defect. Khouri and coworkers were the first who used BMP for custom prefabrication of a small artificial femur head in a rat [5]. Several authors followed with prefabricated bone flaps in small animals without using them as a graft for reconstruction [6–11]. A prefabrication and transplantation in a large animal model was performed by our group in minipigs [12] (Fig. 1). In 10 minipigs an osteoconductive scaffold was placed in a soft tissue pouch inside the Latissimus dorsi muscle [13]. The scaffold consisted of single blocks of xenogenic bone (BioOss®, Geistlich, Wolhusen, Switzerland) which were connected with resorbable threads forming an implant of 4.5 × 2 × 1 cm size. Prior to surgery 600 µg of rhOP-1 in 1.2 ml acetate-mannitol buffer solution was poured over the scaffold and soaked by the material. Bone growth in the blocks was studied by computed tomography (Fig. 2) and histology (Fig. 3). It was found that 6 weeks of prefabrication time are sufficient (Fig. 4).

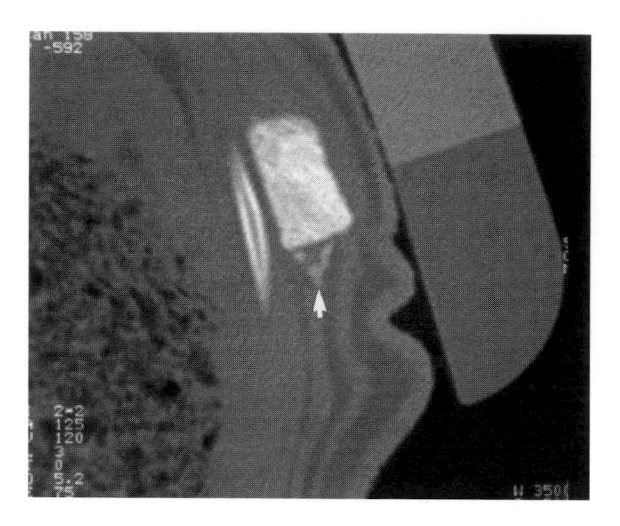

*Figure 2*
*Computed tomography of the thoracic wall with the prefabricated bone graft in the shape of an BioOss® block. There is no fusion with the ribs and only minimal bone overgrowth (arrow). Bone overgrowth was planimetrically assessed in subsequent CT sections (2.4% of the total graft volume).*

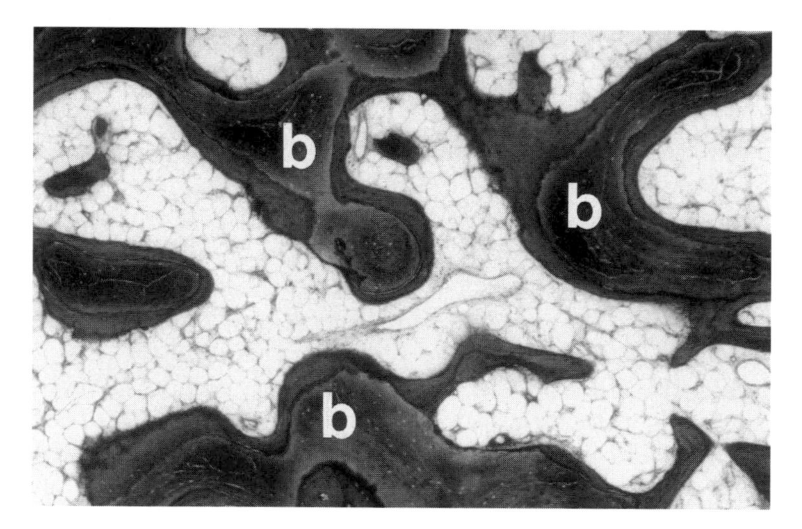


*Figure 3*
*Prefabricated vascularized bone graft: Histology 12 weeks after implantation of the newly formed bone which formed a thin continuous layer on the scaffold of the BioOss® trabeculae (b). Between the bone trabeculae in the interconnecting spaces usually a central arteriole and a complete bone marrow cell population was observed (arrows) (non decalcified, ground and polished section, Toluidine blue × 60).*

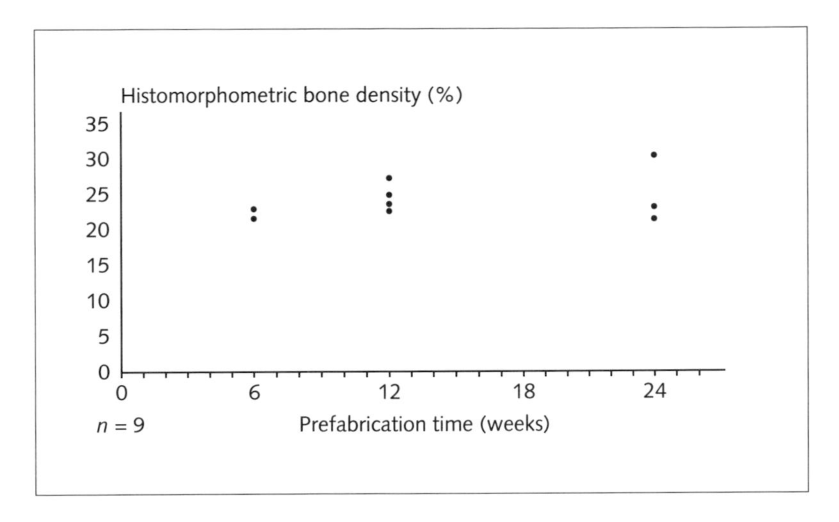


*Figure 4*
*Prefabricated vascularized bone graft: A prefabrication time of 6 weeks is sufficient. No increase of bone density after 6 weeks.*

Vascularisation in the grafts was studied by macro- and microangiography indicating a good vascularization on the microscopic level in areas of bone growth (Figs. 5, 6).

In a subsequent study in minipigs a dose dependency of the parameters blood vessel density and bone density was observed (Figs. 7, 8). The highest best values were obtained with the dosage of 1000 μg rhOP-1 in a gram of carrier (xenogenic bone particles) [14]. In a subsequent study such prefabricated grafts of 4.5 × 2 × 1 cm size were used to treat mandibular defects in minipigs [15]. The grafts were harvested and grafted to a mandibular defect at the angle of the mandible in Göttingen miniature pigs. The defect was created in the mandibular angle using an epiperiosteal preparation and resection of the periosteum (Figs. 9, 10). The newly formed bone was stable enough to be fixed in the defects with conventional titanium miniplates and screws (Fig. 11). Graft perfusion was restored by anastomosis with the facial vessels using a microsurgical technique. An identical defect of the contralateral side served as a control group and was treated by directly applied xenogenic bone scaffold and 600 μg rhOP-1. The first result of the study was that grafted prefabricated vascularized bone stayed viable. The continuous viability of large parts of the bone marrow was demonstrated by tracer uptake in bone scintigraphy (Fig. 12) and secondly shown in histology. Bone apposition in several areas was not interrupted by the transplantation process as proved by continuous polychromatic fluorescent labeling.

As a second result it was possible to restore the mandible with a prefabricated bone graft which was designed to fit into a certain mandibular defect (Fig. 13). His-

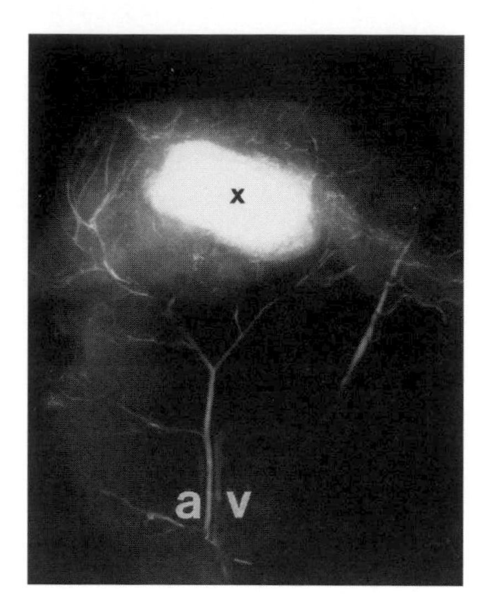


*Figure 5*
*Prefabricated vascularized bone graft: Angiography of the latissimus dorsi flap containing the prefabricated bone graft (×). The thoracodorsal artery and vein (a, v) continuously branch in to the graft.*

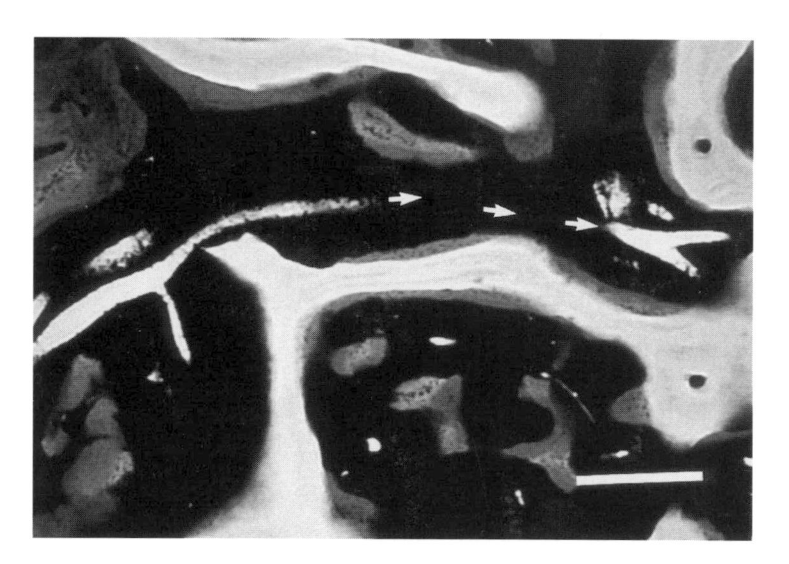

*Figure 6*
*Prefabricated vascularized bone graft: Microangiography of the regenerated bone within the BioOss block. The white BioOss trabeculae are lined by newly formed bone (gray, less mineralized). Usually one newly formed artery (arrows) is found in every pore between the trabeculae of the BioOss (microradiography/microangiography, bar equals 100 μm).*

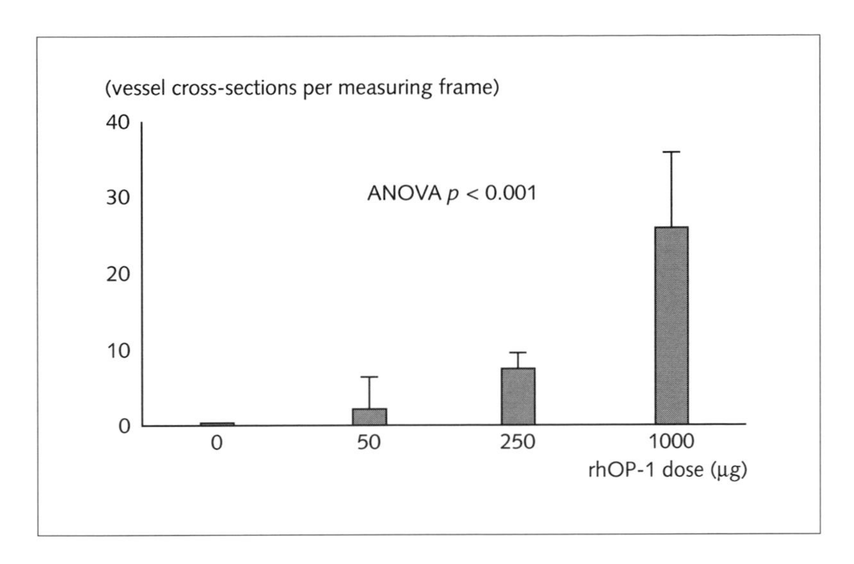


*Figure 7*
*Prefabricated vascularized bone graft: The density of blood vessel in the prefabricated graft depends on rhOP-1 dosage.*

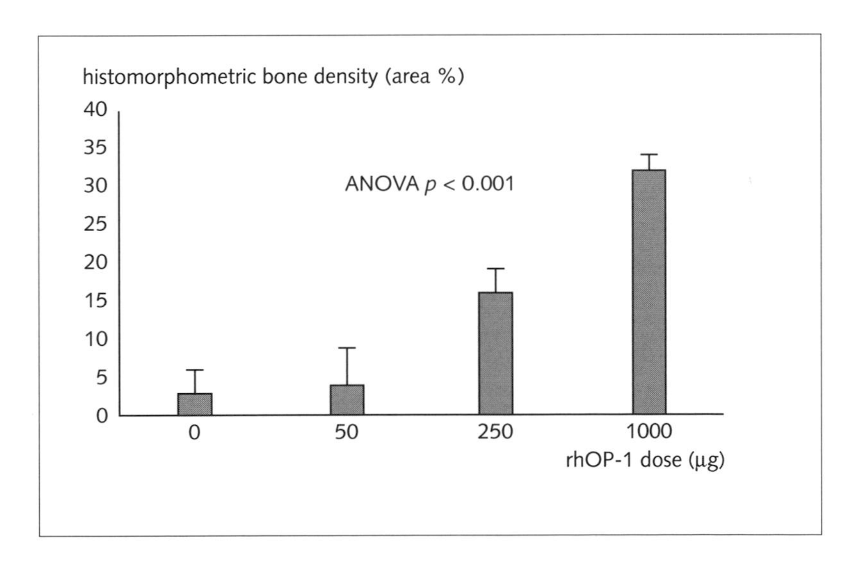


*Figure 8*
*Prefabricated vascularized bone graft: The bone density of the prefabricated graft depends on rhOP-1 dosage.*

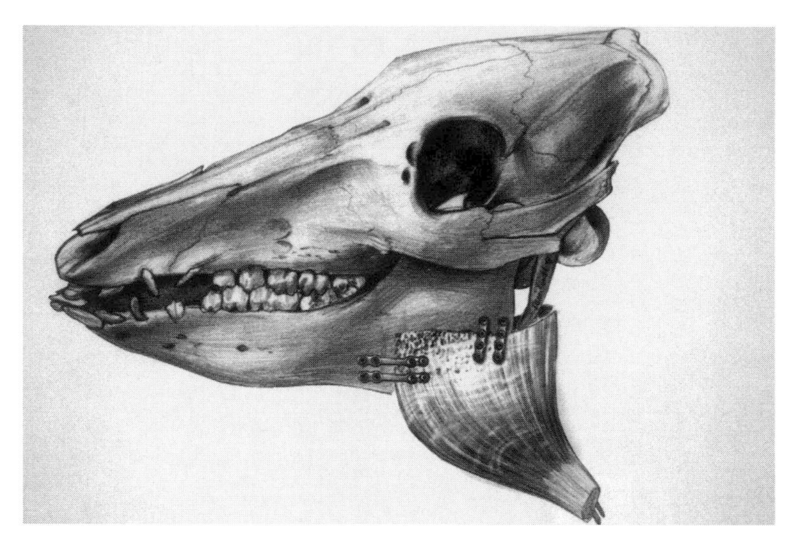

*Figure 9*
*Test side: the defect in the mandibular angle of an miniature pig according to Schmelzeisen et al. [52], modified by Shirota et al. [53], was treated by a vascularized prefabricated bone graft fixed with miniplates.*

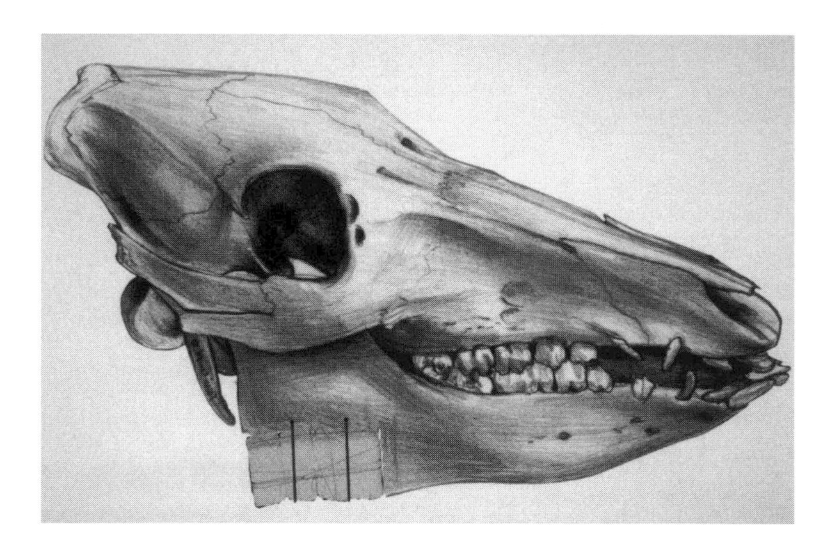

*Figure 10*
*Control side: an identical defect on the contralateral side of the same animal treated with 4 blocks of xenogenic bone and directly applied 600 μg rhOP-1. The BioOss blocks were fixed with resorbable sutures to the residual bone.*

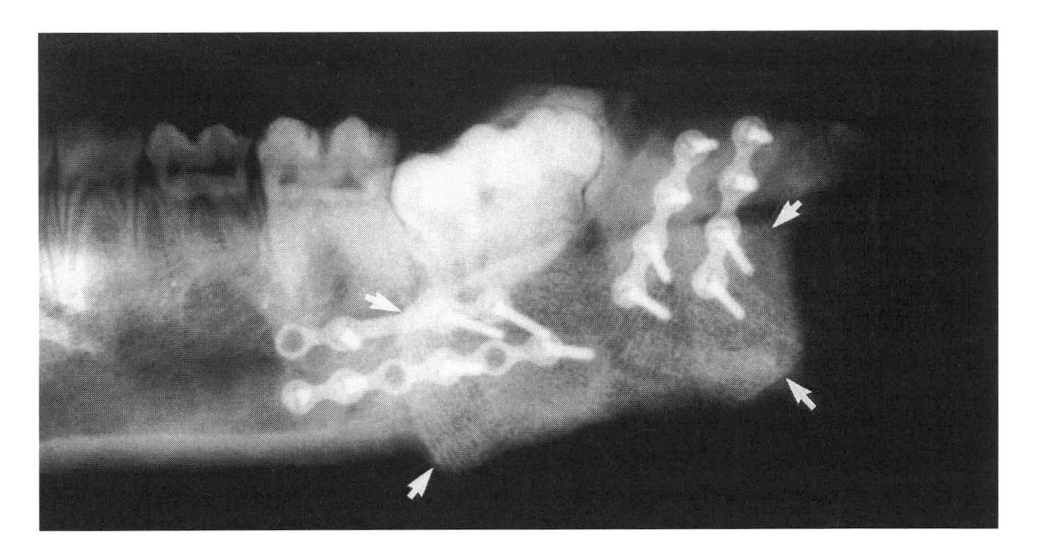

*Figure 11*
*Prefabricated vascularized bone graft: Plain radiograph of the prefabricated bone flap fixed with titanium miniplates in the defect (arrows).*

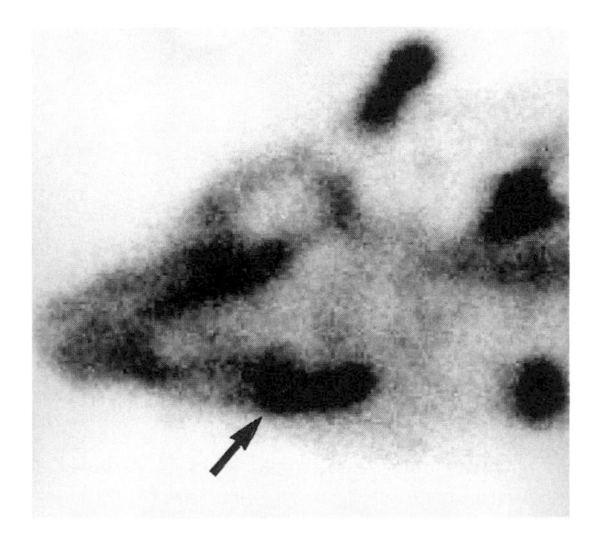

*Fig. 12*
*Prefabricated vascularized bone graft: Planar Tc99m-bone scintigraphy 7 days after transplantation demonstrates vitality and perfusion of the graft (arrow). The remaining spots of tracer accumulation are, clockwise, the contralateral side (DirOP-1), the ear vein with the site of injection and both thyroid lobes.*

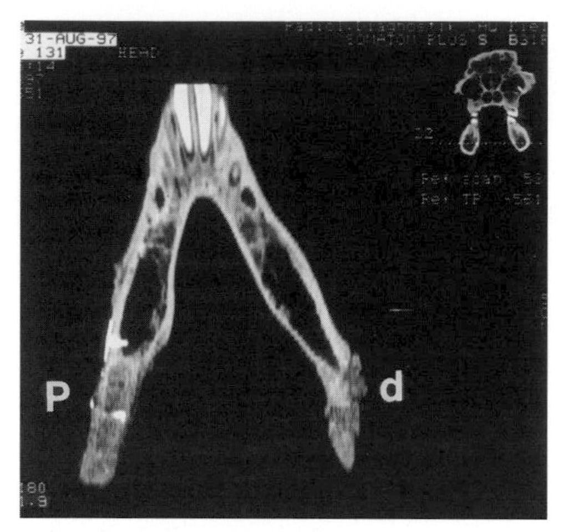


*Figure 13*
*Prefabricated vascularized bone graft: Computed tomography 3 months after surgery. The transversal data reconstruction of the mandibular arch shows the reconstructed area and the residual mandible. The regenerated bone on the test side (prefabricated graft = p) matches the contour of the resected mandible. On the control side after direct application of rhOP-1 (= d) the volume is deficient and bone has grown less controlled.*

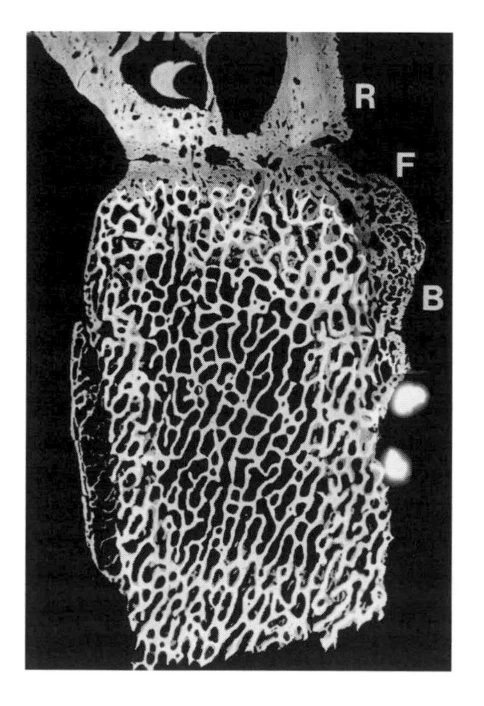


*Figure 14*
*Prefabricated vascularized bone graft: Vertical histological section through the graft with residual mandible (R), fusion zone (F) and BioOss block (B). Bone has developed in every quadrant of the xenogenic bone mineral scaffold and minimal bone overgrowth is present (Microradiography, digital slide composition, non decalcified, bar equals 3 mm).*

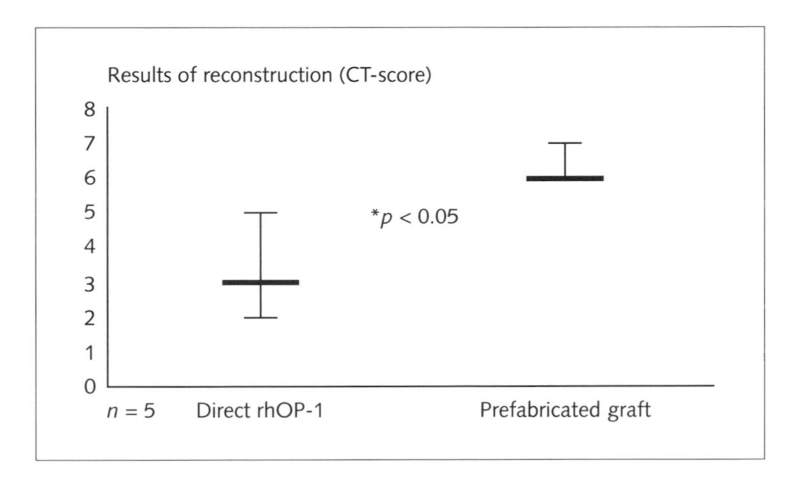


*Figure 15*
*Quantification of the quality of the skeletal reconstruction after computed tomographic examination by independent examiners using a numerical score (median and first and third quartile, U-Test, $\alpha = 0.05$). The reconstruction with the prefabricated graft received higher scores.*

tologically it was observed that the growth of the newly formed bone was controlled by the osteoconductive scaffold which was filled with viable bone (Fig. 14). Bone overgrowth was noted in only 2.3% of the volume. In CT scans 3 months postoperatively a good restoration of the mandibular contour was observed and the regenerated bone showed good volume constancy, suitable for instance for the insertion of dental impants. An independent rating of CT scans with a numerical score system revealed a significantly better reconstructive result than with the directly applied material on the contralateral side (Fig. 15).

In conclusion the prefabrication technique is likely to open new possibilities in reconstructive surgery. The technology seems ready for clinical use once recombinant BMP are approved for the clinical use. Further studies have to focus on technologies for custom shaping of individual parts for skeletal reconstruction.

## Implantology

### Aims for the use of BMP in implantology

It has been shown that long-term success of any implant under function depends on the achievement of direct bony anchorage [16]. Thus, the two basic aims of the use of growth factors and BMP in implant dentistry are to increase bone implant con-

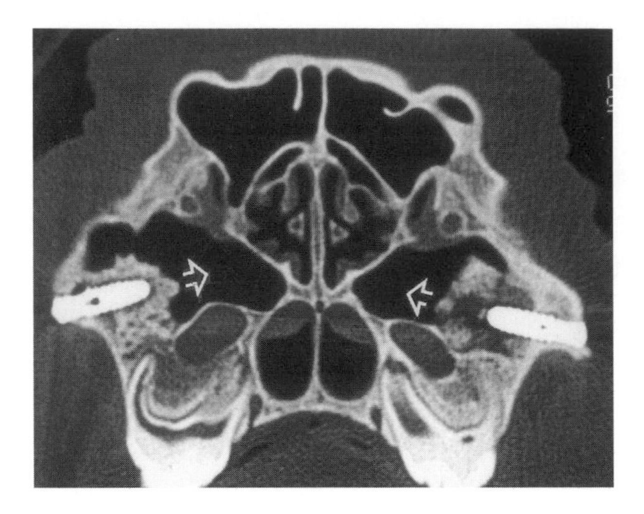

*Figure 16*
*Sinus lift with rhOP-1: Axial CT-scan of miniature pig. Two implants are inserted from a laterocaudal direction into the augmented maxillary sinus area (⇨).*

tact (BIC) and to achieve a faster osseous integration, compared to standard clinical healing times of 3 to 6 months today. Furthermore, there is increasing evidence that in the near future BMP will support or even replace autogenous bone grafting in augmentation of bone deficient sites. A future prospective for the use of BMP may be to increase the quality of bone surrounding the implant and to reosseointegrate an implant after bone loss through periimplant infection (perimplantitis).

## Growths factors in implantology

A mixture of growth factors (PDGF/IGF-1) in a carboxymethylcellulose gel as a carrier was used in a few studies in implantology [17, 18] with some success. However, these studies have not been pursued later. A natural source of PDGF, platelet rich plasma (PRP), has been in demonstrated to be useful to support the healing of bone grafts [19]. This method has been recognized by many dental practitioners. However, no relevant data concerning PRP have been published in implant dentistry yet.

## Enhancement and acceleration of BIC

Bone morphogenetic proteins (BMP) have been reported to enhance osseous contact of dental implants. Some of these studies used naturally-sourced bovine BMP prepa-

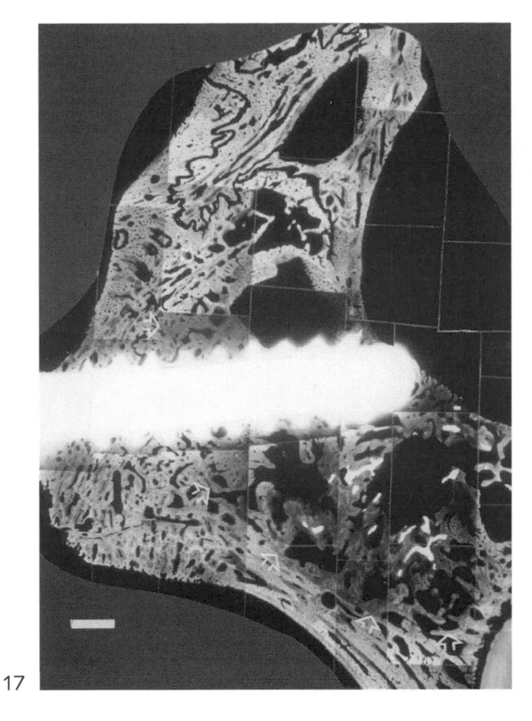
17

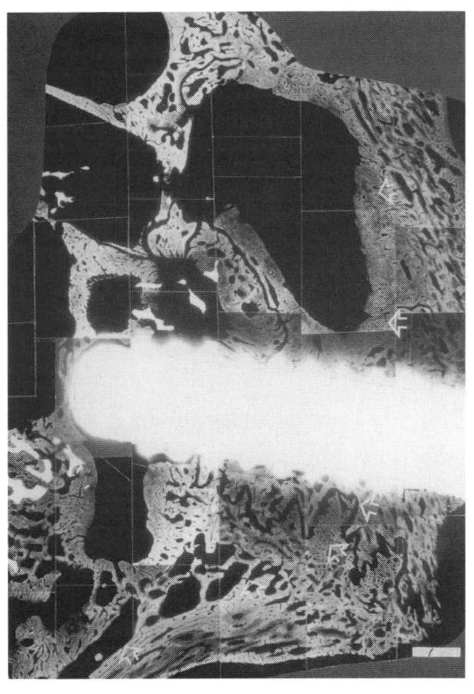
18

*Figure 17*
*Sinus lift with rhOP-1: Histology, frontal section of right maxillary sinus, augmented with rhOP-1 and BioOss with previous (⇨) and new sinus floor (rhOP-1 group) (microradiography in composite slides technique, bar equals 1000 μm).*

*Figure 18*
*Sinus lift with rhOP-1: Frontal section of a maxillary sinus augmented with BioOss alone with previous (⇨) and new sinus floor (control group) (microradiography in composite slides technique, bar equals 1000 μm).*

rations in a canine mandibular site using a descriptive evaluation [20–22]. RhBMP-2 was used in an *in vitro* assay demonstrating a stimulation of osteoblastic cells on a titanium surface [23]. In a canine study rhOP-1 induced new bone and enhanced osseous contact of HA-coated implants (BIC 80%) in combination with bone derived type I collagen in fresh extraction sites in the mandible [24]. Our group observed 80% BIC with rhOP-1 and BioOss® compared to 32% with BioOss® alone in regenerated bone in a sinus augmentation study [25] (Figs. 16–18). Eighty percent BIC is a noticeable value since a 60% BIC in mandibular bone is a representative value for a titanium implant [26]. Attempts have also been made to increase BIC by modifying the surface structure of the implants [27–29] or using HA

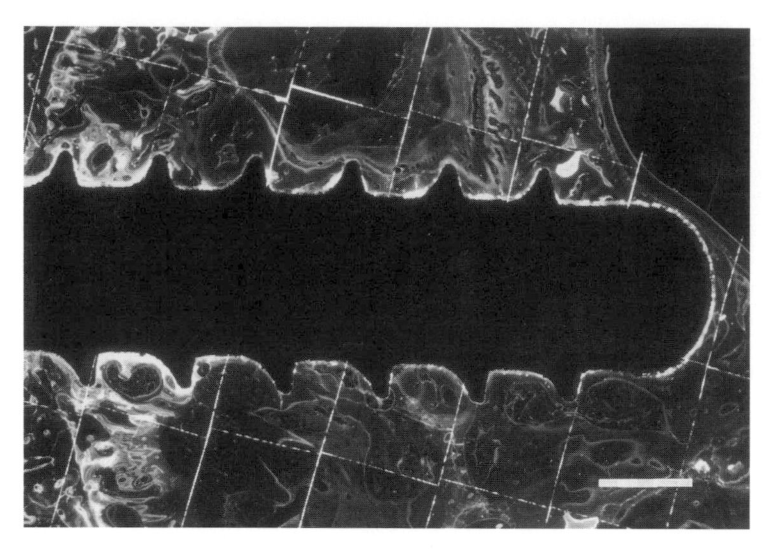

*Figure 19*
*Sinus lift with rhOP-1: Frontal section of the maxillary sinus (rhOP-1 group) with previous (⇨) and new sinus floor. Fluorescent calcified material on the implant surface after polychromatic intravital labelling (fluorescence microscopy in composite slides technique, bar equals 1000 μm).*

*Figure 20*
*Sinus lift with rhOP-1: Frontal section of maxillary sinus (control group) with previous (⇨) and new sinus floor. No fluorescent layer on the implant surface (fluorescence microscopy in composite slides technique, bar equals 1000 μm).*

coatings [30]. Although the studies are not easily compared due to different animal models, experimental periods and surface characteristics, it has to be emphasized, that none of the studies achieved an osseous integration as high as the 80%. It can be concluded that rhOP-1 enhanced BIC.

BMP have also been reported to accelerate bone formation around the implant in naturally derived [31] and recombinant form. In a minipig sinus augmentation study of our group, deposition of calcified material occurred on the implant surface after 2–3 weeks on the rhOP-1 side and after 8–9 weeks in the controls as monitored by polychromatic labelling [24] (Figs. 19, 20). In conclusion, BMP can accelerate BIC formation. Clincal studies will have to elucidate whether clinically this may lead to earlier loading of implants and reduced recommendations for healing time which actually is 6 months in regenerated bone. Further research and development studies are required on biological improvement of dental implants especially on BMP coating.

## Sinus augmentation

From a biomechanical point of view it is useful to distinguish between inlay and onlay augmentations. Maxillary sinus augmentation is an inlay type of augmentation where the augmentation material is put relatively protected into a cavity with excellent contact to residual bone. The procedure is required when implants are planned in the edentulous parts of the lateral upper jaw where protrusion of the maxillary sinus led to an internal reduction of the height of alveolar bone. Sinus augmentation is a clinically very frequently used procedure.

BMP have been applied successfully in preclinical studies on sinus augmentation. In a study utilizing rhBMP-2 and collagen sponges for a maxillary sinus floor augmentation in goats [32] bone growth was observed in the sinus floors. In a primate study rhOP-1 on collagen carrier induced bone, but augmentation with BioOss resulted in a better augmentative effect [33, 34]. Implants were not installed in those studies. In a sinus augmentation study in miniature pigs (Figs. 21–23) using 420 µg rhOP-1 in 1 ml acetate-mannitol buffer solution with 3 ml xenogenic bone mineral (BioOss) as a carrier with simultaneous insertion of dental implants our group reported a successful augmentation over the top of the simultaneously installed implants on the rhOP-1 side and on the control side after 6 months (Figs. 20, 21). In a subsequent study of our group in the same animal model, less BIC and augmentation height were observed with collagen carrier, compared to xenogenic bone or beta-tricalciumphosphate (Cerasorb®, Curasan, Kleinostheim, Germany) (Figs. 24, 25) [35]. The results confirmed the results of Margolin and coworkers [32] and support the view that for augmentation in the sinus the osteoconductive carrier alone was better than soft collagen carrier and rhOP-1. However, osteoconduction takes time (6 months or more) and the role of the BMP in this situation can be the

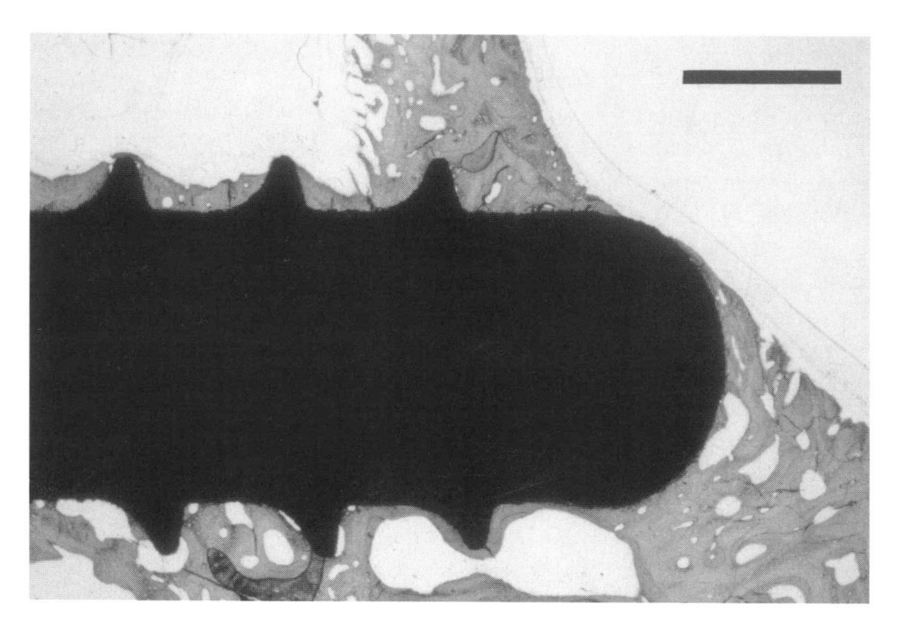

*Figure 21*
*Sinus lift with rhOP-1: Newly formed bone covers the implant surface in the augmented area (rhOP-1 group) (toluidine blue, bar equals 1000 μm).*

acceleration and predictability of ossification. This was confirmed in our study by polychromatic labeling which revealed ossification on the implant surface as early as 3 weeks after implantation in contrast to the osteoconductive control where ossification on the implant occurred after 9 weeks (Figs. 19, 20). As mentioned above, a predictable and significant increase in BIC was observed with simultaneous installation of the dental implant. The fact that implants should be placed simultaneously with the osteoinductive proteins should be emphasized. In a site containing bone morphogenetic proteins the implant is placed into the osteoinductive environment of the developing osteoprogenitor cells. Those cells interact with extracellular matrix and surfaces in their environment [36] and it is well known that the structure of the newly formed bone is influenced by the geometry of the environment [37]. Thus it may be hypothesized , that in implants placed secondarily to bone augmentation with BMP the BIC rates would not be enhanced. In fact in a second stage implantation study using rhBMP-2 on a collagen carrier in sinus augmentation in primates the bone to implant contact was with 41.4% not enhanced compared to the controls [38].

Human studies, as far as they are available, show inconsistent results. A small series of three human patient cases of sinus augmentation with rhOP-1 is reported in the literature [39–41]. The results range from good bone growth in one patient

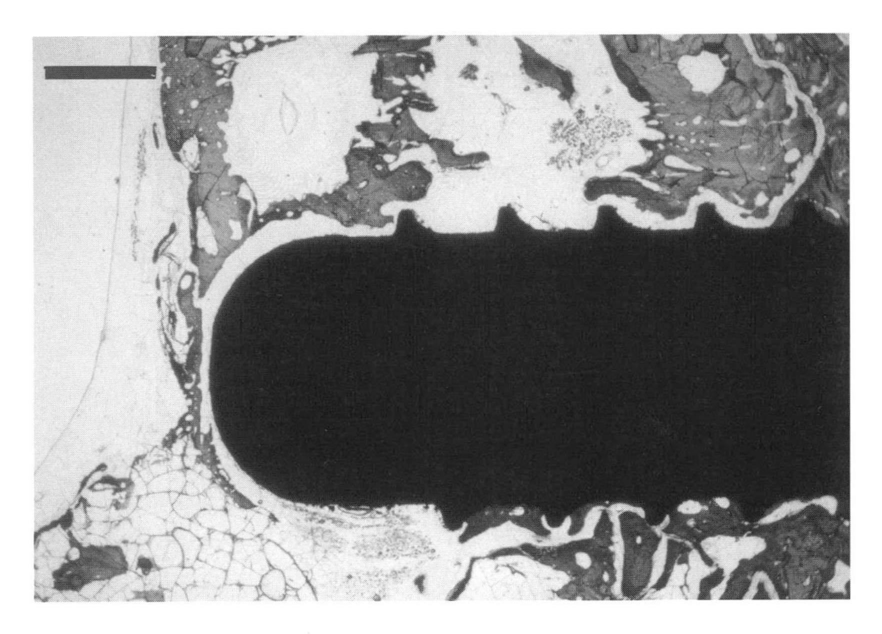

*Figure 22*
*Sinus lift with rhOP-1: Compared to rhOP-1 side sparse bone contact of the implant surface in the augmented sinus area (control group) (toluidine blue, bar equals 1000 μm).*

to absence of bone and persistent swelling in another patient. It was discussed that these inconsistent results may be attributed to the type of carrier used in the study (bone collagen). In a larger series [42] using rhBMP-2 and collagen sponge carrier grossly good augmentative results but not always predictable augmentation height was reported.

In conclusion from animal and human studies for augmentative sinus procedures a mineral osteoconductive carrier seems to be more suitable than soft collagen products and the role of BMP seems to be improving the predictability and speed of ossification and to enhance BIC in cases of primary implant installation.

## Ridge augmentation

Alveolar ridge augmentation in implant dentistry is indicated when an edentulous part of the alveolar ridge has partially lost height and/or width due to ridge atrophy following tooth extraction. Ridge augmentation is an onlay type augmentation where the augmentation material is placed on top of the bone surface or into very shallow defects, where it has only limited contact to the residual bone. In this situation mechanical load (occlusal load and soft tissue pressure) acts towards the

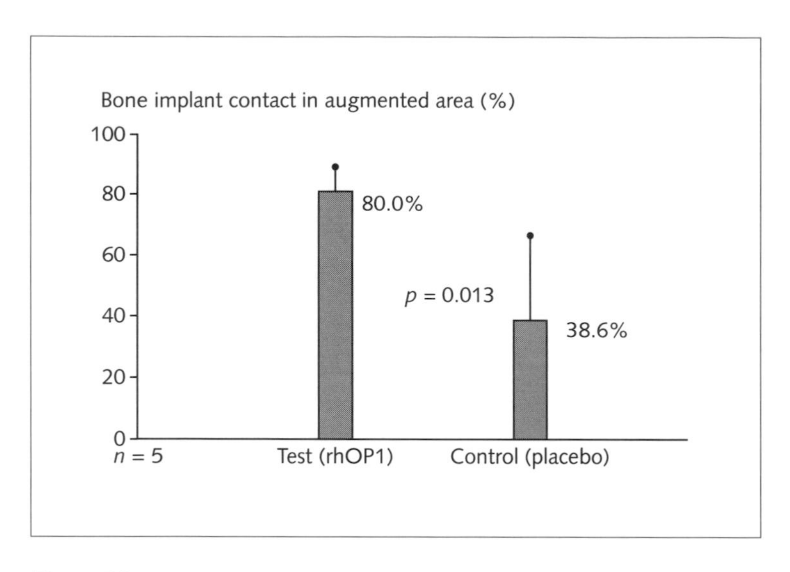


*Figure 23*
*Sinus lift with rhOP-1: Bone implant contact (BIC) in the augmented area of rhOP-1 and control side (mean value and SD, t-test).*

graft and the suture line. Furthermore the augmentation is situated just below the suture line and is in higher risk of bacterial contamination and wound healing problems. This is more pronounced in vertical than in horizontal ridge augmentation.

Several studies dealt with BMP in ridge augmentation. In a basic study of our group using seven different carrier materials in mandibular augmentation in the rat it was confirmed that mineralized calciumphosphate carriers result in a more predictable bone augmentation than collagens and that the different ostoconductivity of carrier materials influences structure of the newly formed bone (Figs. 26, 27) [37, 43]. In a canine study comparing periimplant defects in the mandible treated with and without rhBMP-2 on collagen sponge carrier significant differences to the controls were noted after 12 weeks, but not after 4 weeks by radiographic evaluation [44]. In another canine study with rhBMP-2 and collagen sponge a bone augmentation was observed. However, a low BIC of only 29.1% was reported after 16 weeks in regenerated bone [45]. A subsequent study could demonstrate that using a mixture of the collagen carrier with hydroxyapatite the results significantly improved. Thus, the conclusions of sinus augmentation have to be repeated. All data support the use of a mineralized osteoconductive carriers in augmentations.

Clinical studies on ridge augmentation are sparse and of preliminary character [46].

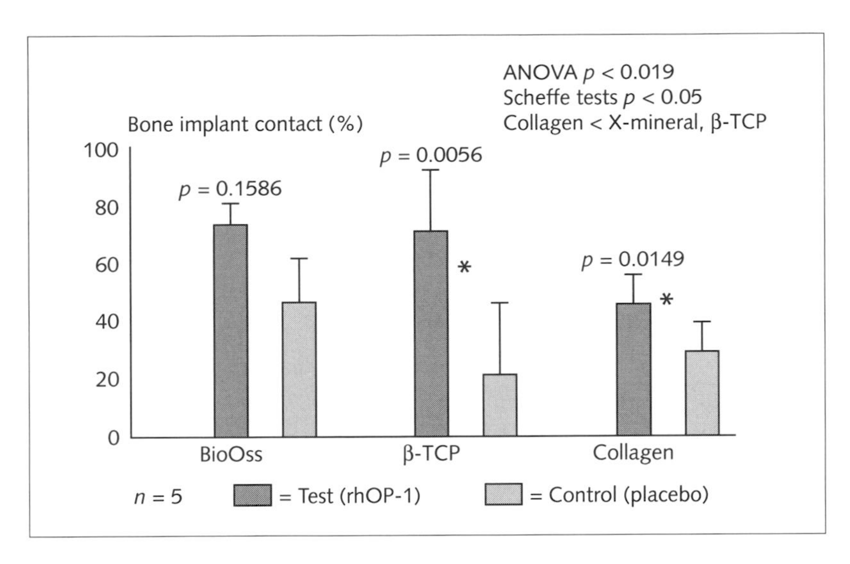


*Figure 24*
*Comparison of three carrier materials in sinus lift with rhOP-1: Xenogenic bone (slow resorption by osteoclasts within years), beta-tricalciumphosphate (spontaneous solubility within months under body conditions) and bone collagen (fast enzymatic resorption in the body within weeks). Better BIC (bone implant contact) in rhOP-1 sites compared to carrier alone. Better BIC for the mineralised carriers compared to collagen on the rhOP-1 sides (mean value and SD, ANOVA and Scheffé-test).*

## Reosseointegration and improvement of bone quality

An investigation on the use of a growth- or differentiation factors for improvement of the local bone quality for example in type IV bone has not been reported yet. Reosseointegration after infection was observed with rhBMP-2 in a primate study [47]. This field remains to be an open question, although hypothetically this seems to be a reasonable field of research.

## Other fields of maxillofacial reconstruction

There are plenty of indications for bone grafting in the craniomaxillofacial field. Cranial defects were successfully restored with Osteogenin as it is required in pediatric and adult craniofacial surgery [48]. RhBMP-2 was successfully applied with a collagen sponge carrier in a cleft palate defect in a monkey study [49] and with polylactide beads carrier in a dogs study [50].

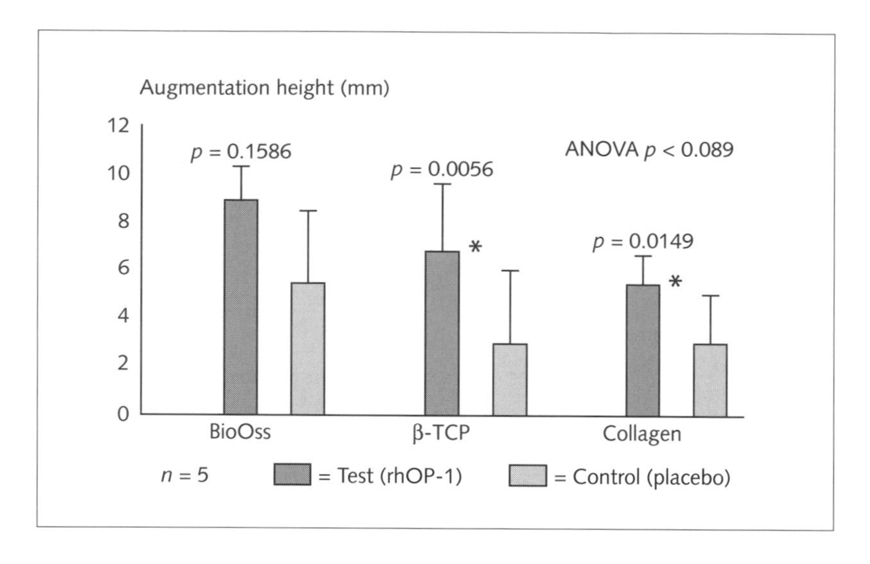


*Figure 25*
*Comparison of three carrier materials in Sinus lift with rhOP-1: Xenogenic bone (slow resorption by osteoclasts within years), beta-tricalciumphosphate (spontaneous solubility within months under body conditions) and bone collagen (fast enzymatic resorption in the body within weeks). Better augmentation height in rhOP-1 sites compared to carrier alone. Better augmentation height for the mineralised carriers compared to collagen on the rhOP-1 sides (mean value and SD, ANOVA and Scheffé-test).*

## Conclusion

The question of carrier materials for rhBMP may be more important in craniofacial surgery than in other fields of reconstructive surgery. Volume and shape of the regenerated bone is important either in continuity reconstruction as in augmentations. A proven way to control the osteoinductive process is to use an osteoconductive scaffold for the induced bone cells [51]. The induced osteoprogenitors will adhere along the surface of this substratum and start matrix production in a controlled fashion. This theoretical principle has proven in many of the reviewed studies. Porous hydroxyapatite as well as porous beta-tricalciumphosphate has been demonstrated to be suitable as delivery agent, as space-keeping material, as well as osteoconductive scaffold for the bone cells. Further studies are required in the field of delivery materials.

As far as preclinical evaluation in animal studies can predict clinical conditions, recombinant BMP may have the ability to replace autogenous bone in most maxillofacial applications.

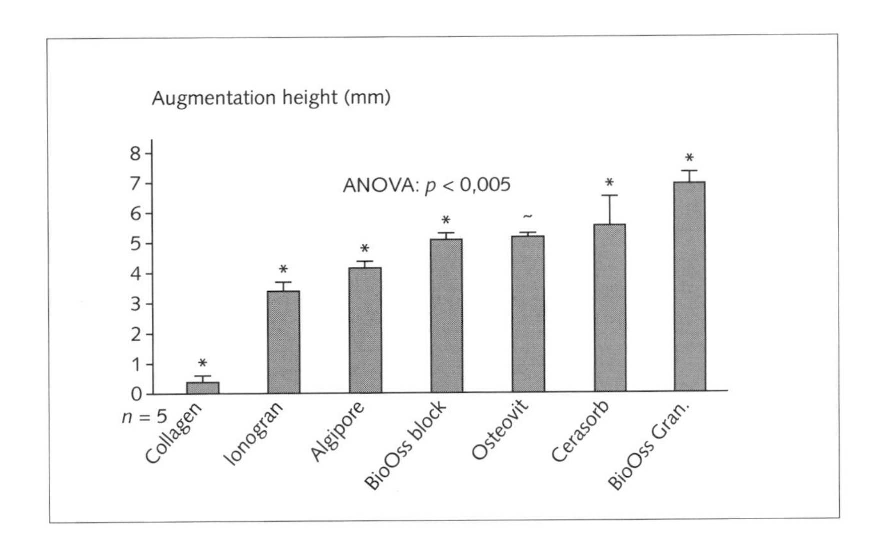


*Figure 26*
*Mandibular augmentation in the rat using seven different carrier materials for 50 µg rhOP-1. The achieved height of augmentation differs significantly with the carrier materials (mean value and SD, ANOVA, Scheffé-test).*

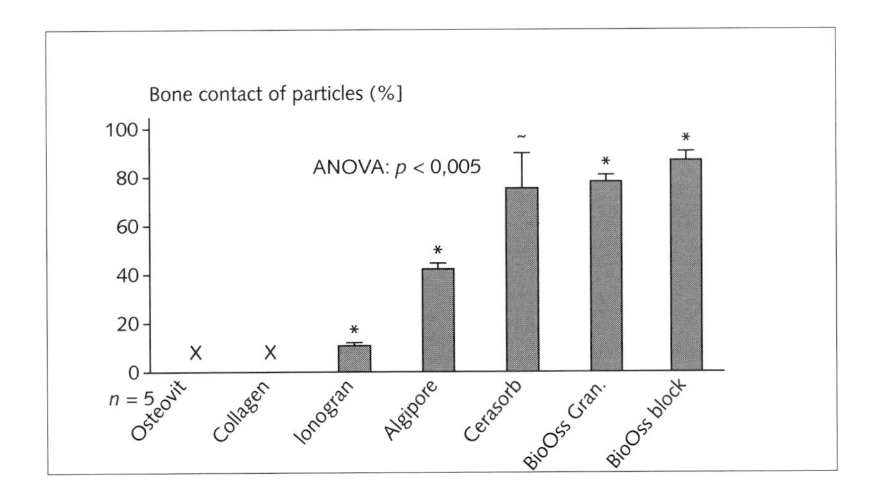


*Figure 27*
*Mandibular augmentation in the rat using seven different carrier materials for 50 µg rhOP-1. The structure and particle contact of the bone differs significantly with the carrier materials. The natural bone mineral demonstrates almost 100% coverage with newly formed bone and has the highest osteoconductivity in this model (mean value and SD, ANOVA, Scheffé-test).*

## References

1 Toriumi DM, Kotler HS, Luxemberg DP, Holtrop ME, Wang EA (1991) Mandibular reconstruction with an recombinant bone-inducing factor. *Arch Otolaryngol Head Neck Surg* 117: 1101–1112
2 Toriumi DM, O'Grady K, Horlbeck DM, Desai D, Turek TJ, Wozney J (1999) Mandibular reconstruction using bone morphogenetic protein 2: long-term follow-up in a canine model. *Laryngoscope* 109: 1481–1489
3 Boyne PJ (1996) Animal studies of application of rhBMP-2 in maxillofacial reconstruction. *Bone* 19 (Suppl) 1: 83–92
4 Khouri RK, Brown DM, Koudsi B, Deune EG, Gilula LA, Cooley BC, Reddi AH (1996) Repair of calvarial defects with flap tissue: role of bone morphogenetic proteins and competent responding tissues. *Plast Reconstr Surg* 98: 103–109
5 Khouri RK, Koudsi B, Reddi H (1991) Tissue transformation into bone *in vivo*. A potential practical application. *JAMA* 266: 1953–1955
6 Viljanen VV, Gao TJ, Lindholm TS (1997) Producing vascularized bone by heterotopic bone induction and guided tissue regeneration: a silicone membrane-isolated latissimus dorsi island flap in a rat model. *Reconstr Microsurg* 13: 207–214
7 Mizumoto S, Inada Y, Weiland AJ (1993) Fabrication of vascularized bone grafts using ceramic chambers. *J Reconstr Microsurg* 1993: 441–449
8 Cavadas PC, Bonanad E, Baena-Montilla P, Vera-Sempere FJ (1996) Prefabrication of a free flap for tracheal reconstruction: an experimental study. Preliminary report. *Plast Reconstr Surg* 98: 1052–1062
9 Casabona F, Martin I, Muraglia A, Berrino P, Santi P, Cancedda R, Quarto R (1998) Prefabricated engineered bone flaps: an experimental model of tissue reconstruction in plastic surgery. *Plast Reconstr Surg* 101: 577–581
10 Levine J P, Bradley J, Turk AE, Ricci JL, Benedict JJ, Steiner G, Longaker MT, McCarthy JG (1997) Bone morphogenetic protein promotes vascularization and osteoinduction in preformed hydroxyapatite in the rabbit. *Ann Plast Surg* 39: 158–168
11 Kusumoto K, Bessho K, Fujimura K, Akioka J, Ogawa Y, Iizuka T (1998) Prefabricated muscle flap including bone induced by recombinant human bone morphogenetic protein-2: an experimental study of ectopic osteoinduction in a rat latissimus dorsi muscle flap. *Br J Plast Surg* 51: 275–280
12 Terheyden H, Jepsen S, Rueger D (1999) Mandibular reconstruction with prefabricated vascularized bone grafts using recombinant human osteogenic protein-1 – a preliminary study. *Int J Oral Maxillofac Surg* 28: 461–463
13 Terheyden H, Knak Ch, Jepsen S, Palmie S, Rueger D (2001) Prefabrication of vascularized bone grafts using recombinant human osteogenic protein-1. Pt. 1 Prefabrication. *Int J Oral Maxillofac Surg* 30: 373–379
14 Terheyden H, Dunsche A, Jepsen S, Menzel C, Rueger D (2000) Prefabrication and microvascular anastomosis of a vascularized bone flap using rhOP-1 – new aspects. *J Craniomaxillofac Surg* (Suppl 3) 28: 22

15 Terheyden H, Warncke P, Jepsen S, Dunsche A, Brenner W, Toth C, Rueger D (2001) Prefabrication of vascularized bone grafts using recombinant human osteogenic protein-1. Pt. 2 Transplantation. *Int J Oral Maxillofac Surg* 30: 469–478
16 Brånemark PI (1983) Osseointegration and its experimental background. *J Prosth Dent* 50: 399–410
17 Lynch SE, Buser D, Hernandez RA, Weber HP, Stich H, Fox CH, Williams RC (1991) Effects of the platelet-derived growth factor/insulin-like growth factor-I combination on bone regeneration around titanium dental implants. Results of a pilot study in beagle dogs. *J Periodontol* 62: 710–716
18 Becker W, Lynch SE, Lekholm U, Becker BE, Caffesse R, Donath K, Sanchez R (1992) A comparison of ePTFE membranes alone or in combination with platelet-derived growth factors and insulin-like growth factor-I or demineralized freeze-dried bone in promoting bone formation around immediate extraction socket implants. *J Periodontol* 63: 929–940
19 Marx RE, Carlson ER, Eichstaedt RM, Schimmele AR, Strauss JE, Georgeff KR (1998) Platelet-rich plasma: Growth factor enhancement for bone grafts. *Oral Surg Oral Med Oral Pathol Oral Radiol Endod* 85: 638–646
20 Yan J, Xiang W, Baolin L, White FH (1994) Early histologic response to titanium implants complexed with bovine bone morphogenetic protein. *J Prosth Dent* 71: 289–294
21 Wang X, Jin Y, Liu B, Zhou S, Yang L, Yang X, White FH (1994) Tissue reactions to titanium implants containing bovine bone morphogenetic protein: a scanning electron microscopic investigation. *Int J Oral Maxillofac Surg* 23: 115–199
22 Wang X, Liu B, Jin Y, Yang X (1993) The effect of bone morphogenetic protein on osseointegration of titanium implants. *J Oral Maxillofac Surg* 51: 647–651
23 Ong JL, Cardenas HL, Cavin R, Carnes DL Jr (1997) Osteoblast responses to BMP-2-treated titanium *in vitro*. *Int J Oral Maxillofac Implants* 12: 649–654
24 Cook SD, Salkeld SL, Rueger DC (1995) Evaluation of recombinant human osteogenic protein-1 (rhOP-1) placed with dental implants in fresh extraction sites. *J Oral Implantol* 21: 281–289
25 Terheyden H, Jepsen S, Möller B, Tucker MM, Rueger DC (1999) Sinus floor augmentation with simultaneous placement of dental implants using a combination of deproteinized bone xenografts and recombinant human osteogenic protein-1. A histometric study in miniature pigs. *Clin Oral Impl Res* 10: 510–521
26 Arvidson K, Bystedt H, Ericsson I (1990) Histometric and ultrastructural studies of tissues surrounding Astra dental implants in dogs. *Int J Oral Maxillofac Impl* 5: 127–134
27 Buser D, Schenk RK, Steinemann S, Fiorellini JP, Fox CH, Stich H (1991) Influence of surface characteristics on bone integration of titanium implants. A histomorphometric study in miniature pigs. *J Biomed Mater Res* 25: 889–902
28 Ericsson I, Johansson CB, Bystedt H, Norton MR (1994) A histomorphometric evaluation of bone-to-implant contact on machine-prepared and roughened titanium dental implants. A pilot study in the dog. *Clin Oral Implants Res* 5: 202–206

29 Gotfredsen K, Wennerberg A, Johansson C, Skovgaard LT, Hjørting-Hansen E (1995) Anchorage of TiO2-blasted, HA-coated, and machined implants: an experimental study with rabbits. *J Biomed Mater Res* 29: 1223–1231

30 Weinlaender M, Kenney EB, Lekovic V, Beumer J 3rd, Moy PK, Lewis S (1992) Histomorphometry of bone apposition around three types of endosseous dental implants. *Int J Oral Maxillofac Implants* 7: 491–496

31 Rutherford RB, Sampath TK, Rueger DC, Taylor TD (1992) Use of bovine osteogenic protein to promote rapid osseointegration of endosseous dental implants. *Int J Oral Maxillofac Implants* 7: 297–301

32 Kirker-Head CA, Nevins M, Palmer R, Nevins ML, Schelling SH (1997) A new animal model for maxillary sinus floor augmentation: evaluation parameters. *Int J Oral Maxillofac Implants* 12: 403–411

33 Margolin MD, Cogan AG, Taylor M, Buck D, McAllister TN, Toth C, McAllister B (1998) Maxillary sinus Augmenation in the non human primate. A comparative radiographic and histologic study between recombinant human osteogenic protein-1 and natural bone mineral. *J Periodontol* 69: 911–919

34 McAllister BS, Margolin MD, Cogan AG, Taylor M, Wollins J (1998) Residual lateral wall defects following sinus grafting with recombinant human osteogenic protein-1 or Bio-Oss in the chimpanzee. *Int J Periodontics Restorative Dent* 18: 227–239

35 Terheyden H, Mueller H, Schulz-Walz JE, Jepsen S, Rueger D (2000) Comparison of three carrier materials for rhOP-1 in sinus augmentation. *J Dent Res* 79: 512

36 Ripamonti U, Reddi AH (1994) Periodontal regeneration: potential role of bone morphogenetic proteins. *J Periodont Res* 29: 225–235

37 Terheyden H, Jepsen S, Vogler S, Tucker MM, Rueger DC (1997) Recombinant human osteogenic protein-1 (rhBMP-7) in the rat mandibular augmenation model: differences in bone mophology are dependent on the type of carrier. *Mund Kiefer Gesichtschir* 1: 272–275

38 Hanisch O, Tatakis DN, Rohrer MD, Wöhrle PS, Wozney JM, Wikesjö UME (1997) Bone formation and osseointegration stimulated by rhBMP-2 following subantral augmentation procedures in nonhuman primates. *Int J Oral Maxillofac Implants* 12: 785–792

39 GroeneveldEH, van-den-Bergh JP, Holzmann P, ten-Bruggenkate CM, Tuinzing DB, Burger EH (1999) Histomorphometrical analysis of bone formed in human maxillary sinus floor elevations grafted with rhOP-1 device, demineralized bone matrix or autogenous bone. Comparison with non-grafted sites in a series of case reports. *Clin Oral Implants Res* 10: 499–509

40 Groenveld HH, van-den-Bergh JP, Holzmann P, ten-Bruggenkate CM, Tuinzing-DB, Burger-EH (1999) Histological observations of a bilateral maxillary sinus floor elevation 6 and 12 months after grafting with osteogenic protein-1 device. *J Clin Periodontol* 26: 841–846

41 van den Bergh JP, ten Bruggenkate CM, Groeneveld EH, Burger EH, Tuinzing DB (2000) Recombinant human bone morphogenetic protein-7 in maxillary sinus floor ele-

vation suregry in 3 patients compared to autogenous bone grafts. A. clinical pilot study. *J Clin Periodontol* 27: 627–636
42 Boyne P, Marx RE, Nevins M, Triplett G, Lazaro E, Lilly LC, Adler M, Nummikowski P (1997) A feasibility study evaluating rhBMP-2/absorbable collagen sponge for maxillary sinus floor augmentation. *Int J Periodont Rest Dent* 17: 11–25
43 Terheyden H, Jepsen S, Vogler S, Tucker M, Rueger DC (1996) Recombinant human osteogenic protein 1 (rhBMP-7) in the rat mandibular augmentation model using different carrier materials. *J Cran Maxillofac Surg* (Suppl 1) 24: 114
44 Cochran D, Nummikoski PV, Jones AA, Makins SR, Turek TJ, Buser D (1997) Radiographic analysis of regenerated bone around endosseous implants in the canine using recombinant human bone morphogenetic protein-2. *Int J Oral Maxillofac Implants* 12: 739–748
45 Sigurdsson TJ, Fu E, Takakis D, Rohrer M, Wikesjö UME (1997) Bone morphogenetic protein-2 for peri-implant bone regeneration and osseointegration. *Clin Oral Impl Res* 8: 375–385
46 Howell TH, Fiorellini J, Jones A, Alder M, Nummikoski P, Lazaro M, Lilly L, Cochran D (1997) A feasibility study evaluating rhBMP-2/absorbable collagen sponge device for local alveolar ridge preservation or augmentation. *Int J Periodontics Restorative Dent* 17: 124–139
47 Hanisch O, Tatakis DN, Boskovic MM, Rohrer MD, Wikesjö UME (1997) Bone formation and reosseointegration in peri-implantitis defects following surgical implantation of rhBMP-2. *Int J Oral Maxillofac Implants* 12: 604–610
48 Ripamonti U. Ma SS. Cunningham NS, Yeates L, Reddi AH (1993) Reconstruction of the bone -bone marrow organ by osteogenin, a bone morphogenetic protein, and demineralized bone matrix in calvarial defects of adult primates. *Plast Reconstr Surg* 91: 27–36
49 Boyne PJ, Nath R, Nakamura A (1998) Human recombinant BMP-2 in osseous reconstruction of simulated cleft palate defects. *Br J Oral Maxillofac Surg* 36: 84–90
50 Mayer M, Hollinger J, Ron E, Wozney J (1996) Maxillary alveolar cleft repair in dogs using recombinant human bone morphogenetic protein-2 and a polymer carrier. *Plast Reconstr Surg* 98: 247–59
51 Ripamonti U, Ma S, Reddi AH (1992) The critical role of geometry of porous hydroxyapatite delivery system in induction of bone by osteogenin, a bone morphogenetic protein. *Matrix* 12: 202–212
52 Schmelzeisen R, Boetel C, Schuberth HJ, Pohlmeyer K (1991) Experimental transplantation of vascularized autologous and allogenic bone grafts for mandibular defects. Anatomical, immunological and surgical basis for vascularized bone transfer in the Göttingen minipig. *Int J Oral Maxillofac Surg* 20: 239–244
53 Shirota T, Schmelzeisen R, Ohno K, Michi KI (1995) Experimental reconstruction of mandibular defects with vascularized iliac bone grafts. *J Oral Maxillofac Surg* 53: 566–571

# Bone morphogenetic proteins in periodontal regeneration

*Søren Jepsen[1] and Hendrik Terheyden[2]*

[1]Department of Restorative Dentistry and Periodontology, University of Kiel, Arnold Heller Str. 16, 24105 Kiel, Germany; [2]Department of Oral and Maxillofacial Surgery, University of Kiel, Arnold Heller Str. 16, 24105 Kiel, Germany

## Introduction

Periodontitis is a chronic inflammatory disease causing breakdown of the periodontal tissues eventually resulting in tooth loss. Successful periodontal reconstruction includes regeneration of a variety of tissues including cementum, periodontal ligament, alveolar bone and gingiva. Wound healing is thought to be regulated by various growth and differentiation factors, such as bone morphogenetic proteins (BMPs), and cytokines. The expression of these biologic mediators following bone and soft tisue injury is thought to regulate the process of repair and/or regeneration. For example, BMPs are known to be expressed during bone repair following fracture [1] and during periodontal wound healing [2]. The rationale for growth factor administration in periodontics is to enhance and/or accelerate the physiological wound healing capacity that may be insufficient to promote a complete healing of the affected structures. Over the past decade numerous *in vitro* and *in vivo* studies have been performed to elucidate the role of growth and differentiation factors in periodontal wound healing. Several of these factors are now available in recombinant form and can be produced in a highly pure form in a large scale production.

This review will describe the effects of bone morphogenetic proteins on periodontal ligament fibroblasts and hard tissue structures cementum and alveolar bone based on available *in vitro* and *in vivo* studies.

## Effects of BMPs on periodontal ligament cells

Factors that possess stimulating effects on the proliferation, migration and collagen matrix synthesis of periodontal ligament (PDL) cells may have the potential to promote new attachment formation. However, at present there is only limited information available regarding effects on PDL cell activity for BMPs. BMP-7 (osteogenic protein-1 = OP-1) was not mitogenic for PDL cells [3], however changed their phenotype by stimulating their alkaline phosphatase activity in a dose- and time-dependent manner. OP-1 failed to induce bone sialoprotein mRNA in PDL cell culture [4].

Bone Morphogenetic Proteins, edited by Slobodan Vukicevic and Kuber T. Sampath

Likewise, BMP-2 and -12 did not show a mitogenic effect on PDL cells [5]. Recombinant human BMP-2 stimulated osteoblast differentiation in human periodontal ligament cells [6]. Inflammatory cytokines such as TNF-$\alpha$ and interleukin-1$\beta$ differentially modulated this stimulatory effect [6]. BMP-2 application to EDTA demineralized dentin surfaces and promoted a significant increase of alkaline phosphatase activity in human PDL cells but no increase in cell number [7]. Future studies exploring the effects of other BMPs on PDL cells would be of great interest.

## Effects of BMPs on bone cells

The main effects of BMPs are to commit undifferentiated pluripotential cells to differentiate into cartilage and bone-forming cells [8–13]. Also, BMPs were shown to regulate growth factor gene expression [14]. They may act synergistically with IGF-1 to stimulate osteoblastic cell differentiation and proliferation [15].

Even though the role of polypeptide growth and differentiation factors on bone formation has been studied extensively, there is insufficient information specifically on alveolar bone.

## Preclinical and clinical studies on periodontal regeneration

### Demineralized freeze dried bone allograft (DFDBA)

The implantation of demineralized freeze dried bone allograft (DFDBA) has a long tradition in periodontics. Since the early publications by Urist [16, 17] periodontists have tried to utilize the osteoinductive factors presumably present in the graft for the stimulation of periodontal bone regeneration. Indeed, BMP-2, -4 und -7 were found in commercially available bone preparations of different bone banks [18]. However, in contrast to fresh preparations the biological activity appeared to be reduced [18] and the ostoinductive properties of different preparations showed a high variability [19]. Moreover, Becker et al. [20], following their investigations on the osteoinductive properties of DFDBA, questioned the rationale for commercially available demineralized bone in periodontics. Instead, they demanded the loading of a carrier matrix with recombinant BMPs of known quality and quantity.

### Natural BMP

#### *Osteogenin*

Bowers et al. [21], in the first and to our knowledge only published clinical trial using BMP for periodontal regeneration in humans evaluated the effect of osteogenin

(BMP-3) extracted from human bone for the healing of intrabony periodontal defects. Osteogenin was delivered in a DFDBA carrier matrix. In 36 defects in eight patients healing proceeded following removal of the crown in a submerged environment and in 50 defects in an additional six patients in a transgingival fashion. Defects treated with either carrier matrix or with non-osseous collagen served as controls. Block biopsies were obtained after 6 months and healing was histologically evaluated. Whereas in the submerged environment the combination of osteogenin/DFDBA was significantly superior to DFDBA the observed differences did not reach statistical significance in the transgingival model, the clinically relevant situation. The least favorable results were obtained with the collagen matrix, with or without the osteogenin. No immunological reactions due to osteogenin were found.

### *BMP-2/BMP-3*

Ripamonti et al. [22] in a pilot study in four monkeys tested the effect of a BMP-extract (bovine bone extracts, containing mostly BMP-2 and BMP-3) in an insoluble collagenous bone matrix (ICBM) for healing of eight surgically created deep mandibular class II furcation defects. Eight contralateral defects treated with the carrier material served as controls. After 2 months there was a significantly enhanced regeneration of cementum, periodontal ligament and bone in BMP/ICBM treated furcations.

Using partially purified bovine BMP incorporated in a fibrous collagen membrane, Kuboki et al. [23] demonstrated periodontal regeneration in class II furcation defects in three monkeys after 12 weeks.

## Recombinant BMPs

### *rhBMP-2*

Sigurdsson et al. [24] applied recombinant human BMP-2 (rhBMP-2) in a carrier consisting of resorbable PLGA-microparticles using the supraalveolar defect model in six beagle dogs. Reconstructive surgery included application of test substance on the test side and of the carrier on the control side. To facilitate protected healing crowns were cut and flaps were sutured above the teeth (submerged model). After 2 months a substantial regeneration of bone (and cementum) was observed in test defects that was significantly superior to control treatment. The incidence of root resorption was less in test sites, the incidence of ankylosis was similar to control treatment.

In a subsequent study by the same group substantial BMP-induced periodontal regeneration could also be observed in the transgingival model [25]. Healing results were significantly influenced by the kind of carrier material (six different carriers for BMP-2 evaluated) that was used.

Kinoshita et al. [26] performed periodontal reconstructive surgery with BMP-2 in a gelatin and polylactic acid polyglycolide acid copolymer carrier in ligature induced circumferential periodontal defects in six beagle dogs. Histometric evaluation after 3 months demonstrated significantly more new bone and cementum formation with no signs of ankylosis as compared to carrier alone.

King et al. [27] studied the effects of rhBMP-2 in a rat fenestration defect model. Following 10 days of healing significant bone formation and 100% more cementum formation was noted as compared to controls. However, after 38 days complete healing was found on both sides, leading the authors to the conclusion that in this model BMP-2 would accelerate bone and cementum formation during early wound healing.

In a subsequent study using the same model King and Hughes [28] investigated the influence of occlusal loading on rhBMP-2 induced bone and cementum formation. Hypofunction and BMP-2 increased the development of transient ankylosis. Occlusal loading enhanced BMP-2 induced cementogenesis.

Wikesjö et al. (1999) [29] evaluated the effect of rhBMP-2 concentration on periodontal regeneration and associated root resorption and ankylosis in supraalveolar defects in eight beagle dogs. Alveolar bone regeneration amounted to 86–96% and cementum to 6–8% of defect height, respectively. Root resorption and ankylosis was seen in all rhBMP-2 treated teeth. They concluded that within the selected concentrations there appeared to be no meaningful differences in regeneration of bone and cementum and no significant differences in the incidence of root resorption and ankylosis.

### *rhBMP-7/OP-1*

Ripamonti et al. [30] evaluated the effects of rhBMP-7 (OP-1) on healing of class II mandibular furcation defects. A total of six defects in three baboons received BMP-7 at a concentration of either 0.1 or 0.5 µg/mg collagen matrix carrier. No bone formation was observed, however substantial new cementum formation was observed. The authors concluded that BMP-7 at the given concentrations stimulated the cementoblast phenotype.

Jepsen et al. [31] demonstrated the possibility of substantial bone regeneration and new cementum formation in class II furcations of four non-human primates (*Macaca fascicularis*) by using higher concentrations of rhBMP-7 (2.5 µg/mg). Bone fill, as determined histologically and volumetrically during surgical reentry, amounted to 84 and 83%, respectively.

Giannobile et al. [32] evaluated different concentrations of rhBMP-7/OP-1 in a dose study in 18 beagle dogs. At a dose of 7.5 µg/mg collagen carrier a significant stimulation for all wound healing parameters was found that was statistically different from either vehicle or surgery-alone sites. No significant increase in root ankylosis was found.

The formation of not only bone but also of a new attachment apparatus following administration of BMPs is difficult to explain. It can be speculated that following the initiation of the wound healing cascade by BMPs, other cytokines and/or growth factors stimulate the differentiation of cells to other non-osseous periodontal phenotypes, since direct mitogenic effects of BMP on periodontal ligament cells appear unlikely. Future research, including BMP receptor studies in periodontal tissues, will hopefully help to better understand the molecular mechanisms of BMP modulated periodontal wound healing.

In summary, there is strong evidence from different preclinical models that rhBMP-2 and -7 can stimulate periodontal regeneration. Human clinical trials are in progress to determine the safety and efficacy of recombinant morphogenetic proteins for periodontal reconstruction with the first results being anticipated in the year 2002.

## Open questions and future perspectives

The success of tissue regeneration by bone morphogenetic proteins depends on the development of suitable delivery systems for these factors to their target cells. Much research has been performed to find optimal carriers for BMP application. The development of suitable delivery systems presents an important step for clinical growth factor therapy. Although a carrier matrix is not a prerequisite for BMP induced bone formation [33] it presents multiple advantages [34] by immobilizing the protein in the target area. The carrier matrix not only defines the shape of the resulting bone, but allows smaller amounts of BMP to be active by retaining it until induction has occurred. An ideal carrier should bind the active protein and protect it against unspecific proteolysis. It should be biocompatible, non-immunogenic and biodegradable and not interfere with the wound healing process [35, 36]. It should facilitate rapid vascular invasion [37] to enable contact between progenitor cells and the rhBMP bound to the carrier. A bone collagen matrix is the natural carrier for BMP, however, when using organic xenogenic materials or bone allografts the risk of disease transmission cannot be ruled out [38]. In this regard, resorbable synthetic materials such as polymers or calcium phosphate ceramics might be advantageous. Such alternative synthetic delivery systems have been evaluated in various animal models [24, 26, 36, 39–41].

Sigurdsson et al. [25] evaluated different candidate carriers for rhBMP-2 in a screening study in the supraalveolar defect model in the beagle dog (among others: bovine deproteinized bone mineral, PLGA-microparticles, PLA-granules). They found distinct differences in the amount and quality of the induced bone and cementum dependent on the type of carrier that was utilized. None of the materials appeared to be ideal in all aspects.

In a recent study, Talwar et al. [42] compared the effects of slow and fast degrading gelatin carriers on BMP-2 induced periodontal healing in rats. New cementum formation was promoted by slow release of BMP.

When comparing different carriers for rhBMP-7/OP-1 in the rat mandibular augmentation model, statistically significant differences for the carriers were found with regard to bone density, height of augmentation, bone quality [43, 44] and the time-course of bone induction [45]. Differences in the release kinetics of rhOP-1 from the different biomaterials could partly explain the observed differences [46].

These findings indicate that in the future different delivery systems could be used for different surgical indications. Whereas a soft material that quickly resorbs might be well suited for the fill of periodontal intraosseous or furcation defects, larger circumferential alveolar defects might require a more rigid, slowly resorbable material with higher mechanical stability.

In addition to the question of the ideal delivery system, other problems remain to be solved: What is the biological and therapeutic significance of the existence of multiple forms of BMPs? What is the optimal therapeutic dose? Future research should investigate different doses as well as molecular combinations to develop an activity profile for the different members of the BMP-family. Finally and most important, to confirm the preclinical data in patients with periodontitis, human biopsies as well as the results from randomized controlled clinical studies are needed.

A shortcoming of current delivery methods of growth factors to periodontal wounds is the short half-life of factors at the target site. The use of DNA delivery systems could become an alternative technique for the application of proteins to the wound site. Thus, the goal of gene therapy would be an elevated and sustained growth factor supply (of days instead of a few hours) in the healing wound. The rationale for this approach is based on the observations that BMPs are expressed up to 14 days during tissue injury [1, 2].

A prerequisite is the successful transduction of appropriate target cells. The efficient delivery of genes into cells can either be done *in vitro* or *in vivo*. Ex vivo therapies require transgene expansion from a tissue specimen. *In vivo* gene therapy resulting in higher but transient gene expression has been performed using plasmid DNA to bone wounds [47]. In another approach it was recently reported that human gingival fibroblasts after transduction with a recombinant adenovirus containing the OP-1 gene produced active BMP-7 resulting in bone formation *in vivo* [48, 49].

Much research remains to be done to optimize gene expresssion, maximize the number of transduced cells and to evaluate whether periodontal wound healing can be enhanced by gene transfer.

## Conclusions

A large number of studies, performed over the last ten years, has demonstrated the possibility of periodontal tissue regeneration by bone morphogenetic proteins. There is evidence for the promotion of periodontal wound healing by rhBMP-2 and rhBMP-7 from multiple *in vitro* and preclinical trials. Provided human clinical trials confirm these findings and growth factor therapies receive approval by the health authorities, the therapeutic use of these potent biologics will certainly add to our regenerative clinical strategies. In addition, in the future the development of gene therapy may become a novel approach in growth factor therapy for tissue engineering in periodontics.

## References


1 Nakase T, Nomura S, Hashimoto J, Yoshikawa H, Takaoka K (1994) Transient and localized expression of bone morphogenetic protein-4 during fracture repair. *J Bone Miner Res* 9: 651–659

2 Ivanovski S, Li H, Daley T, Bartold PM (2000) An immunohistochemical study of matrix molecules associated with barrier membrane-mediated periodontal wound healing. *J Periodontol Res* 35: 115–126

3 Rutherford R B, Charette M, Rueger D (1994) Role of osteogenic (bone morphogenetic) protein and platelet-derived growth factor in periodontal wound healing. In: RJ Genco, S Hamada, T Lehner, J McGee, S Mergenhagen (eds): *Molecular pathogenesis of periodontal disease*. American Society for Microbiology Press, Washington, DC, 427–437

4 Cho MI, Zhumabayeva B, Herr Y, Ryan S, El-Ghorab N (1999) Preclinical considerations for use of OP-1 in periodontal regeneration and osseous reconstruction. In: SE Lynch, RJ Genco, RE Marx (eds): *Tissue engineering*. Quintessence Publishing, Chicago, 161–181

5 Nguyen AM, Tran M, Oates T, Alvares O, Cochran DL (1995) Mitogenic responses of human PDL cells to tissue growth factors. *J Dent Res* 74 (Special issue): 251 (Abstr. 1918)

6 Kobayashi M, Takiguchi T, Suzuki R, Yamaguchi, Deguchi K, Shionome M, Miyazawa Y, Nishihara T, Nagumo M, Hasegawa K (1999) Recombinant human bone morphogenetic protein-2 stimulates osteoblastic differentiation in cells isolated from human periodontal ligament. *J Dent Res* 78: 1624–1633

7 Zaman KU, Sugaya T, Kato H (1999) Effect of recombinant human platelet-derived growth factor-BB and bone morphogenetic protein-2 application to demineralized dentin on early periodontal ligament cell response. *J Periodontal Res* 34: 244–250

8 Asahina I, Sampath TK, Nishimura I, Hauschka PV (1993) Human osteogenic protein-

1 induces both chondroblastic and osteoblastic differentiation of osteoprogenitor cells derived from newborn rat calvaria. *J Cell Biol* 123: 921–933

9 Knutsen R, Wergedal J, Sampath K, Baylink DJ, Mohan S (1993) Osteogenic protein-1 stimulates proliferation and differentiation of human bone cells *in vitro*. *Biochem Biophys Res Comm* 194: 1352–1358

10 Reddi AH, Cunningham NS (1993) Initiation and promotion of bone differentiation by bone morphogenetic proteins. *J Bone Miner Res* 8: 499–502

11 Sampath TK, Maliakal JC, Hauschka PV, Jones WK, Sasak H, Tucker RF, White KH, Coughlin JE, Tucker MM, Pang RHL et al (1992) Recombinant human osteogenic protein-1 (hOP-1) induces new bone formation *in vivo* with a specific activity comparable with natural bovine osteogenic protein and stimulates osteoblast proliferation and differentiation *in vitro*. *J Biol Chem* 267: 20352–20362

12 Takiguchi T, Kobayashi M, Suzuki R, Yamaguchi A, Isatsu K, Nishihara T, Nagumo M, Hasegawa K (1998) Recombinant human bone morphogenetic protein-2 stimulates osteoblast differentiation and suppressess matrix-metalloproteinase-1 production in human bone cells isolated from mandibulae. *J Periodontal Res* 33: 476–485

13 Wozney JM (1992) The bone morphogenetic protein family and osteogenesis. *Mol Reprod Dev* 32: 160–167

14 Yeh LCC, Adamo ML, Duan C, Lee JC (1998) Osteogenic protein-1 regulates insulin-like growth factor-I (IGF-I), IGF-II, and IGF-binding protein-5 (IGFBP-5) gene expression in fetal rat calvaria cells by different mechanisms. *J Cell Pysiol* 175: 78–88

15 Yeh LCC, Adamo ML, Olson MS, Lee JC (1997) Osteogenic protein-1 and insulin-like growth factor I synergistically stimulate rat osteoblastic cell differentiation and proliferation. *Endocrinology* 138: 4181–4190

16 Urist MR (1965) Bone: formation by autoinduction. *Science* 150: 893–899

17 Urist MR, Strates BS (1971) Bone morphogenetic protein. *J Dent Res* 50:1392–1406

18 Shigeyama Y, D'Errico JA, Stone R, Somerman MJ (1995) Commercially-prepared allograft material has biological activity *in vitro*. *J Periodontol* 66: 478–487.

19 Schwartz Z, Mellonig JT, Carnes D L, Delafontaine J, Cochran D L, Dean DD, Boyan BD (1996) Ability of commercial demineralized freeze-dried bone allograft to induce new bone formation. *J Periodontol* 67: 918–926

20 Becker W, Urist MR, Tucker LM, Becker BE, Ochsenbein C (1995) Human demineralized freeze-dried bone: inadequate induced bone formation in athymic mice. A preliminary report. *J Periodontol* 66: 822–828.

21 Bowers GM, Felton F, Middleton C, Glynn D, Sharp S, Mellonig J, Corio R, Emerson J, Park S, Suzuki J et al (1991) Histologic comparison of regeneration in human intrabony defects when osteogenin is combined with demineralized freeze-dried bone allograft and with purified bovine collagen. *J Periodontol* 62: 690–702

22 Ripamonti U, Heliotis M, van den Heever B, Reddi AH (1994) Bone morphogenetic proteins induce periodontal regeneration in the baboon (*Papio ursinus*). *J Periodont Res* 68: 761–767

23 Kuboki Y, Sasaki M, Saito A, Takita H, Kato H (1998) Regeneration of periodontal ligament and cementum by BMP-applied tissue engineering. *Eur J Oral Sci* 106: 197–203
24 Sigurdsson TJ, Lee MB, Kubota K, Turek TJ, Wozney JM, Wikesjo UME (1995) Periodontal repair in dogs: Recombinant human bone morphogenetic protein-2 significantly enhances periodontal regeneration. *J Periodontol* 66: 131–138
25 Sigurdsson TJ, Nygaard L, Tatakis DN, Fu E, Turek TJ, Jin L, Wozney J M, Wikesjö UME (1996) Periodontal repair in dogs: Evaluation of rhBMP-2 carriers. *Int J Periodont Rest Dent* 16: 525–537
26 Kinoshita A, Oda S, Takahashi K, Yokota S, Ishikawa I (1997) Periodontal regeneration by application of recombinant human bone morphogenetic protein-2 to horizontal circumferential defects created by experimental periodontitis in beagle dogs. *J Periodontol* 68: 103–109
27 King GN, King N, Cruchley AT, Wozney JM, Hughes FJ (1997) Recombinant human bone morphogenetic protein-2 promotes wound healing in rat periodontal fenestration defects. *J Dent Res* 76: 1460–1470
28 King GN, Hughes FJ (1999) Effects of occlusal loading on ankylosis, bone, and cementum formation during bone morphogenetic protein-2 stimulated periodontal regeneration *in vivo*. *J Periodontol* 70:1125–1135
29 Wikesjö UM, Guglielmoni P, Promsudthi A, Cho KS, Trombelli L, Jin L, Wozney JM (1999) Periodontal repair in dogs: effect of rhBMP-2 concentration on regeneration of alveolar bone and periodontal attachment. *J Clin Periodontol* 26:392–400
30 Ripamonti U, Heliotis M, Rueger DC, Sampath TK (1996) Induction of cementogenesis by recombinant human osteogenic protein-1 (rhOP-1/BMP-7) in the baboon (*Papio ursinus*). *Arch Oral Biol* 41: 121–126
31 Jepsen S, Schmitz B, Richter KD, Rueger D (1998) Recombinant human osteogenic protein-1 significantly enhances osseous regeneration of class II furcations in monkeys. *J Periodontol* 69: 292
32 Giannobile WV, Ryan S, Shih MS, Su DL, Kaplan PL, Chan TCK (1998) Recombinant human osteogenic protein-1 (OP-1) stimulates periodontal wound healing in class III furcation defects. *J Periodont* 69: 129–137
33 Wang EA, Rosen V, D'Alessandro JS, Bauduy M, Cordes P, Harada T (1990) Recombinant human bone morphogenetic protein induces bone formation. *Proc Natl Acad Sci USA* 87: 2220–2224
34 Reddi AH, Wientroub S, Muthukumaran N (1997) Biologic principles of bone induction. *Orthop Clin North Am* 18: 207–212
35 Kenley RA, Yim K, Abrams J, Ron E, Turek T, Marden LJ, Hollinger JO (1993) Biotechnology and bone graft substitutes. *Pharmaceut Res* 10: 1393–1401
36 Miyamoto S, Takaoka K, Okada T, Yoshikawa H, Hashimoto J, Suzuki S, Ono K (1993) Polylactic acid-polyethylene glycol block copolymer. *Clin Orthop* 294: 333–343.
37 Hollinger J, Chaudhari A (1992) Bone regeneration materials for the mandibular and craniofacial complex. *Cells Mater* 2: 143–151
38 Lindhe J, Cortellini P (1997) Chemicals in periodontal regeneration – Consensus report

of session IV, in: NP Lang, T Karring, J Lindhe (eds): *Proceedings of the 2nd European Workshop on Periodontology*. Quintessenz, Berlin, 359–360

39 Ripamonti U, Ma S, von den Heyden B, Reddi AH (1992) Osteogenin, a bone morphogenetic protein, adsorbed on porous hydroxyapatite substrata, induces rapid bone differentiation in calvarial defects of adult primates. *Plast Reconstr Surg* 90: 382–393

40 Ripamonti U, Ma S, Reddi AH (1992) The critical rôle of geometry of porous hydroxyapatite delivery system in induction of bone by osteogenin, a bone morphogentic protein. *Matrix* 12: 202–121

41 Yamazaki Y, Oida S, Akimoto Y, Shioda S (1988) Response of the mouse femoral muscle to an implant of a composite of bone morphogenetic protein and plaster of paris. *Clin Orthop* 234: 240–249

42 Talwar R, di Silvio L, Hughes FJ, King GN (2001) Effects of carrier release kinetics on bone morphogenetic protein-2-induced periodontal regeneration *in vivo*. *J Clin Periodontol* 28: 340–347

43 Jepsen S, Terheyden H, Vogler S, Tucker M, Rueger D (1997) Mandibular augmentation by recombinant human osteogenic protein-1. *J Clin Periodonto*l 24: 870.

44 Terheyden H, Jepsen S, Vogler S, Tucker MM, Rueger DC (1997) Recombinant human osteogenic protein-1 (rhBMP-7) in the rat mandibular augmenation model: differences in bone mophology are dependent on the type of carrier. *Mund-, Kiefer- Gesichtschir* 1: 272–274

45 Terheyden H, Oehlert C, Jepsen S (2000) The time course of bone induction by human osteogenic protein-1 using different delivery systems. *J Clin Periodontol* 27 (Suppl 1): 67

46 Jepsen S, Chang AC, Terheyden H, Rueger D, Tucker M (1999) In-vitro release of recombinant human osteogenic protein-1 (rhOP-1) from different carrier materials. *J Periodontol* 70: 337

47 Fang J, Zhu YY, Smiley E, Bonadio J, Rouleau JP, Goldstein SA, McCauley, Davidson BL, Roessler BJ (1996) Stimulation of new bone formation by direct transfer of osteogenic plasmid genes. *Proc Natl Acad Sci USA* 93: 5753–5758

48 Hill E, Austin C, Rutherford RB (1999) Human oral cells transduced by BMP-7 produce ectopic bone. *J Dent Res* 78 (special issue), IADR abstract 2495

49 Krebsbach PH, Gu K, Franceschi RT, Rutherford RB (2000) Gene-therapy-directed osteogenesis: BMP-7 transduced human fibroblasts form bone *in vivo*. *Human Gene Therapy* 11: 1201–1210

# Osteogenic protein-1 (OP-1) in the repair of bone defects and fractures of long bones: clinical experience

*Lex R. Giltaij*[1], *Andrew Shimmin*[2] *and Gary E. Friedlaender*[3]

[1]Stryker Biotech, Notengaard 13, 3941 LV Doorn, The Netherlands; [2]Melbourne Orthopaedic Group, 33 The Avenue, Windsor Victoria 3181, Australia; [3]Department of Orthopaedics and Rehabilitation, Yale University School of Medicine, P.O. Box 208071, New Haven, CT, 06520-8071, USA

## Introduction

The concept of osteoinductive or bone morphogenetic proteins (BMPs) was first introduced by Urist nearly 40 years ago [1], and by the late 1980's the human cDNA for OP-1 (BMP-7) was cloned [2]. Utilizing recombinant technology, human OP-1 (rhOP-1) was produced and this molecule has demonstrated its capacity to induce bone formation [3, 4]. Subsequently, extensive preclinical and clinical research has confirmed the efficacy as well as safety of OP-1 in the process of bone repair and regeneration [5–8]. This paper will focus on clinical experience with OP-1 in the treatment of nonunions of the appendicular skeleton.

## Preclinical experience

Preclinical studies have demonstrated the ability of OP-l to cause repair of critical sized defects in numerous animal models, including the long bones of rabbits [9], dogs [10] and nonhuman primates [11]. In each circumstance, the resected segmental deficits, implanted with OP-1 Implant (3.5 mg recombinant human OP-1 in 1g type I collagen matrix), regenerated a complete bony bridge. This repair was accomplished with the same or better frequency than observed in the bone autograft controls, and with the same capacity to remodel and reestablish a marrow cavity as seen with autogenous graft (Figs. 1 and 2).

## Clinical experience

### Fibular defect (The Netherlands; clinical trial)

A prospective, randomized and double-blinded clinical trial, recently reported by Geesink and colleagues [5], demonstrated the ability of OP-1 to cause repair of a

Bone Morphogenetic Proteins, edited by Slobodan Vukicevic and Kuber T. Sampath

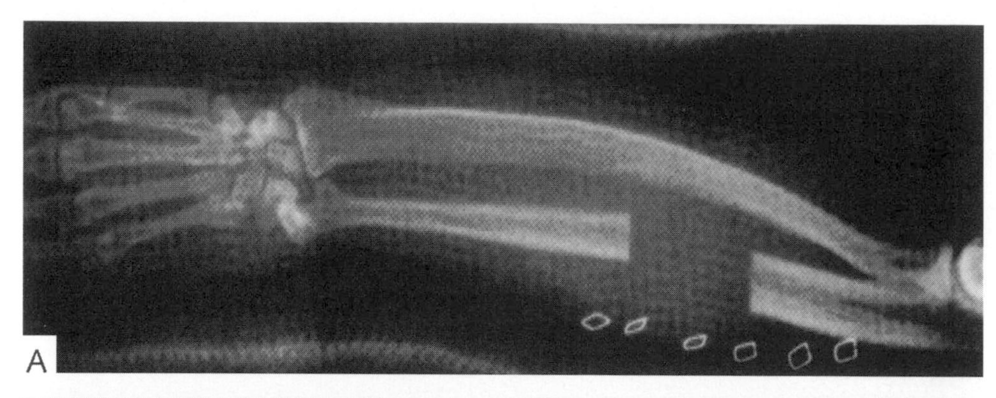


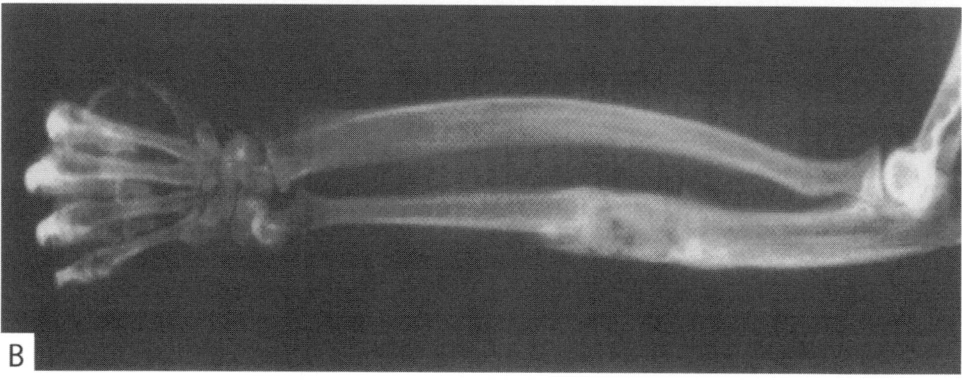


*Figure 1*
*Critical sized defect in monkey ulna, treated with the OP-1 implant. Immediately postoperative (A) and at 20 weeks (B). OP-1 implant is radiolucent.*

critical-sized fibular defect in patients undergoing opening wedge high tibial osteotomy with fibulectomy. In a preliminary study, it was determined that the fibular defect, approximately 1.5 to 2.0 cm in length, would not spontaneously heal. In the subsequent investigation, these segmental defects were implanted with either OP-1 implant or with the matrix alone in a double-blinded fashion. Five of the six patients receiving OP-1 implant bridged their defects by 4 months, as determined by a radiologist blinded to treatment, while none of those patients treated with matrix alone bridged their gap (Fig. 3).

## Tibial nonunions (U.S.; clinical trial)

In a prospective, randomized, controlled clinical study, accomplished under a United States Food and Drug Administration (FDA) approved Investigational Device

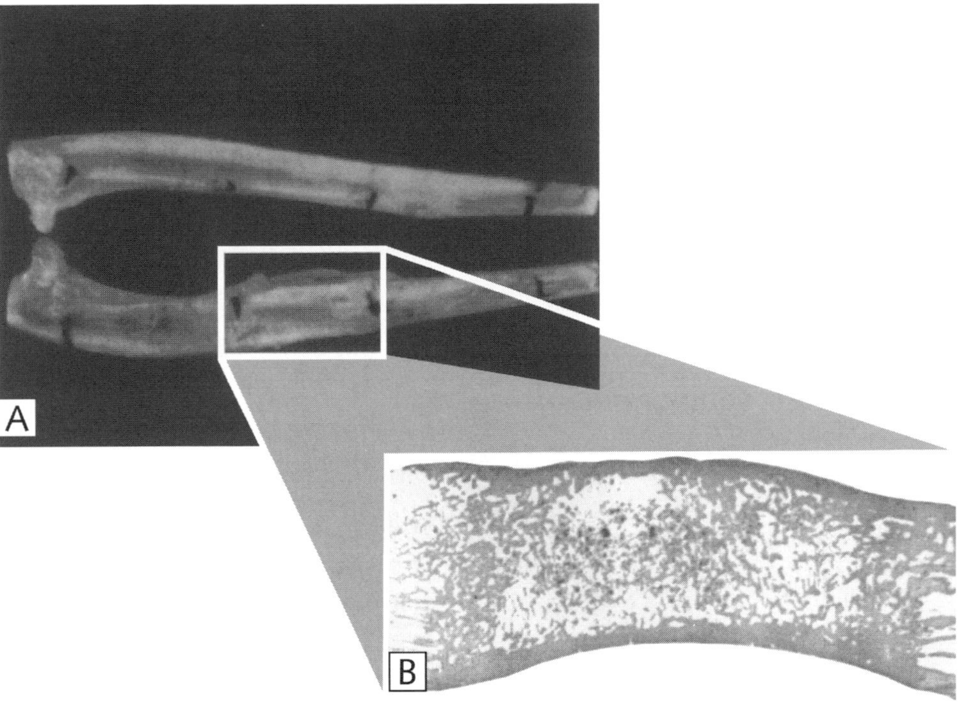


*Figure 2*
*Critical sized defect in monkey ulna, treated with the OP-1 implant. Non-operated contralateral control ulna (A) and the OP-1 implant-treated ulna at 20 weeks (B). Histological specimen shows remodeled mature bone with full cortex and bone marrow cavity [11] (B).*

Exemption (IDE), the safety and effectiveness of OP-1 implant in healing a tibial nonunion was compared with that achieved with bone autograft [8]. The study included 122 patients with 124 tibial nonunions treated at 17 sites within the United States between February, 1992 and August, 1996. The protocol inclusion criteria required that the tibial nonunion in these adults be acquired as the result of trauma, and that the responsible surgeon had determined that treatment would otherwise require intramedullary fixation and bone autograft. Nonunion was defined as the failure to heal the fracture over at least 9 months, and that there was no evidence of healing or surgical intervention within the 3 months prior to investigational treatment.

The demographics of the two groups were similar with the exception of some established risk factors for fracture healing, suggesting a possible bias in favor of the autograft-treated group. For example, the incidence of atrophic nonunion was 41% in the OP-1 implant group compared to 25% in the autograft-treated patients ($p$ =

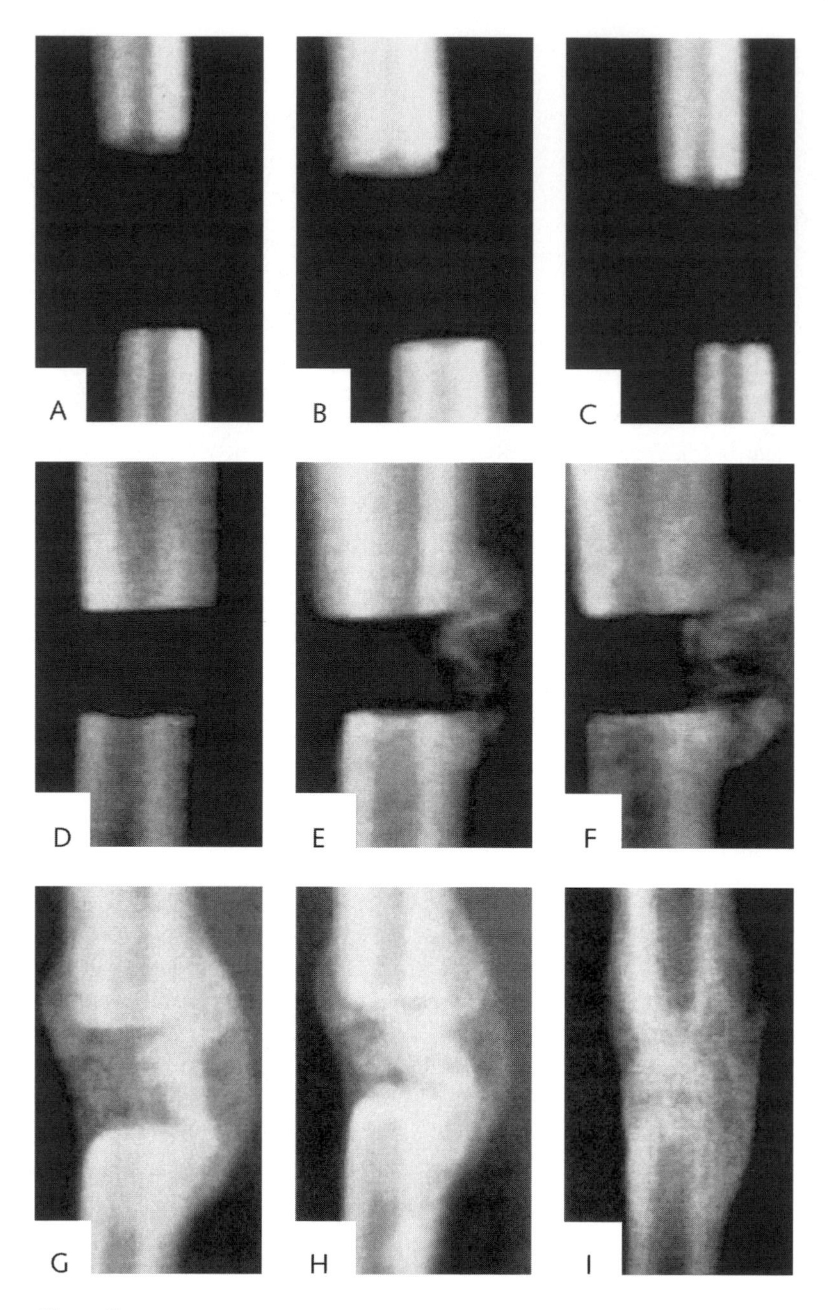


*Figure 3.*
*Radiographs showing a fibula defect treated with collagen matrix alone at 6 weeks (A), 10 weeks (B) and 1 year (C). Radiographs of a fibula defect treated with collagen matrix and BMP-7 at 1 week (D), 6 weeks (E), 10 weeks (F), 4 months (G), 6 months (H) and 1 year (I).*

0.048), and the prevalence of tobacco/nicotine use was 74% in the OP-1 implant patients and 57% in the autograft-treated group ($p$ = 0.057). In addition, more patients in the OP-1 implant group had comminuted fractures at the time of injury (67% vs. 56%, $p$ = 0.212), experienced prior failed autograft procedures (43% vs. 31%, $p$ = 0.177) and previously received intramedullary rods without subsequent success (54% vs. 44%, $p$ = 0.280).

The mean blood loss during the procedure was 345 cc in the autograft-treated patients, which was statistically greater than the 254 cc loss experienced by the OP-1 implant group ($p$ = 0.049). Similarly, hemoglobin and hematocrit levels were significantly lower in the autograft-treated group 1 month following surgery. Furthermore, a significant difference was seen in the incidence of post-operative acute or subacute osteomyelitis at the operative site, which occurred in 21% of the autograft-treated patients and only 3% of those implanted with the OP-1 implant ($p$ = 0.002).

All autograft-treated patients reported post-operative pain at the bone graft donor site. Chronic donor site pain at the 6-month follow-up visit was observed in more than 20% of patients, and 13% continued to complain of pain 12 months following their surgical procedure. Accordingly, the OP-1 Implant-treated patients required less pain medication.

Both treatment groups experienced adverse events, usualy mild or moderate in nature, and these events occurred with comparable frequency in OP-1 implant and autograft-treated patients. No serious adverse events were related to the OP-1 implant or the implanted graft material.

Patients in this study were assessed by both clinical and radiographic criteria. The two groups were compared at 9 months with respect to their ability to fully weight-bear with less than severe pain, the lack of a surgical retreatment of their nonunion as well as physician satisfaction with the patients' repair (Tab. 1). Using these criteria, the outcomes of the two groups were comparable.

Radiographic analysis at 9 months following surgery, by a panel of three musculoskeletal radiologists blinded to treatment, demonstrated bridging of the fracture on at least one view in 75% of the OP-1-treated patients and 84% of those receiving autograft ($p$ = 0.218, an insignificant difference between the groups). A more strict interpretation, requiring bridging on at least three of four views (determined on AP, lateral and two oblique x-rays) demonstrated healing in 62% of the OP-1 implant and 74% of the autograft-treated patients ($p$ = 0.158) at this same time interval (Tab. 1).

The conclusions supported by this study include that the OP-1 implant is a safe and effective treatment modality for tibial nonunions, and comparable to the use of bone autografts. In addition, the OP-1 implant demonstrated a number of safety advantages over autograft bone, including a reduction in the amount of operative blood loss, decreased incidence of osteomyelitis at the surgical site, the elimination of donor site specific complications and pain as well as a decrease in the use of post-operative pain medication.

*Table 1 - Clinical and radiological outcomes at 9 months following treatment*

| Criteria | OP-1 implant | | Autograft | | *p*-Value** |
|---|---|---|---|---|---|
| | n | success | n | success | |
| Full weight-bearing with less than severe pain | 56 | 89% | 55 | 90% | 0.817 |
| Radiographic bridging (in at least one view)* | 47 | 75% | 51 | 84% | 0.218 |
| Radiographic bridging (in at least three views)* | 39 | 62% | 45 | 74% | 0.158 |
| No surgical retreatment* | 60 | 95% | 55 | 90% | 0.276 |
| Physician satisfaction* | 54 | 86% | 55 | 90% | 0.447 |

*based on the number of nonunions rather than the number of patients*
***Chi Square Test where $p > 0.05$ indicates no significant difference between groups*

## Appendicular salvage cases (Australia)

In Australia, 163 consecutive patients were treated with the OP-1 implant between August, 1997 and December, 1999, for a variety of skeletal disorders (Tab. 2). Individual Patient Usage (IPU) approval for compassionate release was obtained in each case from the Therapeutic Goods Administration (TGA) prior to treatment. IPU approval was only obtained for patients having previously failed conventional treatment or who were deemed unsuitable for other standard treatment option; consequently, all of these cases were particularly challenging. Seventy-one surgeons in five states of Australia have contributed to this series of cases, with an average follow-up of 15 months. Since May, 1998, the OP-1 implant has been combined with the excipient, carboxymethycellulose (CMC), to improve handling properties. Forty-four of these cases have been previously reported [7].

Data were collected on standardized forms with clinical outcome being assessed by the treating surgeon. Radiological assessment was performed by one of the authors (AS) and by the treating surgeon. Nonunions were considered to be radiologically healed if continuous bridging was clearly present. Outcomes were considered failures if the patient was unable to return to normal or near normal activities or if they required additional surgical treatment for the same condition.

In many cases, the OP-1 implant was combined with autograft or with other osteoconductive fillers, such as bone allograft or hydroxyapatite preparations (Tab. 3). Forty (35%) of these patients had prior autograft procedures. Most others were either considered ineligible for autograft by the treating surgeon, usually due to concomitant conditions that result in poor bone stock, or had failed customary

*Table 2 - Demographics of Australian patient population*

| Patients | Indications |
|---|---|
| 113 | Nonunions (see Table 5) |
| 18 | Revision arthroplasty |
| 16 | Failed arthrodesis |
| 9 | Bone defects |
| 3 | Peri-prosthetic fracture |
| 1 | Elective osteotomy |
| 1 | Congenital pseudarthrosis |
| 2 | Osteochondral defects |
| 163 | Total |

*Table 3 - Clinical application of OP-1 implant*

| Combined with | No of cases |
|---|---|
| Iliac crest autograft | 57 |
| Local or other autograft | 36 |
| Allograft | 20 |
| Bone marrow aspirate | 1 |
| Osteoconductive fillers | 7 |
| Combinations of above | 6 |
| OP-1 implant ALONE | 36 |
| Total | 163 |

treatment by intramedullary reaming at the time of exchange rodding for ununited femoral and tibial fractures (Fig. 4).

In 46 cases, there was significant pre-existing pathology or illness known to be associated with impaired fracture repair or the biomechanical character of bone, including prior or recent infection, chronic osteomyelitis, rheumatoid arthritis requiring high-dose steroid treatment, severe osteoporosis, osteogenesis imperfecta, fibrous dysplasia and Paget disease (Fig. 5).

The outcomes of 76% of these 163 patients were considered successful by clinical criteria and 69% were successful by a combination of both clinical (Tab. 4) and radiographic criteria. Twelve patients (7%) could not be adequately assessed by both clinical and radiographic parameters. Thirty-nine patients, 24% of these challenging cases, were clinically unsuccessful. In 18 of these 39 cases, significant fac-

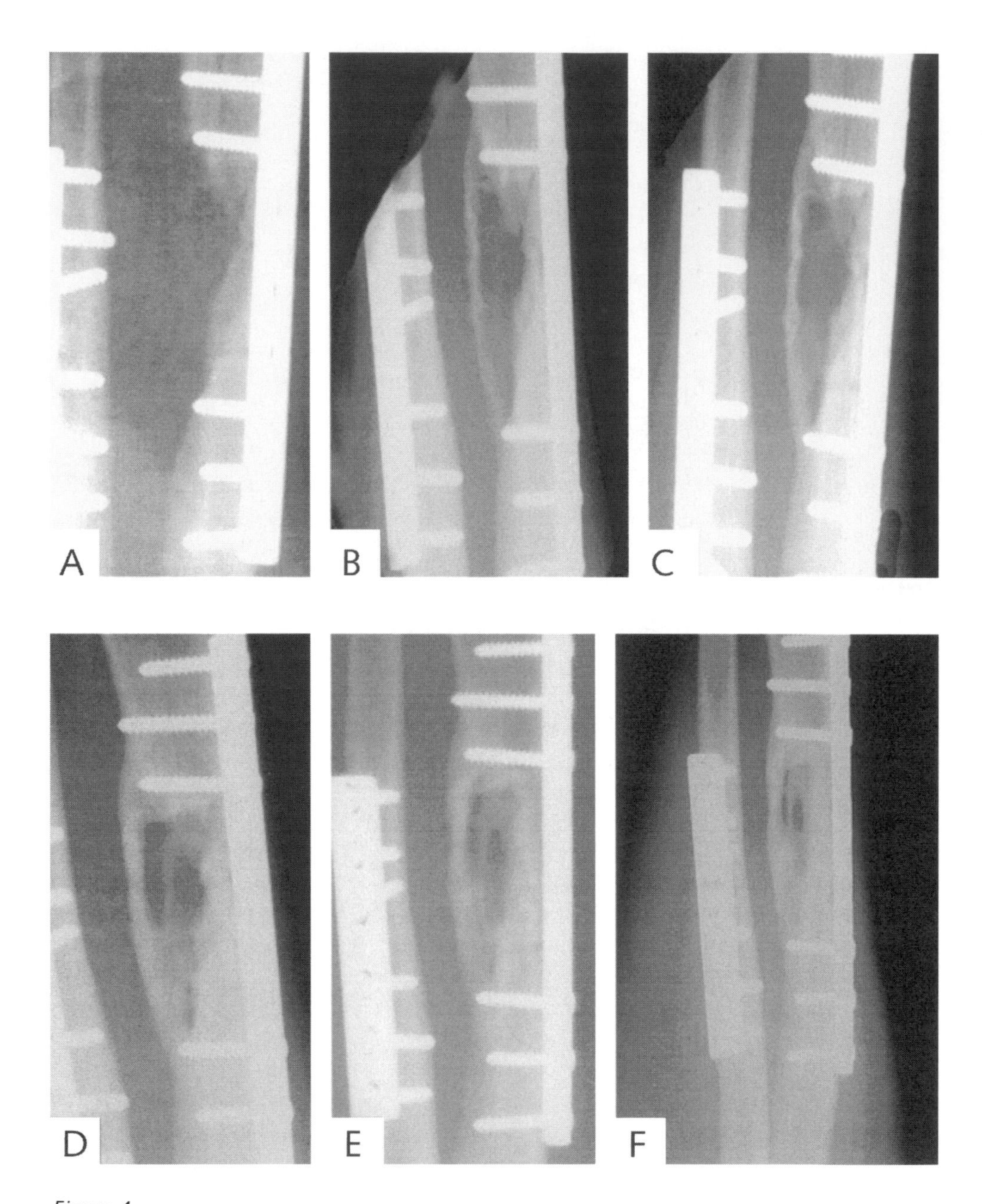


*Figure 4.*
*Radiographs of a 35-year-old male who suffered a comminuted mid-shaft radial fracture in a motor bike accident and did not unite due to a large bone defect. The defect was filled with the OP-1 implant alone. The radiographs show a progression of bone formation from day 0; 5 and 8 weeks (A, B, C) and 6, 22 and 30 months (D, E, F), respectively.*

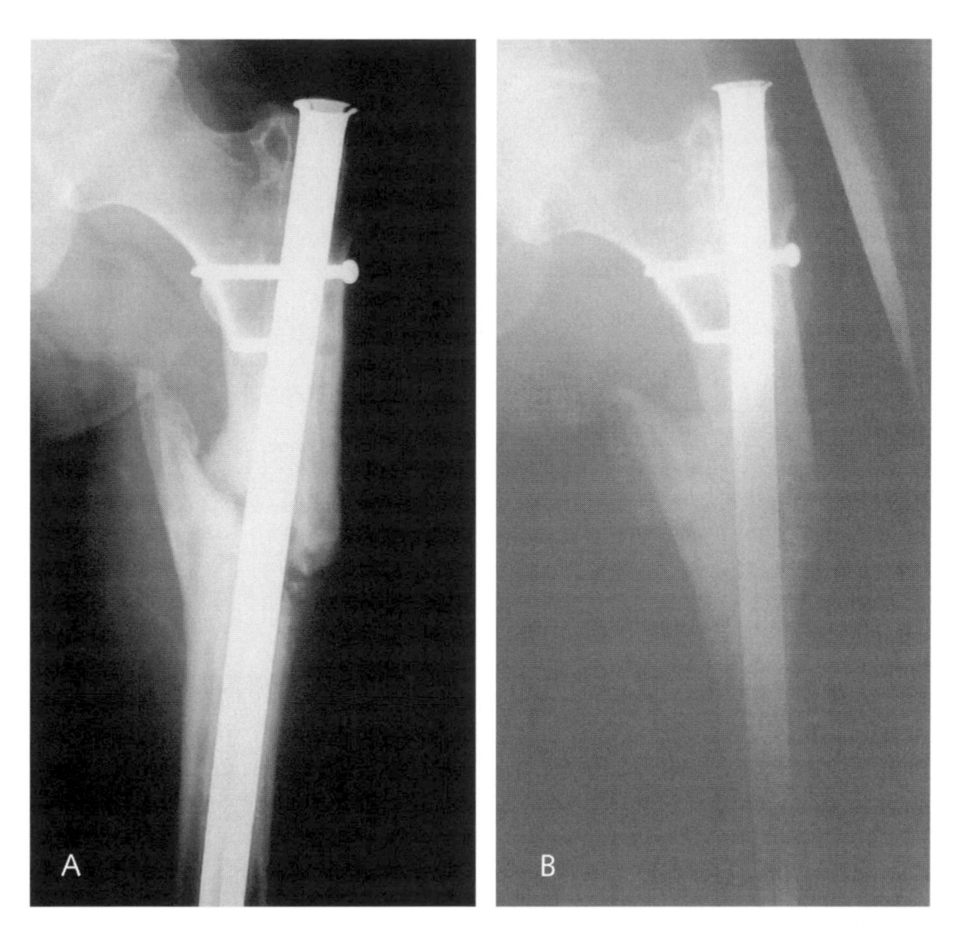


*Figure 5*
*Twenty-seven-year-old male who sustained an open fracture to the proximal third of his femur in a motor vehicle accident 7 years prior to treatment with the OP-1 implant. Subsequent treatments involved management of infection and five attempts at achieving union which included internal fixation with plates and screws, intramedullary nailing (on four occasions) and autografts (on two occasions). Fixation was unaltered at the time of application of OP-1. Union was achieved in 9 months (left x-ray pre-operative, right x-ray at 4 months).*

tors were identified which may contribute to failure, including pre-existing or active infection, sub-optimal internal fixation, the need for early amputation, inadequate local blood supply or severely compromised soft tissue coverage of the fracture site. Several cases were associated with unusually challenging circumstances or the application of unconventional surgical procedures.

*Table 4 - Clinical Results of OP-1 implant application*

| Diagnosis | Total | Failed |
|---|---|---|
| Nonunions | 113 | 28 |
| Arthrodeses | 16 | 3 |
| Revision arthroplasties | 18 | 3 |
| Bone defects | 9 | 3 |
| Peri-prosthetic fractures | 3 | 0 |
| Elective osteotomies | 1 | 0 |
| Congenital pseudarthroses | 1 | 0 |
| Osteochondral defects | 2 | 2 |
| Total | 163 | 39 |

Complications and adverse reactions were uncommon. One patient with a nonunion of the ulna developed a local erythematous reaction following surgery. This reaction resolved with antihistamine treatment and went on to union. There were two cases of deep wound infection (1.4%) and one reported superficial wound infection follow surgery. In addition, 13 patients had prior treatment for osteomyelitis. Reactivation of this infection occurred in seven cases, of which four failed to unite.

Within this series, were 113 patients who sustained fractures following trauma to various long bones and developed nonunions (Tab. 5). These patients had a mean and median of two prior surgical procedures for treatment of their nonunions (range 0 to 12), usually involving exchange rodding with intramedullary reaming or change of plate and screw fixation. Forty (35%) of these patients also had prior bone autograft procedures. This subset of patients was treated with OP-1 an average of 23.3 months following injury (range 1 to 148 months, median 14 months), and the average time of follow-up was 19 months (range 3 to 28 months).

Union was achieved, using clinical criteria, in 79 of these 113 patients (70%). In 6 cases (5%), the patient's clinical outcome could not be adequately assessed due to the presence of reflex sympathetic dystrophy (RSD) or other associated limb fractures. Radiographic union was seen in 74 cases (65%). An additional three patients (2.5%) demonstrated moderate, but incomplete new bone formation and six patients (5%) could not be evaluated radiographically due to obscuring hardware.

As mentioned above, 40 patients had prior autografts. This group had an average of 3.1 (range 1 to 12, median 3) prior surgical procedures for treatment of their nonunions, generally bone grafting, exchange rodding, revision of plate fixation or a combination of these procedures. Treatment with the OP-1 implant occurred at an average of 28 months (range 5 to 84 months, median 19 months) following initial injury.

*Table 5 - Patients with nonunions treated with OP-1 implant*

| **Anatomic sites of nonunions** | |
|---|---|
| Femur | 35 |
| Tibia | 32 |
| Humerus | 12 |
| Radius/ulnae/fibula | 9 |
| Clavicle | 8 |
| Scaphoid | 6 |
| Navicular | 8 |
| Pelvis | 3 |
| Total | 113 |

Clinically, 28 patients (70%) in this subset of appendicular nonunions previously treated with autograft healed following implantation of OP-1; two patients could not be assessed (one lost to follow-up and one with RSD). The average time to union was 5.6 months (range 3 to 15 months). Ten patients (25%) were considered failures by clinical criteria.

Radiographically, 28 of these 40 patients went on to union. One additional patient was forming new bone that was not yet united, and one patient was lost to follow-up. Ten patients failed to unite, three of whom had reactivation of a prior infection at the fracture site, one had Paget disease, one had early failure of fixation and two requested amputation at 12 weeks. One of these patients had a large segmental defect fixed with a cylindrical strut allograft.

This experience with OP-1 in a variety of challenging cases in Australia confirms and strengthens the findings of the U.S. tibial nonunion study and expands the scope of skeletal sites and conditions that have been successfully treated with this osteoinductive molecule. In particular, this study demonstrated that the OP-1 implant induces new bone formation in humans. It is also clear that a successful outcome requires attention to sound surgical principles, including adequate treatment of prior infection, good soft tissue coverage of the fracture, an adequate blood supply to the fracture site and stable internal fixation. Some of these patients were treated with OP-1 alone and others with OP-1 and a variety of additional commonly accepted adjuncts to bone healing, including bone autograft and allograft and osteoconductive materials. These combinations did not appear to detract from the efficacy of OP-1. Finally, OP-1 appears safe, with no significant adverse reactions attributed to the implant.

## Discussion

The nature of bone induction and regeneration is now better understood, and the important roles of a variety of molecular factors are becoming clear. Some BMPs, including OP-1 (BMP-7), have demonstrated their ability in a number of animal models to induce new bone formation and favorably influence the process of bone regeneration and repair. These growth and differentiation factor are capable of causing the recruitment, differentiation and proliferation of osteogenic cell populations.

Recombinant human osteogenic protein-1 has demonstrated both safety and efficacy in the treatment of nonunions of various long bones in humans, building upon substantial preclinical success in a wide variety of animal models. It is important, however, to recognize the need to follow established surgical principles prerequisite to the success of any osteogenic stimulus. This includes the establishment of a bacteriologically clean, viable and well-vascularized surgical site with adequate stabilization of fractures. The usefulness of OP-1 also obviates the need, in many cases, for a bone autograft donor site and its associated morbidity [12]. It is likely that similar enhancement of spinal arthrodesis [13–16], implant fixation [17] and allograft incorporation [18, 19] by implantation of OP-1, as demonstrated in animals, will be confirmed by the growing clinical experience with this evolving technology.

## References


1 Urist MR (1965) Bone: Formation by autoinduction. *Science* 150: 893–899

2 Özkaynack E, Rueger DC, Drier EA, Corbett C, Ridge RJ, Sampath TK, Oppermann H (1990) OP-1 cDNA encodes an osteogenic protein in the TGF-beta family. *EMBO J* 9 (7): 2085–2093

3 Sampath TK, Maliakal JC., Hauschka PV, Jones WK, Sasak H, Tucker RF, White KH, Coughlin JE, Tucker MM, Pang RH et al. Recombinant human osteogenic protein-1 (HOP-1) induces new bone formation *in vivo* with a specific activity comparable with natural bovine osteogenic protein and stimulates osteoblast proliferation and differentiation *in vitro*. *J Biol Chem* 267 (28): 20352–20362

4 Asahina I, Sampath TK and Hauschka PV (1993) Human osteogenic protein–1 induces both chondroblastic and osteoblastic differentiation of osteoprogenitor cells derived from newborn rat calvaria. *J Cell Biol* 123: 921–933

5 Geesink RGT, Hoefnagels NHM, Bulstra SK (1999) Osteogenic activity of OP-1 bone morphogenetic protein (BMP-7) in a human fibular defect. *JBJS (Br)* 81-B: 710–718

6 Cook, SD (1999) Preclinical and clinical evaluation of osteogenic protein-1 (BMP-7) in bony sites. *Orthopedics* 22 (7): 669–671

7 Shimmin A, Ruff S (1999) Clinical use of recombinant human osteogenic protein-1 (rhOP-1). *Bone* 24 (4): 409–431
8 Friedlaender GE, Perry CR, Cole JD, Cook SD, Cierny G, Muschler GF, Zych GA, Calhoun JH, Laforte AJ, Yin S (2001) OP-1 (BMP-7) in the treatment of tibial nonunions. A prospective randomized clinical trial comparing rhOP-1 with fresh bone autograft. *J Bone Joint Surg* 83-A (Suppl 1): 151–158
9 Cook SD, Baffes GC, Wolfe MW, Sampath TK, Rueger DC, Whitecloud TS III (1994) The effect of recombinant human osteogenic protein-1 on healing of large segmental bone defects. *J Bone Joint Surg Am* 76 (6): 827–828
10 Cook SD, Baffes GC, Wolfe MW, Sampath TK, Rueger DC (1994) Recombinant human bone morphogenetic protein-7 induces healing in a canine long-bone segmental defect model. *Clin Orthop* 301: 302–312
11 Cook SD, Wolfe MW, Salkeld SL, Rueger DC (1995) Effect of recombinant human osteogenic protein-1 on healing of segmental defects in non-human primates. *J Bone Joint Surg Am (US)* 77 (5): 734–750
12 Summers BN, Eisenstein SM (1989) Donor site pain from the ilium. A complication of lumbar spine fusion. *J Bone Joint Surg (Brit)* 71: 677–680
13 Cook SD, Dalton JE, Tan EH, Whitecloud III TS and Rueger DC (1994) *In vivo* evaluation of recombinant human osteogenic protein (rhOP-1) implants as a bone graft substitute for spinal fusions. *Spine* 19–15: 1655–1663
14 Cunningham BW, Kanayama M, Parker LM, Weis JC, Sefter JC, Fedder IL, McAfee PC (1999) Osteogenic protein *versus* autologous interbody arthrodesis in the sheep thoracic spine. *Spine* 24 (6): 509–518
15 Grauer JN, Patel TCh, Erulkar JS, Troiano NW, Panjabi MM, Friedlaender GE (2001) Evaluation of OP-1 as a graft substitute for intertransverse process lumbar fusion. *Spine* 26 (5): 127–133
16 Magin MN, Delling G (2001) Improved lumbar vertebral interbody fusion using rhOP-1: a comparison of autogenous bone graft, bovine hydroxylapatite (Bio-Oss) and BMP-7 (rhOP-1) in sheep. *Spine* 26 (5): 469–478
17 Cook SD, Rueger DC (1996) Osteogenic protein-1: biology and applications. *Clin Orthop* 324: 29–38
18 Cook SD, Barrack LB, Santman M, Popich Patron L, Salkeld SL, Whitecloud III TS (2000) Strut allograft healing to the femur with recombinant human osteogenic protein-1. *Clin Orth Rel Res* 381: 47–57
19 Salkeld SL, Patron LP, Barrack RL (2001) Effect of osteogenic protein 1 (OP-1, BMP 7) on the healing of segmental bone defects treated with autograft or allograft bone. *JBJS* 83-A (6): 803–816

# Evaluation of OP-1 in a rabbit model of lumbar fusions

*Tushar Ch. Patel, Jonathan N. Grauer and Jonathan S. Erulkar*

Department of Orthopedics & Rehabilitation, Yale University School of Medicine, P.O. Box 208071, New Haven, CT 06520-8071, USA

## Introduction

Lumbar fusion is a common surgical procedure for which there are multiple approaches and techniques. It has recently been estimated that more than 185 thousand spinal fusions are performed each year in the United States alone. There are different types of lumbar fusion with posterolateral intertransverse fusion using iliac crest autograft being the most common. This type of fusion is used in the treatment of a wide variety of conditions including spondylolisthesis and discogenic disease.

Iliac crest autograft is the most common material utilized in fusion; it is, however, associated with certain limitations and morbidity. The amount of donor bone may be limited due to poor bone quality or previous graft harvest. As a result pseudoarthrosis or nonunion may occur. There has been a reported incidence of pseudoarthrosis of up to 26% [1]. In addition, there are many variables that affect the results of lumbar fusion. For example, smoking has been shown to increase the rate of pseudoarthrosis anywhere from two- to five-fold [2, 3].

In addition to pseudoarthrosis, chronic pain at the iliac crest donor site is a frequently encountered complication of autograft harvest and in fact chronic donor site pain occurs in 25% of all patients undergoing iliac crest harvest [4].

To address the problems of pseudoarthrosis, various adjuncts such as instrumention and electrical stimulation have been suggested as a possible means to enhance the outcome of iliac crest autograft fusion. However, neither of these modalities alleviates the need for an osteo-conductive and/or osteo-inductive agent in order to achieve fusion. For this reason, various bone graft alternatives are being investigated. Allograft may serve in either an osteo-conductive or osteo-inductive role depending on its method of preservation. However, allograft has been shown to have inferior results when compared to autograft in a clinical setting [5].

With the advent of new methods of instrumentation as well as the isolation and purification of various bone growth factors, the surgeon is faced with many surgical options for the spine. Objective comparison of these many variables is impera-

Bone Morphogenetic Proteins, edited by Slobodan Vukicevic and Kuber T. Sampath

tive. Not only must existing fusion modalities be critically evaluated, but novel modalities must also be compared to those already in use. And ultimately, the indications for the various modalities must be defined.

To this end, animal studies are an effective means of addressing the limitations of human *in vitro* and *in vivo* studies. Outcomes of such studies may be extrapolated to human scenarios. Of significance, the more closely the animal models mirror human clinical scenarios, the more confidence that can be placed in such extrapolations. As this implies, models must be designed to address specific clinical questions [6]. Animal size, cost, and ease of care are also issues to be considered.

Multiple animal models have been established to study the spine. One mode of model validation has been to compare the physiologic motions of animal spines to those of human spines. Multi-directional flexibility testing has been used for such determinations. For example, Wilke et al. studied the physiologic motions of sheep and calf spines [7–9]. The resulting values were compared to baseline human values [10]. Wilke et al. concluded that the physiologic motions of the sheep and calf were roughly similar to those of a human, and thus, these animals could reasonably model human spinal kinematics.

Boden and colleagues have developed and extensively published about a spine model using a smaller animal, the New Zealand white rabbit [11–13]. Methodology was developed specifically to study the intertransverse process fusion. The surgical technique used by this group is similar to that used clinically. The observed pseudoarthrosis rate of 33% with autograft alone also mirrors human clinical outcomes. Nonetheless, physiologic biomechanical testing was not performed. The biomechanical testing performed quantified tensile fusion mass strength, but did not evaluate the physiologic effects of local fusion masses. Since its introduction, many other research groups have adopted this model [14–17]. Unlike the sheep and calf, there are significant anatomic differences between the rabbit and human spines. The rabbit has seven lumbar vertebra as compared to human five lumbar vertebra. The rabbit's spine has evolved to facilitate ambulation as a quadruped, as opposed to humans who evolved to facilitate bipedal ambulation. Furthermore, the rabbit is significantly smaller than any animal for which physiologic spine motion has been previously studied. Objective physiologic data of the rabbit spine would thus be useful. In particular, the lower lumbar spine needs to be characterized to further interpret the model of Boden et al., which focuses on the L5-6 intervertebral level. The first portion of this study was performed to evaluate the normal physiologic motion of the rabbit's lumbar spine. The data was then compared to human data as well as other animal studies previously studied in the lumbar spine.

Despite the frequency with which posterolateral fusion is performed, limited information is available regarding the quantitative biomechanical stabilization provided by this technique. That is to say, the stability that can be expected has not been clearly defined. As mentioned above, the posterolateral fusion produced in the New Zealand white rabbit model has been described with histologic analysis and with tensile testing. Although this tensile testing has determined the physical strength and stiffness of both fusion masses themselves, it has not provided information regarding the physiologic multi-directional stabilization afforded by intertransverse process fusion. The hypothesis for the second portion of our study was that successful fusion does not eliminate intervertebral motions. Using the normative rabbit data collected in the first portion of the study, the New Zealand white rabbit model was used to define the physiologic multi-directional stabilization provided by posterolateral lumbar fusion.

More recently, several methodologies have been used to enhance autograft fusion. Recently, more potent osteoinductive agents have been evaluated as potential bone graft alternatives. These studies began with the evaluation of slurries of demineralized bone matrix. Despite encouraging results [18] the content of such preparations was poorly defined and not always reproducible. Subsequent work has focused on individual, well characterized molecules such as bone morphogenetic proteins, or BMPs that have been prepared using molecular biologic techniques.

Recombinant human BMP-2 (Genetics Institute, Cambridge, MA) has been studied in a New Zealand white rabbit intertransverse process fusion model. The surgical technique used with this model was similar to that used clinically, as is the 67% rate of fusion with autograft. The BMP-2 molecule was found to induce fusion more rapidly than autograft bone and with a lower pseudoarthrosis rate [13].

Recombinant human BMP-7, also known as osteogenic protein-1 (OP-1), is the only other BMP currently being commercially developed (Stryker Biotech, Hopkinton, MA). Extensively evaluated with demonstrated efficacy in a wide variety of applications [19, 20], OP-1 has not been thoroughly studied in the lumbar spine.

We hypothesized that OP-1 can be used to induce solid intertransverse process fusion in the rabbit model and act as a substitute for autograft. The third portion of our current study was designed to define the functional radiographic and histologic outcomes of OP-1 induced intertransverse process fusion in the established New Zealand white rabbit model.

As stated earlier, there are also a multiplicity of conditions which can increase the rate of pseudoarthrosis, such as tobacco use. Smoking interferes with bone homeostasis and repair in several ways. It has been demonstrated to decrease bone density in the axial skeleton and to increase parathyroid hormone as well as resistance to calcitonin. Furthermore, nicotine stimulates sympathetic vasoconstriction, which may limit cellular metabolic processes, and there is evidence that nicotine decreases neovascularization [21, 22].

With the New Zealand white rabbit model, the clinical observation that smoking interferes with fusion has been confirmed [23, 24]. Nicotine exposure decreased the rate of autograft fusion from 53–56% down to 0% in the two reported studies to date. These dramatic results were defined by manual palpation and tensile testing. Silcox went on to show that combining autograft with an osteoinductive protein extract produced 100% fusion rate in the rabbit model, even in the presence of nicotine [25]. This suggested that BMPs might offer a method to overcome the inhibitory effects of nicotine on spinal fusion.

It was hypothesized that OP-1 might be used alone as a graft substitute to overcome the inhibitory effect of nicotine on posterolateral lumbar fusions. The purpose of the final portion of our present study was to use the New Zealand white rabbit model to study autograft in OP-1 induced fusions in the presence of systemic nicotine.

## Materials and methods

### Study design

The present study was divided to address four questions. First, the New Zealand white rabbit lumbar spine physiologic biomechanical characteristics were defined using multi-directional flexibility testing. Second, the New Zealand white rabbit was established as a model for posterolateral lumbar fusion, and the biomechanical stability provided by such fusion was defined. Third, OP-1 was evaluated as a substitute for autograft in posterolateral fusion and, finally, the inhibitory effect of nicotine on posterolateral fusion was confirmed, and the ability of OP-1 to overcome that inhibitory effect was evaluated.

For the first part of the study, ten skeletally mature rabbit cadaveric lumbar spines were evaluated using biomechanical flexibility testing. For the subsequent parts of the study, single level intertransverse process fusions were performed at the L5-6 level in 49 New Zealand white rabbits [11]. The rabbits were divided into five groups: (1) autograft, (2) OP-1 with its commercially prepared carrier, (3) carrier alone, (4) autograft in the presence of nicotine, and (5) OP-1 with its carrier in the presence of nicotine. Autograft was harvested from both iliac crests of all animals. As such, autograft was discarded for those animals in non-autograft groups. Animals in the nicotine groups were exposed to systemic nicotine *via* subcutaneous mini-osmotic pumps. Animals were sacrificed 5 weeks after surgery and the success of the fusion was evaluated by multiple testing modalities including manual palpation, plain radiographs and flexibility testing. This protocol was approved by the Yale Animal Care & Use Committee.

### Cadaveric specimens for biomechanic testing

Ten skeletally mature New Zealand white rabbit cadaveric spines were obtained. As noted above, the species has seven lumbar vertebrae. Previous studies have focused on the rabbit L5-6 intervertebral level. It was thus determined to be appropriate to study this level as well as one level above and one level below. Osteo-ligamentous L4-7 specimens were harvested *en bloc*. Specimens were dissected of all soft tissues except for ligaments and joint capsules. Specimens were stored at –20° C wrapped in saline moistened gauze and sealed in double plastic bags until testing was performed. Such storage conditions have been shown not to affect the outcome of standard biomechanical testing [26]. Biomechanical flexibility testing is described later in this section.

## Posterolateral fusions

Adult New Zealand white rabbits weighing approximately 4.5–5 kg were housed at an established animal facility for a minimum of 1 week prior to surgery to allow acclimatization. Preoperative radiographs were obtained to rule out underlying pathology.

Surgical anesthesia was achieved with subcutaneous injection of Acepromazine (0.75 mg/kg) followed by Ketamine (15 mg/kg) and Xylazine (2.5 mg/kg). The rabbits were then intubated and isofluorane inhalation was used to maintain anesthesia. Enrofloxacin (5–10 mg/kg SC) was given subcutaneously immediately prior to surgery. The rabbits were shaved, positioned, and prepped in a standard surgical fashion. A dorsal midline incision was made in the lumbar region. The L5 and L6 transverse processes were identified and exposed through two paramedian fascial incisions. These levels were identified intraoperatively by referencing from the sacrum with manual palpation. Autograft was recovered from all animals, regardless of the experimental group to which they would be assigned. This was done to expose all animals to the same operative stresses. Both iliac crests were exposed through separate fascial incisions and approximately 2–3 $cm^3$ cortico-cancellous graft was obtained. The crest sites were then irrigated, packed with gel foam, and closed.

Attention was returned to the fusion beds. After irrigation, the transverse processes were decorticated with a power burr. The transverse process shavings produced by decortication were left in the lateral gutters in all cases. One of three graft materials was used. The grafting materials were: (1) approximately 1–1.5 $cm^3$ of the recovered autograft per side, (2) 0.3 grams of Bovine Type I collagen matrix and 77 mg of carboxymethylcellulose per side (the commercially developed carrier for OP-1), or (3) the above carrier with 1.2 mL of OP-1 per side. This quantity of OP-1 was based upon previous studies [12, 20]. This was considered to be an appropriate volume for the fusion bed.

For those rabbits in the nicotine portion of the study, nicotine pumps were then implanted subcutaneously in the interscapular region. These mini-osmotic pumps (Alzet, Palo Alto, CA) delivered 4.5 μgrams/kg/min of nicotine at a rate of 2.5 μl/h. This dosing was based on earlier rabbit studies which were able to achieve serum nicotine levels in the range of 10–70 Ng/ml [21, 23, 25] which is comparable to those of a human smoking 20–30 cigarettes per day [27–29].

Once the graft material was placed and the incisions were closed, the rabbit was extubated. Postoperative radiographs were taken to confirm the level of fusion. Buprenex (0.04 mg/kg bid) and Enrofloxacin (5 mg/kg qd) were given subcutaneously for 2 days.

## Postoperative animal care

The rabbits were then individually housed for 5 weeks in cages that were approximately 0.9 × 1.2 m in size. Daily rounds insured that the animals were moving all extremities, posturing well, and feeding appropriately.

Serum levels of nicotine and its primary metabolite, cotinine, were monitored initially and with weekly subsequent blood samplings of those animals implanted with nicotine pumps. Serum samples were collected, stored at -20°C, and later analyzed at an independent commercial laboratory.

A follow-up of 5 weeks was chosen because fusions have been shown to be distinguishable from nonunions by this time [11]. Rabbits were given calcein 1 and 11 days prior to sacrifice as a fluorescent marker of mineralization for lateral histologic examination. The rabbits were sacrificed with a sedating dose of subcutaneous Ketamine followed by a lethal dose of intravenous Pentobarbital.

## Evaluation of specimens

The fusion masses of postoperative specimens were characterized and compared with manual, radiographic, biomechanical, and histologic evaluations. As stated previously, ten non-operated cadaveric specimens were tested using biomechanical flexibility testing.

## Manual palpation testing

Manual palpation has been thought of as an accurate indicator of successful lumbar fusion. In the clinical setting, the spine may be evaluated by direct manual palpation and surgical exploration to determine whether or not pseudoarthrosis exists. Due to clinical limitations of other methodologies, this is widely considered the definitive method for determination of fusion in the clinical setting. In an analogous manner, two independent observers manually evaluated the rabbit lumbar spines immediately after sacrifice. The L5 and L6 vertebra were manipulated with forces small enough not to produce gross trauma, but great enough to evaluate for gross intervertebral motion. Specimens were determined to be fused when no significant motion was noted by either observer.

## Radiologic evaluation

PA and lateral radiographs were taken to evaluate the fusion masses. Films were reviewed in a blinded fashion with fusion defined as calcification bridging from one transverse process to the next.

## Specimen preparation

The superior (L4) and inferior (L7) vertebra were potted in resin mounts with the L5-L6 intervertebral discs oriented in the horizontal position. Screws were placed in the border vertebra for additional fixation in the resin mounts. Bolts were also imbedded in each mount to allow fixation of the lower vertebra to the testing table to apply pure moments to the upper vertebra *via* a headpiece.

The upper and lower mounts were fitted with Plexiglas motion detection flags on the lateral aspect of the specimen. L5 and L6 were fitted with similar flags attached to the vertebral bodies *via* pairs of 0.062 inch k-wires. Each flag was equipped with three non-co-linear inferred light emitting diodes designed for detection by an opto-electronic motion measure system (Optotrak, Northern Digital, Waterloo, Ontario, Canada). Radiographs were taken of each specimen to insure that no underlying abnormalities or injuries were present.

## Three dimensional flexibility testing

The specimens were kept moist with normal saline throughout the flexibility testing as previously established and described in human specimens [30, 31]. Human specimens were loaded to a maximum of 10 N-m in the studies referenced earlier. It was, however, determined appropriate to decrease the testing moment applied to the rabbit spines in a body mass proportional fashion. Thus, a maximum moment of 0.27 N-m was selected for testing.

Further validation of this selected testing moment was obtained from preliminary reproducibility experiments. The loading protocol involved loading in a stepwise fashion to the maximum load. Each step (0.00, 0.09, 0.18, and 0.27 N-m) was sequentially applied for 30 s to allow visco elastic relaxation. A total of three load/unload cycles was performed for each motion study and data was gathered from the final loading cycle. This protocol had been established to minimize air due to the effects of creep.

### Histologic analysis

Histologic analysis was then performed to evaluate the maturity of bone induced by OP-1 as compared to that induced by autograft or carrier alone. This included an assessment of callus constituents: bone, cartilage, and fibrous tissue. Immediately after biomechanical testing, the L5-L6 spine segments were isolated and divided along the mid-sagittal plane. Each half specimen was prepared for either decalcified or undecalcified sectioning.

## Results

### Baseline cadaveric spines

Using the flexibility testing protocols as previously described we found that a significant portion of the motion for each direction of applied moment was due to the neutral zone with a gradual increase in displacement with subsequent loading up to range of motion with the application of 0.27 N-m. Flexion and extension were studied for independent study parameters. Lateral bending and torque were expected to be symmetric due to the symmetry of the lumbar spine. The relative differences with this parallel data are comparable to that of the reported human data.

The three levels tested have roughly similar range of motion and neutral zone parameters. There is a trend toward increased flexion and decreased lateral bending moving caudal through the levels tested. The greatest motion for each level tested was in flexion with lesser motion in extension and lateral bending, and the least amount of motion with torque. As such, flexion was used as the basis for comparison.

### Surgical complication rates

Out of the forty-nine rabbits receiving surgical fusion, 10 were excluded (24%): five due to subclinical deep infections discovered at the time of sacrifice, four due to anesthetic related complications and one due to sciatic nerve decompression from the iliac crest harvest site. This complication rate is comparable to previous studies using this model (20%) [11]. Of the remaining 39 rabbits, eight each were in the autograft, OP-1, carrier alone, and nicotine exposed autograft groups. Seven rabbits were in the nicotine exposed OP-1 group.

### Autograft fusion spines

By manual palpation, five of the eight rabbits had solid fusion. There were no differences in opinion between the two observers regarding the fusion status of the specimens. Radiographically, fusion masses were clearly visualized. However, as all specimens were interpreted to have some trabecular bridging, all radiographs were read as fused. In other words, pseudoarthrosis was not noted by radiographic evaluation at the 5-week timepoint.

Baseline flexibility data of the L5-6 level of non-operative rabbit spines was based on a group of 10 animals of similar age and mass to those of the current study. The range of motion of the fused specimens was significantly decreased from that of baseline non-operative specimens in flexion (81%), extension (61%) and right and left lateral bending (67% and 83% respectively). Right and left axial rotations, which had significantly smaller baseline values than the other motions, were without change.

The specimens determined to be unfused by manual palpation were similarly studied biomechanically. This group consisted of three specimens. In comparison to baseline non-operative flexibility data, the unfused specimens had a decrease in flexion range of motion of 51%. In flexion, the range of motion of fused specimens had an additional decrease of 63% from the unfused specimens. Thus, the pseudoarthrosis specimens represented a distinct intermediate stability between the baseline and fused specimens.

Similar to range of motion, the neutral zone of the fused specimens was significantly decreased from that of baseline non-operated specimens in flexion (85%), extension (65%), and left lateral bending (88%). In comparison to baseline non-operative flexibility data the unfused specimens had a decrease in flexion neutral zone of 50%. In flexion, the neutral zone of fused specimens had an additional decrease of 71% from the unfused specimens.

## OP-1 fusion spines

By manual palpation five of the eight autograft rabbits fused (63%), none of the carrier alone rabbits fused (0%), and all of the OP-1 rabbits fused (100%). Both autograft and OP-1 fusion rates, as determined by manual palpation, were significantly different from the carrier alone group, but were not significantly different from each other. Radiographically, all of the autograft specimens were thought to be fused with three unfused specimens incorrectly assessed by this approach. Some of the eight carrier alone specimens were correctly determined to be unfused, but two were incorrectly thought to be fused. Seven of the eight OP-1 specimens were correctly determined to be fused, but one was incorrectly thought to be fused. Overall, the radiographs were 92% sensitive and 55% specific for determining fusion with a positive predictive value of 71% and negative predictive value of 86%.

The findings of biomechanical testing further characterized the fusion masses. Based on findings from cadaveric rabbit spines, flexion was determined to be the best indicator for fusion as it was the direction of greatest motion for the rabbit lumbar spine. Of the autograft specimens, the five that were fused by manual palpation had 2.3° of flexion. Conversely, those that were unfused by manual palpation had 6.3° of flexion. The OP-1 specimens which were fused by manual palpation had 0.8° of flexion. The carrier alone specimens, which were unfused by manual palpation, had 6.3° of flexion. The differences in flexion range of motion between the three groups was significant using one way ANOVA analysis. Not surprisingly there was little difference between the flexion range of motion of the unfused autograft specimens and the carrier alone specimens. In addition, the OP-1 specimens had significantly less flexion than fused autograft specimens.

Histologic sections were analyzed using several staining preparations. Toluidine blue staining highlighted the regions of calcification. Calcified islands were seen in the autograft fusion masses corresponding to the original grafting material. Essentially, no calcified material was seen in the carrier alone fusion masses. Conversely, bridging calcification was clearly seen in the OP-1 fusion masses. Higher magnification toluidine blue and hematoxylin used in staining further defined the fusion masses with the autograft fusion masses characterized predominantly by cartilaginous tissue and small amounts of fibrous tissue between bone graft fragments. The intertransverse region of the carrier alone specimens demonstrated moderate fibrous tissue and remnants of the reabsorbing collagen-based carrier. Despite endochondral bone formation around the decorticated surfaces of the transverse process, no intertransverse callus was seen. There was also no significant inflammatory reaction appreciated.

OP-1 induced fusion masses were characterized by a cortical rim of woven bone surrounding trabecular bone. While small amounts of cartilaginous material were present, the OP-1 fusion masses were predominantly maturing bone with high magnification revealing significant osteoblast activity. Calcein fluorescent staining confirmed active mineralization fronts in the OP-1 specimens. This was present to a lesser extent in the autograft specimens and was negligible in the carrier alone specimens.

## Nicotine exposed fusion spines

Weekly nicotine and cotinine levels were determined by gas chromatography. The average nicotine value for each timepoint studied was within the target range of 10–70 Ng/ml. No clinical signs of nicotine toxicity were noted.

By manual palpation, two of the eight nicotine exposed autograft rabbits fused (25%). This is less than the five of eight autograft fusions in rabbits not exposed to nicotine (63%). These results were consistent with the inhibitory effect of nicotine on fusion. Of note, the two nicotine exposed autograft rabbits that were fused at 5 weeks had nicotine levels within the range of the other unfused rabbits.

By manual palpation, all of the nicotine exposed OP-1 rabbits fused (100%). This fusion rate is comparable to the 100% fusion rate of OP-1 rabbits not exposed to nicotine. In comparing fusion rates of the two nicotine exposed groups, OP-1 specimens had a significantly higher fusion rate than autograft specimens (Chi Squared Analysis).

Radiographically, five of seven nicotine exposed OP-1 rabbits were determined to be fused. Thus, two of the fused nicotine exposed OP-1 specimens were misinterpreted to be unfused. Of the nicotine exposed autograft rabbits three of the six unfused specimens were interpreted to be unfused. One of the two nicotine exposed autograft specimens that fused was interpreted to be fused. Overall, radiographs were 67% sensitive and 50% specific for determining fusion with the 67% positive predictive value and 50% negative predictive value.

The results of biomechanical testing correlated well with those of manual palpation. Of the nicotine exposed autograft specimens the six that were unfused by manual palpation had 4.2° of flexion. Conversely, those that were fused by manual palpation had significantly less flexion. The seven nicotine exposed OP-1 specimens, which were all fused by manual palpation, had 0.6° of flexion.

The differences in flexion range of motion between autograft and OP-1 groups with and without nicotine were significant using one way ANOVA analysis. In addition, there was little difference between the flexion data of the OP-1 group with nicotine and the OP-1 group without nicotine.

Histologic sections were analyzed in a similar fashion to that previously described. Calcified islands corresponding to the original graft material characterized the nicotine exposed autograft specimens. Calcified bridging was clearly seen in the nicotine exposed OP-1 group. The fusion masses of this latter group were notable for a bony cortical rim with central trabecular bone. Upon higher magnification, nicotine exposed autograft fusion masses, particularly in the unfused specimens, were characterized by minimal amounts of cartilaginous and fibrous tissue between bone graft fragments. The nicotine exposed OP-1 fusion masses were characterized by a maturing bony callus. Higher magnification of the OP-1 fusion masses revealed significant osteoblast activity and substantial osteoid formation indicative of newly forming bone. Calcein fluorescent staining confirmed active mineralization fronts in the OP-1 specimens. Fluorescent staining was negligible around the islands of bone graft found in the autograft group.

## Discussion

The New Zealand white rabbit has been used as a spine model in looking at the effectiveness of posterolateral lumbar fusions. Although the biomechanical properties of the fusion masses themselves have been studied, the baseline and resulting alterations in physiologic motion have not been established. This is important, as solid fusion masses do not necessarily eliminate inter-body motion.

The primary purpose of the first portion of this study was to determine baseline biomechanical flexibility parameters of the New Zealand white rabbit lumbar spine.

The presented range of motion and neutral zone data can be used as normative values to which future experiments can be compared.

The secondary purpose of this portion of the study was to determine the physiologic motion of the rabbit lumbar spine to that of the human lumbar spine. The data from our physiologic range of motion study is remarkably similar between the rabbit and the human. In fact, the average difference in range of motion between the two species at the three lowest intervertebral levels was only 2.42°.

The purpose of the second portion of this study was to use the New Zealand white rabbit model to perform *in vitro* characterizations of *in vivo* fusions using the techniques of manual palpation, radiography, and biomechanical multi directional flexibility testing. By manual palpation, we found a 35% pseudoarthrosis rate with posterolateral fusion using autologous iliac crest bone graft. This approximates the previously reported pseudoarthrosis rate of 33% with this model.

While radiography revealed fusion masses, the technique was not useful in identifying pseudoarthrosis. This is consistent with previous studies that have found a limited role for plain radiographs in defining fusion [32].

Physiologic biomechanical flexibility testing offers a precise method to characterize the changes in physiologic motion the result from spinal fusion. In the current study, posterolateral fusion led to a significant stabilization of the L5-6 motion segment with significant range of motion decreases in flexion, extension, and lateral bending of 61–83%. Interestingly, the changes in neutral zone closely mirrored the changes in range of motion and remained a relatively constant percentage of the range of motion.

These findings suggest that successful fusion significantly limited, but did not eliminate, intervertebral motion at the time of point studied. Certainly, further studies elucidating the contributing factors to fusion flexibility are indicated. Nevertheless, the findings of this study should remind the clinician that the primary goal of fusion surgery is spinal stabilization sufficient to eliminate pain and not necessarily to completely eliminate motion.

Unfortunately, the correlation between biomechanical stabilization and pain relief is a difficult one to study. In regards to this question of how much stability is adequate, our pseudoarthrosis specimens are of interest. The flexibility of these specimens, intermediate between those of fused and baseline non-operative specimens was thought to be secondary to scarring produced by surgical exposure. The scarring effect is consistent with the findings of previous animal studies [33]. Whether this decrease in flexibility would have limited clinical symptoms cannot be determined from this animal model.

To gain further perspective on the decrease in flexibility produced by biologic fusions, the present results can be compared to time zero cadaveric instrumentation studies of lumbar fusion. Panjabi and colleagues evaluated the kinematic effects of several spinal fixation devices in human cadaveric spines using flexibility testing similar to that used in the present experiment [33]. Of the posterior fixation constructs tested, hook and rod constructs lead to an approximately 15–70% flexion stabilization with pedicle screw constructs leading to an approximately 65-80% flexion stabilization. In other words, these posterior constructs lead to a time zero stabilization only slightly less than that observed in the present study with biologic posterolateral fusion (81%).

Bone morphogenetic proteins are currently being evaluated as potential substitutes for iliac crest autograft in a wide variety of clinical situations. The purpose of the third portion of this study was to evaluate OP-1 as a bone graft substitute in posterolateral fusion using the New Zealand white rabbit model.

Biomechanical flexibility testing revealed five of eight of the autograft rabbits to be fused. This fusion rate was consistent with previous reports and the histologic evaluation of these fusion masses showed an immature combination of bone and cartilage.

OP-1 induced fusion in all eight of the treated rabbits. This is higher than that seen with autograft and is consistent with the fusion rate described with BMP-2 [13]. While the fusion rates with OP-1 as determined by manual testing were not significantly different from autograft fusion rates, biomechanical testing revealed that OP-1 fusions were more stable than the time matched autograft fusions. Histologically, the OP-1 induced fusion masses were characterized by predominantly remodeling bone that was more mature with that associated with autograft. These data suggest that the fusion process was occurring more rapidly with OP-1 than with autograft.

Conversely, the carrier alone did not induce any fusion. The carrier is an important component of any potential bone graft alternative as it distributes the osteoinductive agent while keeping it in the desired location. In this case, the carrier was clearly not responsible for the osteogenic response. Of note, the Bovine Type I collagen matrix/carboxymethylceloulose carrier was free of any significant inflammatory response.

OP-1 appears to be an effective bone graft alternative for intertransverse process spinal fusion in the New Zealand white rabbit model.

The final portion of this study was performed to evaluate autograft in OP-1 induced posterolateral fusions that were exposed to systemic nicotine. It has previously been shown that nicotine inhibits posterolateral autograft lumbar fusion. The present study showed a similar decrease in autograft fusion rate from 63% to 25% with the introduction of systemic nicotine. As observed in prior studies, nicotine appeared to retard or preclude a successful bony healing process at the histologic level.

In the third portion of this study OP-1 had been shown to induce 100% posterolateral lumbar fusion in the rabbit model in the absence of nicotine exposure. This 100% rate of fusion was shown to persist in the presence of systemic nicotine. The ability of OP-1 to induce fusion was demonstrated with manual and biomechanical testing. Histologically, maturing bony callus with a cortical rim was seen in the OP-1 study group despite the presence of nicotine.

As only one timepoint was evaluated in the study no significant delay in bony repair could be determined for the nicotine exposed OP-1-induced fusion masses. However, there may have been an initial delay in healing which was not evident later in the healing process.

Further, it is possible that additional nicotine exposed autograft specimens may have gone on to fusion with additional time. We were unable to say if nicotine delays or prevents a proportion of posterolateral spine fusion. Nevertheless, it is clear that OP-1 is able to induce more mature fusion masses more rapidly than autograft at the 5 week timepoint studied in this model. In addition, the success of OP-1 to achieve such fusions without the use of autograft implies that the morbidity associated with autograft harvest may be avoided in the future.

Currently, studies are underway to characterize the molecular mechanism of action of OP-1. It has been suggested that more than one growth factor may be necessary in the human clinical setting to achieve a successful fusion. Studies have been undertaken to characterize the influence of an individual bone morphogetic protein on the expression of bone morphogenetic proteins that occur in the natural cascade of bony healing as well as the expression of various autologous growth factors involved in bony healing such as vascular endothelial growth factor, basic fibroblast growth factor, etc.

Overall, OP-1 appears able to overcome the inhibitory effects of nicotine on spinal fusion. While the role of OP-1 in the clinical setting remains to be defined, the final portion of the study suggests that OP-1 may be beneficial in the smoking patient in whom autograft may not provide reliable posterolateral lumbar fusion.

## References

1 Steinmann JC, Herkowitz HN (1992) Pseudoarthrosis of the spine. *Clin Orthop* 284: 80–90

2 Blumenthal SL, Baker J, Dossett A, Selby DK (1986) The role of anterior lumbar fusion for internal disc disruption. *Spine* 13: 566–569

3 Brown CW, Orme TJ, Richardson HD (1986) The rate of pseudoarthrosis (surgical nonunion) in patients who are smokers and patients who are nonsmokers: A comparison study. *Spine* 11: 942–943

4 Summers BN, Eisenstein SM (1989) Donor site pain from the ilium. A complication of lumbar spine fusion. *J Bone Joint Surg {Br}* 71: 677–680

5 Jorgenson SS, Lowe TG, France J, Sabin J (1994) A prospective analysis of autograft versus allograft in posterolateral lumbar fusion in the same patient. A minimum of 1-year follow-up in 144 patients. *Spine* 19: 2048–2053

6 Panjabi MM (1998) Cervical spine models for biomechanical research. *Spine* 24: 2684–2700

7 Wilke HJ, Kettler A, Claes LE (1997) Are sheep spines a valid biomechanical model for human spines? *Spine* 22: 2365–2374
8 Wilke HJ, Krischak S, Claes L (1996) Biomechanical comparison of calf and human spines. *J Orthop Res* 14: 500–503
9 Wilke HJ, Krischak ST, Wenger KH, Claes LE (1997) Load-displacement properties of the thoracolumbar calf spine: experimental results and comparison to known human data. *Eur Spine J* 6: 129–137
10 Yamamoto I, Panjabi MM, Crisco T, Oxland T (1989) Three dimensional movements of the whole lumbar spine and lumbosacral joint. *Spine* 11: 1256–1260
11 Feiertag MA, Boden SD, Schimandle JH, Norman JT (1996) A rabbit model for nonunion of lumbar intertransverse process spine arthrodesis. *Spine* 21: 27–31
12 Schimandle JH, Boden SD (1994) Spine Update: The use of animal models to study spinal fusion. *Spine* 19: 1998–2006
13 Schimandle JH, boden SD, Hutton WC (1995) Experimental spinal fusions with recombinant human bone morphogetic protein-2 (rhBMP-2). *Spine* 20: 1326–1337
14 Curylo LJ, Johnstone B, Petersilge CA, Janicki JA, Yoo JU (1999) Augmentation of spinal arthrodesis with autologous bone marrow in a rabbit posterolateral spine fusion model. *Spine* 24: 434–439
15 Itoh H, Ebara S, Kamimura M, Tateiwa Y, Kinoshita T, Yuzawa Y, Takaoka K (1999) Experimental spinal fusion with use of recombinant human bone morphogenetic protein 2. *Spine* 24: 1402–1405
16 Minamide A, Tamaki T, Kawami M, hashizume H, Yoshida M, Sakata R (1999) Experimental spinal fusion using sintered bovine bone coated with Type I collagen and recombinant human bone morphogenetic protein-2. *Spine* 24: 1863–1872
17 Tay BK, Le AX, Heilman M, Lotz J, Branford DS (1998) Use of collagenhydroxyapatite matrix in spinal fusion: a rabbit model. *Spine* 23: 22762281
18 Lindholm TS, Ragni P, Lindholm TC (1988) Tesponse of bone marrow stroma calls to demineralized cortical bone matrix in experimental spine fusions in rabbits. *Clin Orthop* 230: 296–302
19 Cook SD, Rueger DC (1996) Osteogenic protein-1: biology and applications. *Clin Orthop* 325: 29–38
20 Cook SD, Salkeld SL, Brinker MR, Wolfe MW, Rueger DC (1998) Use of osteoinductive biomaterial (rhOP-1) in healing large segmental bone defects. *J Orthop Trauma* 12: 402–412
21 Daftari TK, Whitesides TE, Heller JG, Goodrich AC, McCarey BE, Hutton WC (1994) Nicotine on the revascularization of bone graft: an experimental study in rabbits. *Spine* 19: 904–911
22 Reibel GD, Boden SD, Whitesides TE, Hutton WC (1994) The effects of nicotine on incorporation of cancellous bone graft in an animal model. *Spine* 20: 2198–2202
23 Silcox DH, Daftari T, Boden Sd, Schimandle JH, Hutton WC, Whitesides TE (1995) The effect of nicotine on spinal fusions. *Spine* 20: 1549–1553

24 Wing KJ, Fisher CG, O'Connell JX, and Wing PC (2000) Stopping nicotine exposure before surgery; the effect on spinal fusion in a rabbit model. *Spine* 25: 30–34
25 Silcox DH, Boden SD, Schimandle JH, Johnson P, Whitesides TE, Hutton WC (1998) Reversing the inhibitory effect of nicotine on spinal fusion using an osteoinductive protein extract. *Spine* 23: 291–297
26 Panjabi MM, Krag M, Summers D, Videman T (1985) Biomechanical time-tolerance of fresh cadaveric human spine specimens. *J Orthop Res* 3: 292–300
27 Benowitz NL, and Jacob P (1984) Daily intake of nicotine during cigarette smoking. *Clin Pharmacol* 35: 499–504
28 Benowitz NL (1986) Clinical pharmacology of nicotine. *Ann Rev Med* 37: 2132
29 Isaac PF, Rand ML (1969) Blood levels of nicotine and physiological effects after inhalation of tobacco smoker. *Eur J Pharmacology* 8: 269–284
30 Panjabi MM, Abumi K, Duranceau J, Crisco JJ (1988) Biomechanical evaluation of spinal fixation devices: II. Stability provided by eight internal fixation devices. *Spine* 13: 1135–1140
31 Panjabi MM, Kifune M, Liu W, Arand M, Vasavada A, Oxland TR (1998) Graded thoracolumbar spinal injuries: development of multidirectional instability. *Eur Spine J* 7: 332–339
32 Kant AP, Daum, WJ, Dean SM, Uchida T (1995) Evaluation of lumbar spine fusion: Plain radiographs versus direct surgical exploration and observation. *Spine* 20: 2313–2317
33 Panjabi MM, Pelker RR, Crisco JJ, Thibodeau L, and Yamaoto I (1988) Biomechanics of healing of posterior cervical spinal injuries in a canine model. *Spine* 13: 803–807

## *Acknowledgements*

All data and contents of this paper have been previously accepted for publication:
1 Grauer JN, Erulkar JS, Patel TCh, Panjabi MM (2000) Biomechanical evaluation of the New Zealand white rabbit lumbar model. *Eur Spine J* 9: 250–255
2 Erulkar JS, Grauer JN, Patel TCh, Panjabi MM (2001) Kinematic analysis of posterolateral fusions in a New Zealand white rabbit model. *Spine* 26: 1125–1130
3 Grauer JN, Patel TCh, Erulkar JS, Troiano NW, Panjabi MM, Friedlaender GE (2001) Evaluation of OP-1 as a graft substitute for posterolateral lumbar fusion. *Spine* 26: 127–133
4 Patel TCh, Erulkar JS, Grauer JN, Troiano NW, Panjabi MM, Friedlaender GE (2001) OP-1 overcomes the inhibitory effect of nicotine on posterolateral lumbar fusion. *Spine* 26: 1656–1661

OP-1 and financial support for this study was provided by Stryker Biotech (Hopkinton, MA).

# Bone morphogenetic proteins and the synovial joints

*Frank P. Luyten, Rik Lories, Dirk De Valck, Cosimo De Bari and Francesco Dell'Accio*

Laboratory of Skeletal Development and Joint Disorders, Department of Rheumatology, University Hospitals Leuven, Herestraat 49, B-3000 Leuven, Belgium

## Bone morphogenetic proteins and joint development

### Morphological events of joint formation

Bone morphogenetic proteins (BMPs) are involved in a broad array of morphogenetic processes. These span from the specification of the dorso-ventral body axis to patterning, organogenesis and differentiation of most tissues. Nevertheless, the initial discovery of BMPs as protein preparations that induce ectopically and *in vivo* a cascade of endochondral bone formation in rats, has strongly stimulated the study of their role in the development of the skeleton and in the patterning of the synovial joints [1–3]. In addition, with their remarkable cartilage and bone morphogenetic activity, BMPs represent an attractive therapeutic option for skeletal and joint disorders. Indeed, growing scientific evidence supports the concept that tissue repair and regeneration recapitulates to a certain extent the process of tissue formation during embryonic development. Taking advantage of the expanding knowledge in the field of developmental biology to define potential new targeted therapeutic approaches, the role of BMPs in the development of the skeleton and in particular in the patterning and differentiation of joint tissues becomes increasingly clinically relevant.

Joint development has been extensively studied in a variety of animal species including human [4–12], chick [13–18], mouse [19, 20], and rat [21]. As the molecular cascades driving organogenesis and tissue specification are highly conserved across species, with some precaution, one can integrate the data available from those different animal species into a common scheme.

The appendicular skeleton develops from a primitive avascular, densely packed cellular mesenchyme derived from the lateral plate mesoderm [22–24]. Limb outgrowth is proceeding in a proximal-distal fashion, in the forelimbs earlier than in the hindlimbs. The condensation of mesenchymal cells leads to the formation of uninterrupted rod-like structures called anlagen. Subsequently, within the condensations, cells undergo chondrocytic differentiation to form cartilaginous templates

Bone Morphogenetic Proteins, edited by Slobodan Vukicevic and Kuber T. Sampath

surrounded by a sheath of spindle-shaped cells, the perichondrium. In the middle of each skeletal element, chondrocytes mature toward hypertrophy to be replaced by bone tissue in a process called endochondral ossification.

Synovial joints form through a process of segmentation of the skeletal elements. In the region of the prospective joint, a narrow zone of mesenchymal cells does not undergo cartilage differentiation [19] and forms a so-called joint interzone. Morphologically [21], the joint interzone represents the first evidence of joint formation. The interzone, at 12 days *post coitum* (dpc), is constituted by a few layers of a morphologically homogeneous elongated cell type. By 15 dpc, the interzone differentiates into three distinguishable layers. Two chondrogenic, perichondrium-like dense layers covering the articulating surfaces of the cartilage elements contain flattened elongated cells at the articular side and rounded, chondrocyte-like cells at the cartilaginous side. One layer of a loose cellular tissue with a sparse cell population and enlarged intercellular spaces is in between the two chondrogenic layers. The dense zones further differentiate into articular cartilage at both ends of the future joint. After a phase of vascular invasion that selectively involves the peripheral part of the interzone, the one that will give rise to the capsulo-synovial apparatus, a cavitation process takes place in the central loose layer of the interzone. Joint cavitation starts with the appearance of small clefts within the interzone, which eventually coalesce to form the synovial cavity. In rats, joint cavitation is seen first in proximal joints at 16 dpc and is completed in distal joints by 20 dpc. Peri- and intra-articular joint associated structures such as joint capsule, menisci, and ligaments differentiate from the mesenchymal cells surrounding the interzone and from the cells constituting the interzone, respectively [21]. In contrast to mammals, in the avian embryo some joint interzones form after the entire mesenchymal condensation underwent cartilage differentiation. This happens presumably by invasion of mesenchymal cells from the perichondrium or by de-differentiation of chondrocytes at the site of interzone formation [25, 26]. In addition, the avian joint is somewhat different from the mammalian joint, since the articular surface is covered by a perichondrium-like fibrocartilage layer, the articular cap, which is absent in mammals [14], with the exception of the temporomandibular joint.

## Molecular signaling in joint formation

From the molecular point of view, joint development consists of two main critical phases: joint patterning, with the specification of the site where a joint will form within a mesenchymal condensation, and tissue differentiation. After the joint interzone has been established, it further differentiates into three layers: two external layers that will give rise to the articular cartilage, while the middle one will undergo a process of cavitation. Subsequently, with an articulated sequence of differentiation events, the subchondral bone and the articular cartilage will differentiate fully.

There is compelling evidence for a role of BMPs in both specification and tissue differentiation in joint development.

The identification of the signal(s) responsible for the determination of the site of joint formation is still a challenge. Hints to address this point come from genetic studies, transgenic models, and natural mutations in which joint formation is disrupted. One of the best documented candidates to play a role in joint determination is cartilage derived morphogenetic protein-1/growth/differentiation factor-5 (CDMP-1/GDF-5) [27–30], a bone morphogenetic protein (BMP) family member. In developing mouse limbs, *cdmp1/gdf5* is expressed in the perichondrium and in every interzone 24–36 h before its morphological appearance [28]. In naturally occurring loss-of-function mutations in the *cdmp1/gdf5* gene in mice (brachypodism) [29] and in humans (Hunter-Thompson chondrodysplasia) [31] the distal elements of the appendicular skeleton develop poorly and a specific subset of joints does not form. Although *cdmp1/gdf5* is expressed in all joint interzones early in limb development, only a subset of joints is affected by *cdmp1/gdf5* null mutations, indicating that other molecules, possibly other BMPs, can compensate the *cdmp1/gdf5* function. This hypothesis finds support in the phenotype of another spontaneous mutation of the *Cdmp1/Gdf5* gene in humans, Grebe chondrodysplasia (OMIM 200700) [32]. In contrast to the Hunter-Thompson variant, in which CDMP-1/GDF-5 protein is presumably absent as the result of a frameshift mutation in the mature region [31], the Grebe chondrodysplasia [33] is associated with a point mutation in the *Cdmp1/Gdf-5* gene. This mutation results in a protein that is not secreted, is inactive *in vitro* and can form non-functional heterodimers with other BMP family members thereby probably preventing their secretion [33]. *In vitro* studies suggest that this mutation generates a molecule that can apparently behave as a dominant negative for a number of other BMP family members. Therefore, this phenotype is much more severe than the Hunter-Thompson, and proper morphogenesis of the entire appendicular skeleton is disrupted, but interestingly still in a proximo-distal fashion. These studies provide support to the intriguing hypothesis that the morphogenesis of different skeletal elements is regulated by different BMP family members, as a result of gene duplications within the BMP family, followed by gain and loss of specific regulatory elements [1]. This would explain the complexity of the skeletal system of evolutionary higher species.

Disruption of joint formation is obtained in a number of different experimental models. *Bmp7/op1* is highly expressed in the differentiating perichondrium of chick limb cartilages at stages 29–34 HH (Hamburger/Hamilton) [34], with characteristic interruptions in the zones of future joint formation [35]. Implantation of BMP-7/OP-1 soaked beads at these stages in the joint region disrupts joint formation [35]. Thus it has been suggested that BMP-7/OP-1 would act as an inhibitory factor for joint formation, preventing joints from forming in non-physiological sites, and that the discontinuities in its expression in the perichondrium would have a permissive role [35]. In contrast to *bmp7/op1*, *bmp2* transcripts exhibit linear domains of expression in the joint interzones over the same developmental stages [35]. *Bmp2* has been also detected with a similar pattern in mice as early as at stage 13.5 dpc, and its expression becomes prominent at stage 15.5 dpc [36]. Overexpression of *bmp2* and *bmp4* by retroviral vectors, also disrupts joint formation [37]. The correct patterning of the appendicular skeleton and the joint formation process is likely to require an interplay of different signaling molecules tightly restricted in their activity and specific expression domains. A fine balance of BMPs may play a pivotal role in joint identity.

BMP signaling is regulated in many ways: at the extracellular level by several binding molecules (e.g. noggin, chordin and DAN/gremlins), at the receptor level by alternative expression of different receptors, and at the intracellular level by both cytosolic proteins including *smads*, and nuclear proteins such as smad-interacting proteins, and finally by several transcriptional regulators at the DNA level [38–41]. Noggin (encoded by the *nog* gene) is a secreted molecule that physically interacts with BMP family members and inhibits their activity [42]. It is expressed in developing murine limbs in the condensing mesenchyme and in immature chondrocytes [43]. Its expression pattern and *in vitro/in vivo* function suggest that its developmental role is to establish boundaries of BMP activity. In noggin deficient mice the resulting excess of BMP activity leads to enlarged appendicular skeletal elements and failure to form joints [43]. This skeletal phenotype closely resembles that of *cdmp1/gdf5* overexpression [44–46]. The absence of joints is likely due to failure of joint formation rather than joint fusion, since the *cdmp1/gdf5* expression domain is disrupted in $nog^{-/-}$ mice, while the expression of other BMPs such as *bmp2*, *bmp4*, *bmp5*, and *bmp6* is unaffected [43]. While heterozygous $nog^{+/-}$ mice appear to be normal, dominant missense mutations in a highly conserved region of the *Nog* gene have been identified in five independent families that segregate proximal symphalangism (SYM1; OMIM 185800) and one dominant missense mutation in a family segregating multiple synostoses syndrome (SYNS1; OMIM 186500) [47]. The principal feature of both syndromes is joint fusion. The mechanism by which these mutations alter the noggin function and cause the phenotypes is not known. Functional haplo-insufficiency is one potential mechanism, as has previously been suggested for *cdmp1/gdf5* mutations in Brachydactyly C families [48]. Alternatively, different mutations may impair the ability of the peptide to bind a subset of TGF-β family members, accounting for the differences in the two syndromes and between the families [47]. These data also suggest that the requirement of noggin for joint morphogenesis may vary between species.

At the receptor level, BMP signaling is regulated by the expression of different BMP receptors [38, 39, 41]. Alk6/BMPR-IB type I BMP receptor is expressed early throughout the prechondrogenic mesenchymal condensations and its expression pattern becomes later restricted to a narrow domain flanking both distally and proximally that of *cdmp1/gdf5* in the joint interzones (Fig. 1). Although *alk6* is expressed in all the skeletal elements, *alk6*-deficient mice display only limited skele-

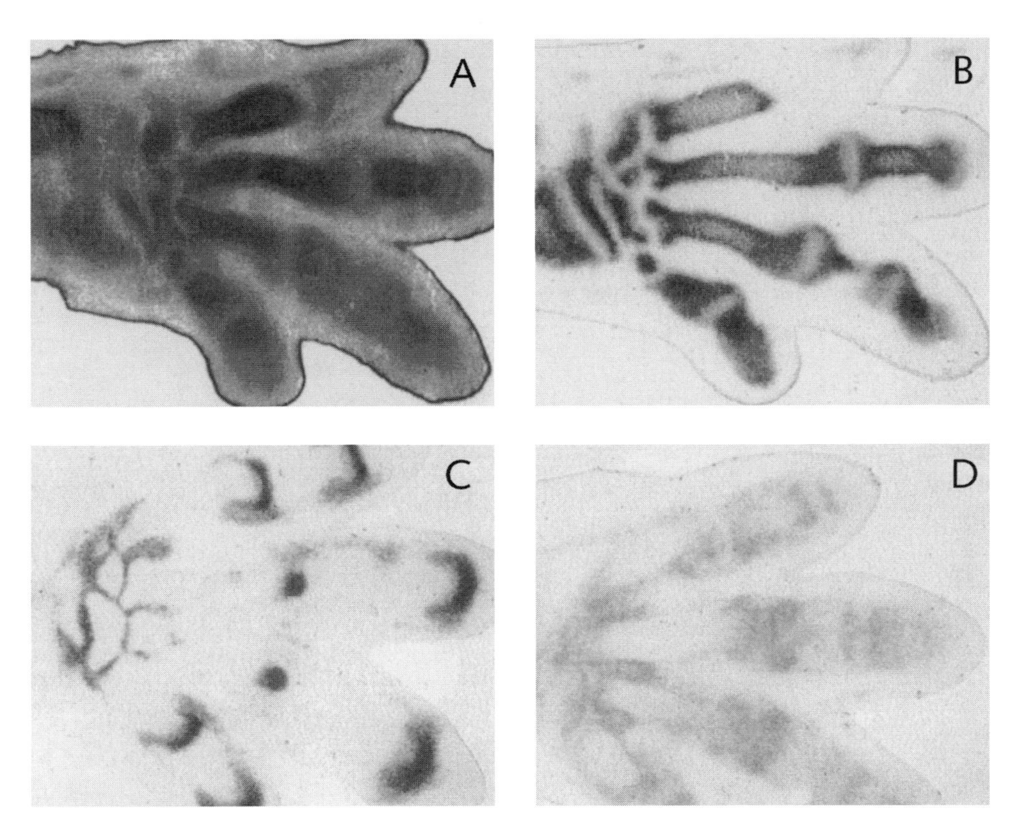


*Figure 1*
*Gene expression pattern of selected BMP signal transduction components during joint morphogenesis. Cryosectioned 14.5 dpc mouse forelimbs were stained with toluidine blue (A) or analyzed by* in situ *hybridization with digoxigenin-labeled cRNA probes for* Col2a1 *(B),* Gdf5/Cdmp1 *(C) and* Bmpr-Ib/Alk6 *(D).*

tal deformities, lacking both the first and the second phalanges [49,50]. This phenotype is overlapping, but not identical to that of *cdmp1/gdf5* deficient mice since, in contrast to *cdmp1/gdf5*$^{bpJ-/-}$ mice, the metacarpal elements are of normal length and articulate directly to a normal distal phalanx. The double homozygous *alk6*$^{-/-}$ *cdmp1/gdf5*$^{bpJ-/-}$, however, resemble more closely the *cdmp1/gdf5*$^{bpJ-/-}$, again with subtle differences [49, 50]. These genetic data, taken together with *in vitro* studies [51, 52] seem to indicate that they function within the same pathway and that their absence can be compensated by other signaling pathways in most skeletal elements. The discrete differences between the phenotypes described indicate that *cdmp1/gdf5* signals prevalently, but not exclusively, through the *alk6* receptor, as well as *alk6* does not transduce only *cdmp1/gdf5* signaling.

Little is known about the molecules upstream of *cdmp-1/gdf-5*. Possible candidates are Hox genes, a family of transcription factors, which are thought to control the positional information of skeletal elements [53]. Indeed, mutations of *hoxa* and *hoxd* genes cause fusion of carpal joints [54–56]. Recently, the characterization of regulatory elements of *gdf5* has been described [57, 58]. The knowledge and availability of these elements should allow further analysis of signaling pathways critically involved in the joint formation process using genetic approaches.

## The process of joint cavity formation

Various mechanisms have been proposed to unravel the molecular basis of cavitation in synovial joints. To date, the factors considered as being involved in the cavitation process are fetal movements [59–64], programmed cell death (PCD) [65–67], and selective secretion and turnover of ECM components [14, 68].

The role of movement in joint cavitation is controversial. The observation that synovial joints fail to develop in immobilized chick embryos [59, 61, 62] has led to hypothesize that mechanical disruption of intercellular matrix could occur under forces generated by muscle activity. However, in myogenin deficient mice, which do not develop contracting skeletal muscles, joint cavitation takes place normally [69].

During mammalian morphogenesis, PCD is an essential mechanism to eliminate selectively cell populations and accomplish histogenesis and organogenesis. In the rat embryo, PCD has been observed histologically within the interzone before cavitation [21]. It has been suggested that cells with chondrogenic potential would be eliminated in this way from the interzone, thus preventing cartilage differentiation [18, 65]. Another mechanism envisages synthesis and deposition of large amounts of hyaluronic acid (HA) as a mechanical factor to separate the opposing joint surfaces [14, 68, 70]. This theory is corroborated by the histochemical localization of free HA at the chick metatarsophalangeal joint interzones concomitant with the first signs of cavitation at stage 37 HH [68], and confirmed by the local increased activity of uridine diphosphoglucose dehydrogenase and HA synthase, enzymes involved in HA synthesis [70]. The swelling pressure of the HA is assumed to physically separate the cells thereby inducing joint cavitation, to increase and maintain the cavity, and prevent secondary fusion across the joint space [68, 70].

More recently, PCD has been described to occur within joint interzones of developing digits in mouse fetuses between 13 and 14 dpc, thus shortly before cavitation starts (14–15 dpc) [67]. These data have been confirmed also in the chick embryo at stages 33–35 HH [71]. *Cdmp1/gdf5* and *bmp2*, expressed in the joint interzone within the same time window, are good candidates in mediating this process, since BMPs have been shown to induce apoptosis in mesenchymal cells at certain sites and stages during development [35, 72, 73]. In *alk6*$^{-/-}$ mice, as a secondary event, *cdmp1/gdf5* is overexpressed with an expanded expression domain [49]. This expression domain overlaps with an area of intense cell death [49]. These data seem to indicate that *cdmp1/gdf5* stimulates chondrogenesis and cartilage growth through the *alk6* receptor, while triggering apoptosis in the absence of *alk6*, therefore through a different receptor. Since the *alk6* expression domains are flanking the narrow stripe of *cdmp1/gdf5* expression at the joint interzones, a role of *cdmp1/gdf5* in inducing apoptotic events associated with joint cavitation is likely.

Finally, it is important to mention that the combined genetic and experimental evidence clearly establish the existence of a signaling center in the joint interzone, directly or indirectly, orchestrating limb growth. For instance, loss of function of *cdmp1/gdf5* results in delayed chondrogenesis and shorter limbs. Overexpression of the same polypeptide modulates dose dependently the size of the limbs and epiphysis, both in the chick and mouse model [45, 46, 74].

## BMP signaling in postnatal synovial joints

### BMP signaling in articular cartilage

In the last decade, our understanding of the molecular events leading to joint formation has been rapidly expanding. However, the whole picture is still far from being drawn. The set of molecules known to be involved has not been completed yet. In addition, information of how these molecules interact with each other and orchestrate the processes of skeletal and joint morphogenesis and tissue differentiation is limited.

Even more limited is our knowledge and data about molecular signaling in postnatal joints. There is some evidence that nature may utilize postnatally the same signaling pathways for comparable roles and functions during development. In other words, the molecular events that regulate tissue differentiation and organogenesis during development may also be involved postnatally in tissue homeostasis and repair. For example, BMPs and hedgehog proteins, critically involved in the formation of cartilage and bone during embryogenesis, are also expressed in fracture healing and distraction osteogenesis [75, 76]. It is conceivable that at least some of the molecules herein discussed in the context of joint development have also a role in the maintenance of joint tissues, and in the processes of tissue repair and regeneration.

An example comes from GDF-5/CDMP-1. This molecule, which during development is strongly associated with the initiation of the joint interzone [3], is also present in normal human adult articular cartilage [77]. Its expression, as determined by immunohistochemistry, is mostly restricted to the superficial cartilage in normal joints, while in osteoarthritic cartilage its expression domain is extending to damaged areas [77]. These data suggest a possible role for GDF-5/CDMP-1 in the homeostasis of normal cartilage, as well as in repair processes. Accordingly, recombinant

GDF-5/CDMP-1 increases proteoglycan biosynthetic activity in adult articular cartilage that has been partially matrix-depleted by mild trypsin treatment [77].

The effects of GDF-5/CDMP-1 on articular chondrocytes may not be limited to a stimulation of matrix synthesis. A 30-min incubation of adult swine articular chondrocytes with recombinant GDF-5/CDMP-1 at a final concentration of 100 ng/ml resulted not only in enhanced matrix deposition, but also in an increased cell number when injected as a cell suspension intramuscularly in nude mice. The wet weight of the implant of hyaline-like cartilage recovered after 3 weeks was two- to three-fold higher. In addition, the cartilage tissue stained more intensely with safranin O as compared with the untreated control (Fig. 2). GDF-5/CDMP-1, therefore, may be implicated in the proliferation and metabolic activity of articular chondrocytes.

A recent study demonstrated the presence of BMP-7/OP-1 in normal adult human articular cartilage, as determined by *in situ* hybridization, Western blotting, and immunohistochemistry [78]. BMP-7/OP-1 mRNA was found in the superficial and middle layers of the cartilage, whereas in the deep layer levels of expression were very low. The topographic distribution of the protein within the tissue was quite interesting as revealed by immunostaining performed using two different antibodies, one recognizing the active mature form, and the other reacting with the inactive pro-form. Mature BMP-7/OP-1 was found predominantly in the superficial and middle layers of the tissue, whereas pro-BMP-7/OP-1 was predominantly detected in the deep layer of the cartilage [78]. The distinct localization of pro- and mature forms of BMP-7/OP-1 suggests that the processing of pro-BMP-7/OP-1 into mature BMP-7/OP-1 may occur primarily in the superficial chondrocytes. The detection of BMP receptors type IA and IB, and type II in normal human articular cartilage [79], further corroborates a possible autocrine/paracrine function for BMPs in the maintenance and repair of the articular surface.

Cartilage morphogenesis is critical for both bone and joint morphogenesis. Articular cartilage and growth plate cartilage are biologically distinct. In contrast to the articular chondrocytes, the transient chondrocytes in the growth plate determine the longitudinal and circumferential growth of the cartilage skeletal elements, which are replaced by bone through a process called endochondral ossification. BMP-2/4 and BMP–7/OP-1, and BMP receptors (BMPR-IA, BMPR-IB, and BMPR-II), and their intracellular signaling transducers Smads have been detected immunohistochemically in the epiphyseal plate of growing rats [80, 81]. Their temporal and spatial expression pattern suggests a morphogenic role for BMPs in the multistep cascade of endochondral ossification in the epiphyseal growth plate.

Conversely, articular cartilage is stable throughout life, being resistant to vascular invasion and endochondral ossification. Factors responsible for the maintenance of articular cartilage include TGF-β superfamily signaling molecules. The occurrence of osteoarthritis in adult mice with tissue specific overexpression of a dominant negative TGF-β type II receptor [82] would support this concept.

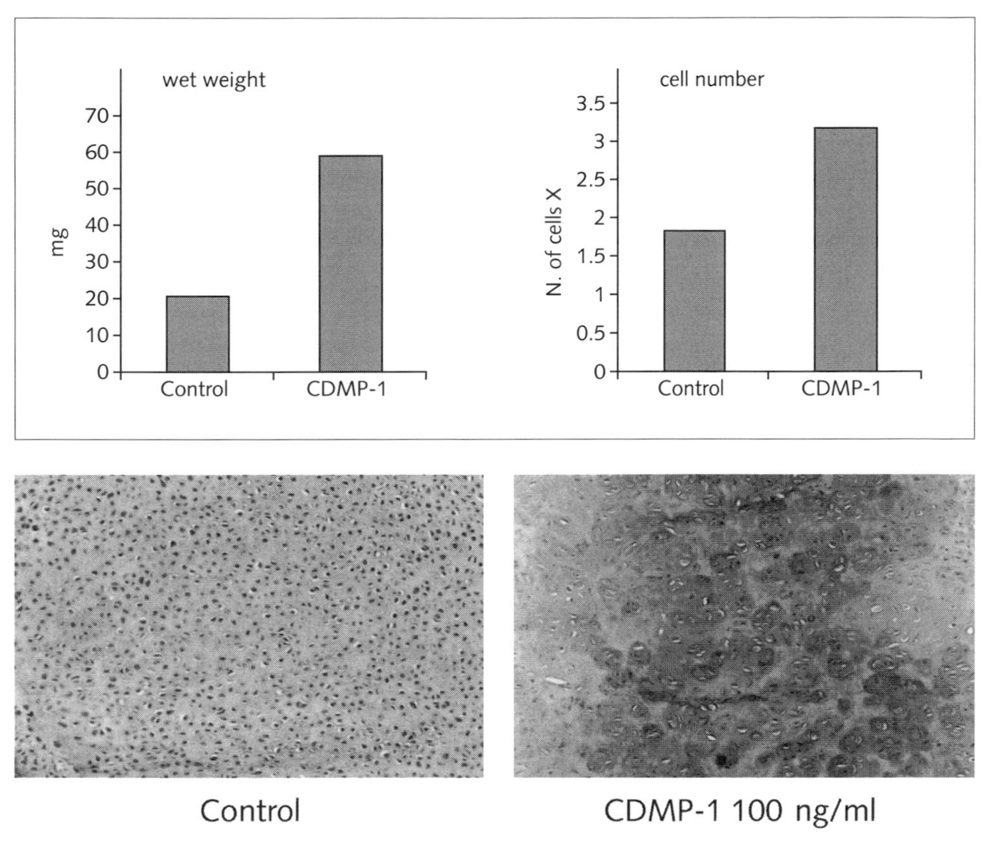


*Figure 2*
*Treatment with GDF-5/CDMP-1 enhances the capacity of articular chondrocytes to organize cartilage tissue* in vivo. *Swine articular chondrocytes from metatarsal joints were treated with 100 ng/ml CDMP1 or with control medium, washed and injected intramuscularly into nude mice. Three weeks later the samples were weighed, and either submitted to histological analysis (safranin O staining) or digested in 0.2% crude collagenase in DMEM for cell count.*

We have determined by semiquantitative RT-PCR the expression of BMPs and related receptors by articular chondrocytes, isolated from normal adult human knee cartilage. BMP-2, -4, and -6, as well as GDF-5/CDMP1 were expressed by freshly isolated cell populations (Fig. 3). We have found a correlation between the BMP expression profile and the phenotype of chondrocytes during *in vitro* expansion. While passaging, chondrocytes are known to undergo a derangement/rearrangement of their phenotypic traits, a phenomenon commonly called de-differentiation [83]. The expression levels of BMP-2 and -6 were downregulated during passaging in parallel with cartilage matrix proteins such as collagen type II (Fig. 3) [84]. These find-

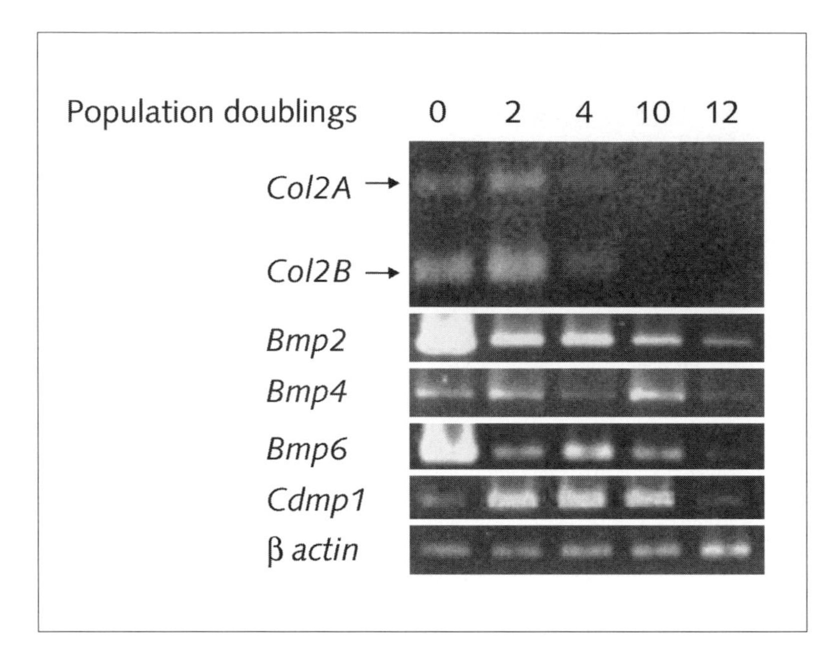


*Figure 3*
*Modulation of the expression of some BMPs during chondrocyte expansion* in vitro.
*Human adult articular chondrocytes lose their phenotypic traits during* in vitro *expansion.* Bmp2 *and* Bmp6 *mRNA levels decrease throughout passaging, paralleling the downregulation of both collagen type IIA and type IIB mRNAs.*

ings underscore the potential role of BMPs in the maintenance of the chondrocyte phenotype. A variety of *in vitro* models have provided evidence that BMPs promote chondrogenesis [85–87], enhance cartilage matrix synthesis [88–90], and support re-expression of the cartilage phenotype [91, 92].

## BMP signaling in postnatal joint associated tissues

The joint is a complex organ that encompasses different tissues, i.e. cartilage, subchondral bone, menisci, and the capsulo-ligamentous apparatus. The synovial membrane lines the inner surface of the joint capsule and covers most intra-articular structures except for the cartilage.

Increasing evidence supports the hypothesis that multipotent stem cells are available postnatally in different organs and tissues. These cells could contribute to postnatal growth and participate in tissue homeostasis by replacing differentiated cells lost to physiological turnover, injury, and senescence. A hypothetical role for BMPs in adult tissues can be the maintenance and recruitment of a pool of progenitor cells for tissue homeostasis and regeneration. Fine balances of BMPs would be required for either the maintenance of this cell population in a quiescent phenotype, or their activation and commitment to a specific lineage.

Mesenchymal stem cells (MSCs) have the potential to differentiate into lineages of mesenchymal tissues, including cartilage, bone, fat, and muscle. Isolation and characterization of MSCs from bone marrow [93] and periosteum [94, 95] have been described. We have identified a population of multipotent MSCs derived from adult human synovial membrane. These cells possess *in vitro* high self-renewal capacity with limited senescence. Under appropriate culture conditions, expanded synovial membrane-derived MSCs can be induced to differentiate *in vitro* toward chondrogenesis, osteogenesis, myogenesis, and adipogenesis [96]. As determined by RT-PCR, synovial membrane-derived MSCs express all BMP receptors and many BMPs (Tab. 1). In recently described studies, no BMP receptors of any type were detected in normal synovial membrane by immunostaining [79]. This apparent discrepancy may have different explanations. The cell isolation technique and the subsequent expansion of this selected cell subpopulation can enrich in BMP receptor expressing cells. In addition, cells in culture are exposed to an environment that is different from the native tissue, and therefore change their molecular phenotype.

Progenitor cells present in the synovial membrane could be responsible for the cartilaginous metaplasia observed in synovial chondromatosis, characterized by the formation of cartilage nodules within the SM [97, 98]. Although TGF-β1 has been detected in synovial chondromatosis [99], its role in the pathogenesis of this process remains unclear.

Multiple injections of TGF-β1 into normal murine knee joints cause the formation of "osteophytes," which have been described as being of periosteal origin [100–102]. The chondro-osteogenic potential of periosteum is known [94, 95, 103]. However, progenitor cells present in the SM might contribute to the process of osteophyte formation observed in TGF-β1 injected joints.

Repair processes require not only tissue regeneration as a re-creation of destroyed cells and extracellular matrix, but also the maintenance of tissue architecture and appropriate relationships between different tissues. TGF-β superfamily members including BMPs/CDMPs are good candidates for the orchestration of these regenerative processes. As morphogens, they would be involved in the coordination of different events such as positional information, patterning, and they could participate in the regulation of the proliferation rate and the progress in the differentiation cascade and maturation process. GDF-5, -6, or -7 appear to be able to induce neotendon/ligament formation when implanted at ectopic sites *in vivo* [104], suggesting that they can influence progenitor cells to differentiate along a tendon/ligament pathway. Implantation of GDF-5 or -6 on collagen sponges has been reported to enhance tendon healing in rats [105]. The elucidation of the functions of mor-

*Table 1 - Expression of BMPs/CDMPs and receptors by human synovial membrane-derived mesenchymal stem cells, as determined by RT-PCR.*

| **Receptors** | | **BMPs/CDMPs** | |
|---|---|---|---|
| ALK1 | + | BMP2 | + |
| ALK2 | + | BMP3 | – |
| ALK3 | + | BMP4 | + |
| ALK4 | + | BMP5 | – |
| ALK5 | + | BMP6 | + |
| ALK6 | + | BMP7/OP1 | – |
| BMPR2 | + | GDF5/CDMP1 | + |
| | | GDF6/CDMP2 | – |
| | | TGF-β1 | + |
| | | TGF-β2 | + |
| | | TGF-β3 | + |

phogens including the BMPs/CDMPs will lead to the identification of additional therapeutic targets and novel tissue engineering protocols to enhance and control repair processes in joint disorders, thereby possibly delaying or limiting major surgery.

## Bone morphogenetic proteins in joint disease

Very limited data have been reported on the potential role of BMPs in joint disease. However, given their well-documented functions in bone and joint development, as well as their potential contribution to joint tissue homeostasis, it seems likely that these molecules also have a role in different diseases affecting the joint. They may influence the disease process itself, or be involved in eventual repair processes taking place as a response to injury.

As for every "organ," different types of disease can affect the synovial joint: degenerative disease, inflammatory and auto-immune disorders, infectious diseases, metabolic diseases as well as benign and malignant tumors.

### TGF-β/BMP signaling in degenerative joint diseases

Osteoarthritis (OA) is a common disorder, occurring mostly in middle and older aged persons, characterized by articular cartilage destruction and subchondral bone remodeling, leading to loss of joint function, and increasing disability. Although several risk factors have been recognized, such as obesity, familial history, skeletal malformations and trauma, the precise pathological events causing the disease and associated with disease progression are not yet clear. The key features appear to be subchondral bone sclerosis, potentially changing the weight-bearing properties and therefore the internal mechanics and dynamics of the joints, together with localized articular cartilage damage. However, the complete picture is far more complex. The whole joint organ is involved. The presence of new bone formation at the joint margins, so-called osteophytes, suggests repair efforts which are either insufficient, or poorly coordinated, since they do not result in repair of the damaged tissue with preserved function. In OA models several stages of the disease have been described each with different characteristics of the cartilage, bone, synovium and their extracellular matrices [106, 107]. The early stage of the degenerative process is characterized by hypertrophy of the articular cartilage with a net increase in matrix synthesis and content. This phase, occurring before macroscopic cartilage damage can be demonstrated, is followed by a phase with net matrix loss by depletion of matrix components, resulting in focal damage and loss of function. In the late phase it is suggested that the release of matrix components and particles from the cartilage lead to synovial activation and inflammation, including the secretion of inflammatory cytokines such as IL-1 and TNFα. The resulting cytokine imbalance further enhances protease and matrix metalloproteinase (MMP) synthesis, stimulation of cyclo-oxygenase and further damage of joint tissues. The complex interactions between these signaling molecules, effector enzymes and different cell populations involved, are likely to be influenced by the presence of growth and differentiation factors such as BMPs, not only in the hypertrophic phase but also in the later stages.

Some evidence regarding the role of TGF-β superfamily signaling in skeletal and joint diseases has been obtained in genetic mouse models. Skeletal tissue-specific overexpression of a truncated, kinase deficient TGF-β type II receptor, acting as a dominant-negative effectively neutralizing TGF-β signaling, results in skeletal malformations. They include progressive skeletal degeneration after birth, leading to kyphoscoliosis, and stiff and torqued joints in heterozygous mice by the age of 4 to 8 months [108]. Strikingly, the histological changes resemble those seen in osteoarthritis. The first signs of joint degeneration are seen in 4 weeks: patches of the articular surface appear denuded and an increase in hypertrophic chondrocytes is seen in the deeper layers of the articular cartilage. In 6-month-old mice, articular cartilages are fibrillated and disorganized. Chondrocytes are organized in clusters, there is an increased number of hypertrophic chondrocytes and a disruption of the tidemark, and bone replaces articular cartilage. Osteophytes can be recognized as outgrowths of chondroid tissue in the articular margins undergoing enchondral bone formation. Proteoglycan synthesis, as shown by Safranin O staining, is decreased in "osteoarthritic" transgenic mice. Type X collagen, normally characteristic of non-proliferating hypertrophic chondrocytes, is expressed in the joints of older transgenic mice, localized to fibrillated articular cartilage, osteophytes and cartilage growing in the joint space as can also be seen in human osteoarthritis [109]. A similar phenotype is apparently found in mice deficient in Smad3, a TGF-β receptor smad [110]. Smad3$^{-/-}$ homozygotes (knock out mice) display skeletal abnormalities, including inwardly turned paws, kyphosis of the spine, osteopetrosis and abnormal ossification of the joints. In 6 months many mutant mice developed an osteoarthritis-like disease, characterized by progressive loss of articular cartilage, surface fibrillation, formation of large osteophytes, upregulation of type X collagen and decreased proteoglycan synthesis. The presence of osteoarthritic changes in a model, in which TGF-β signaling is impaired, suggests that TGF-β is important for the maintenance of tissue integrity, and that the balance between TGF-β and BMP signaling influences joint homeostasis.

Using joint injections, Van Den Berg et al. have extensively studied the *in vivo* effects of TGF-βs and BMPs on cartilage metabolism, and potential interactions with IL-1. BMP2 strongly enhances proteoglycan synthesis after injection in the knee joint of normal mice [111]. The effect, however, is short as compared to the effect of TGF-β1 injection. After TGF-β1 injection, proteoglycan synthesis rises slower and less high but the response is maintained for 20 days. This is probably due to stimulation of endogenous TGF-β or BMP production and/or upregulation of receptors. Remarkably, TGF-β1 counteracts the IL-1 induced suppression of proteoglycan synthesis whereas BMP-2 does not [112, 113]. However, the relative dose of TGF-β used in these experiments (as compared to the amounts used in other settings) is higher than that of BMPs. The effect and the counterbalance of TGF-β and IL-1 are only seen in articular cartilage, but not at the joint margin where osteophytes are formed. TGF-β probably induces cartilage formation from the periost, as has been demonstrated in an *in vitro* model [95] and this process seems not to be influenced by IL-1 in the *in vivo* mouse model. On the other hand, mRNA for BMP-2 as well as BMP-7/OP-1 has been detected in the growing osteophyte ([110, 114] and F.P. Luyten et al., unpublished observations). CDMP-1 and CDMP-2 have been detected in osteoarthritic and normal cartilage, and are able to promote cartilage matrix recovery after enzymatic depletion *in vitro*, with restoration or maintenance of the normal phenotype thus pointing to a potentially important repair mechanism [115]. Recent data by Chubinskaya et al. show the presence of BMP-7/OP1 in human articular cartilage [116]. The expression patterns in normal and OA cartilage are strikingly different. Protein expression analysis by immunohistochemistry and Western blotting shows the presence of mature OP1 in the superficial layer of normal articular cartilage and non-active pro-peptide in the deeper layers. In OA samples where the superficial layer is destroyed, no mature OP1 is detected, the propeptide is, however, present. OP1 expression by RT-PCR is clearly increased in the superficial layer in normal cartilage. However, in OA the deeper layers show an increased OP1 expression. These results suggest an important role for OP1 in tissue maintenance in the superficial layer. However, in OA the chondrocytes of the deeper layer do not seem capable of post-translational modification of the propeptide into the mature bioactive protein, in spite of the upregulation of transcription. Therefore, potential repair mechanisms by OP1 may be impaired.

## BMPs and inflammatory joint disease

Although many systemic inflammatory disorders can also involve the synovial joints, most forms of chronic arthritis can be categorized into two distinct groups: rheumatoid arthritis (RA) and the spondylarthropathies (SpA), the latter consisting of ankylosing spondylitis, psoriatic arthropathy, enteropathic SpA, reactive arthritis (such as Reiter's syndrome) and undifferentiated SpA. It is remarkable that although most of the key inflammatory mediators such as TNF-α and IL-1 have been found within the synovium and the synovial fluid in both disease groups, and at least some of the destructive mechanisms appear to be driven by the same molecular players, the pathological endpoints are strikingly different. RA is mostly characterized by periarticular osteoporosis, extensive cartilage and bone destruction and no appreciable repair efforts. The SpAs mostly have no periarticular osteoporosis, often less destruction and remarkable "repair," not seldom seemingly "overdoing" it, and leading to bony bridging of the joint cavity and ankylosis. Many of these presumed repair processes morphologically closely resemble bone and cartilage formation during development and, therefore, a role of BMPs and BMP signaling can be expected. It is noteworthy that Braun et al. detected by *in situ* hybridization expression of TGF-β2 in biopsies from the sacroiliac joints of patients with ankylosing spondylitis [117]. Investigations in this field are relatively new and largely unexplored so far. Most data on joint pathology have come from samples obtained at joint replacement surgery, and therefore only representing severe and end-stage disease. However, the development of needle arthroscopy as a diagnostic tool in daily rheumatology practice, and the availability of biopsies at distinct stages of the disease, is rapidly increasing our knowledge of the pathology and the molecular players involved.

We have set out to study the potential role of BMPs in inflammatory disorders by studying potential effects of BMPs on the immune system, comparing their function with TGF-βs, members of the same superfamily and well-established immune regulators. The chemotactic potential of some BMPs has already been demonstrated [118]. By RT-PCR we identified the presence of BMP receptor and signaling molecules mRNA in immune cells, including freshly isolated PBMCs, T-cell and monocytic cell lines (R. Lories et al., unpublished). The presence of these receptors and the proposed role of BMPs in hematopoiesis [119] do suggest that BMPs can be partners in immune processes in a way that still has to be elucidated.

## BMPs and infectious arthritis

Bacterial joint infection is probably the most destructive and rapidly progressive pathological process within the organ. Septic arthritis is either caused by a contiguous process or by bacteremia in the subsynovial vessels from a distant focus. Some bacteria preferentially localize within the joint. Bacterial products such as endotoxin, cell fragments, immune complexes, and bacterial opsonisation cause an extensive inflammatory reaction from the innate as well as from the acquired immune system including the production of TNF-$\alpha$ and IL-1, activation of proteolytic enzymes and MMPs, antibody production and generation of effector and memory T-cells. Moreover, phagocytosis by neutrophils causes autolysis thereby releasing lysosymic tissue-destructive enzymes within the joint cavity. Bacterial products are also capable of inducing chondrocyte proteinases which often subsist even after the bacteria have been cleared by the host immune system. Infection also leads to activation of the subsynovial endothelial cells, resulting in thrombosis and ischemia. It should therefore not be surprising that BMPs may be involved in either modulation of the reaction or in a failing attempt to repair the occurring damage. We were able to detect by Western blot BMP-4, CDMP-1 and CDMP-2 in the synovial fluid of patients with septic arthritis (Fig. 4). However, it has not been clear yet which cells and tissues are responsible for the BMP release into the fluid. BMP release can be caused by upregulation of BMP-production as part of a repair effort, but it can also be explained by the release of BMPs previously trapped in the articular cartilage matrix. These preliminary observations provide sufficient impetus to further investigate the potential role of BMPs in infectious joint disease.

## BMPs and skeletal and joint tumors

Joint tumors are rare disorders. BMPs may be important in growth and differentiation of some types, since the embryological and growth cascade are often partially recapitulated. It is obvious that in tumors containing bone and chondroid tissue, these growth factors could be involved.

However, few groups have studied BMP biology in these disorders to date and the available data are often based on scattered observations. Most research in this field has been done by Yoshikawa et al. [119–122]. Osteosarcomas, not necessarily joint-associated, were analyzed for ectopic bone formation, as a way to measure the BMP activity, by implanting the lyophilized fraction of the tumor in a nude mice model. Not only did the BMP-activity containing tumors have some distinct radiological and pathological properties, they also showed a higher resistance to doxorubicin-metothrexate chemotherapy, and a higher tendency to metastasize [119, 120, 122]. Subsequently, BMP-2 or BMP-4 were demonstrated immunohistochemi-

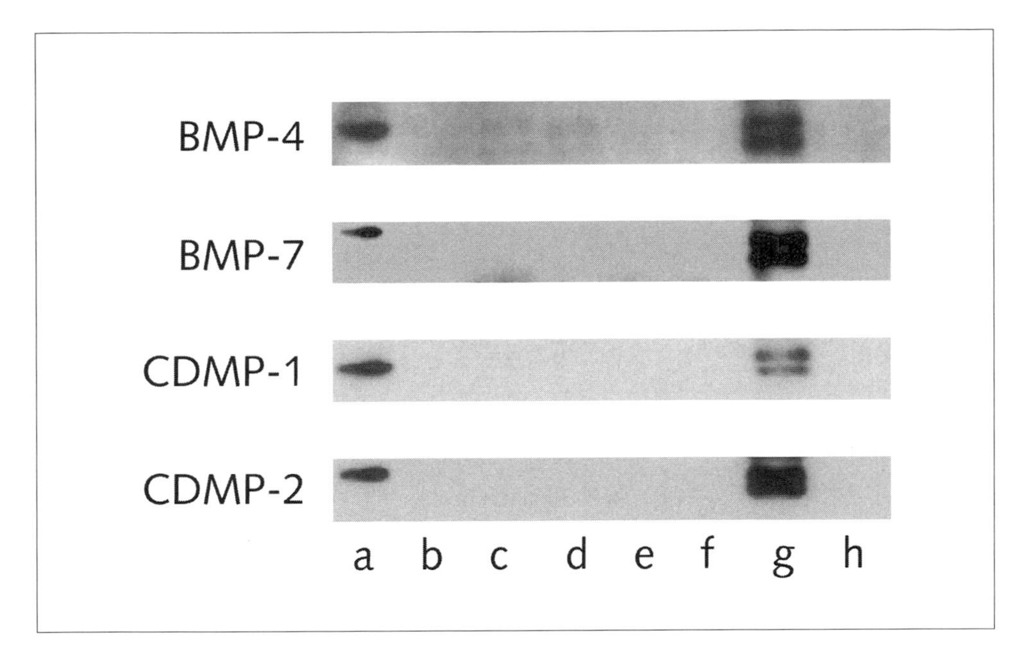


*Figure 4*
*Western blot of BMP-4, BMP-7, CDMP-1 and CDMP-2 in synovial fluid. Growth factors in 1% hyaluronidase treated synovial fluid were concentrated by heparin sepharose binding in 8 M urea, 10 mM Tris, 150 mM NaCl (pH 7.3), washed with 8 M urea 10 mM Tris 3 M NaCl and precipitated with ice cold trichloroacetic acid 30% (w/v); the resulting binding protein pellet was redissolved in 8 M urea 0.05 M Tris and subsequently run on reducing SDS-PAGE gels. Western blots were performed with polyclonal anti-BMP-4, anti-BMP-7, anti-CDMP-1 or anti-CDMP-2 antibodies [114], then incubated with biotinylated secondary antibody and analyzed with peroxidase/luminol staining. Lanes (a) 10 ng of human recombinant protein (b) patient with undifferentiated spondylarthropathy (c) patient with chondrocalcinosis (d) patient with rheumatoid arthritis (e) patient with mono-arthritis of unknown etiology (f) patient with chondrocalcinosis (g) patient with septic arthritis (h) patient with rheumatoid arthritis.*

cally in osteosarcomas, except in nine chondroblastic subtypes, in malignant fibrous histiocytomas (MFH) and in several sarcomas, but not in synovial, rhabdomyo- and fibrosarcoma. However, the sensitivity of the technique can be questioned since no BMP has been detected either in any normal human tissue, or in a 16-week-old human fetus [121].

Guo et al. studied BMP expression in 36 osteosarcomas, six Ewing's sarcoma, 20 synovial sarcomas and 20 chondrosarcomas by RT-PCR [123]. BMP-2 and BMP-4 mRNAs were detected in almost all sarcomas, BMP-6 in 22 osteosarcomas and seven chondrosarcomas. BMP type II receptor was found in 25 osteosarcomas, eight chondrosarcomas, four Ewing's sarcomas and 15 synovial sarcomas. The expression of the type II receptor correlates with metastasis in osteosarcoma and synovial sarcomas. Recently, a new series has been reported [124] in which nine out of 11 osteosarcomas showed expression of BMPs and BMP-receptors by immunohistochemistry. The two negatives were again osteosarcomas of the chondroblastic type; eight out of 10 malignant fibrous histiocytomas also showed BMP staining, but no receptor staining, thus providing a possible explanation for the non-ossification of malignant fibrous histiocytomas.

## References

1 Kingsley DM (1994) What do BMPs do in mammals? Clues from the mouse short-ear mutation. *Trends Genet* 10: 16–21
2 Luyten FP (1997) A scientific basis for the biologic regeneration of synovial joints. Oral *Surg Oral Med Oral Pathol Oral Radiol Endod* 83: 167–169
3 Dell'Accio F, De Bari C, Luyten FP (1999) Molecular basis of joint development. *Jpn J Rheumatol* 9: 17–29
4 Bernays A (1878) Die Entwicklungsgeschichte des Kniegelenkes des Menschen, mit Bemerkungen über die Gelenke im Allgemeinen. *Morphologiesches Jahrbuch* 4: 403–446
5 Andersen H, Bro-Rasmussen F (1961) Histochemical studies on the histogenesis of the joints in human fetuses with special references to the development of the joint cavities in the hand and foot. *Am J Anat* 108: 111–122
6 Andersen H (1962) Histochemical studies on the histogenesis of the human elbow joint. *Acta Anatomica* 51: 50–68
7 Andersen H (1962) Histochemical studies of the development of the human hip joint. *Acta Anatomica* 48: 258–292
8 Andersen H (1963) Histochemistry and development of the human shoulder and acromio- clavicular joint with particular reference to the early development of the clavicle. *Acta Anatomica* 55: 124–165
9 Gardner E, Gray DJ, O'Rahilly R (1959) The prenatal development of the skeleton and joints of the human foot. *J Bone Joint Surg Am* 41A: 847–876
10 Haines RW (1947) The development of joints. *J Anat* 81: 33–55
11 Andersen H (1961) Histochemical studies on the histogenesis of the knee joint and superior tibio-fibular joint in human foetuses. *J Anat* 46: 274–303
12 Merida Velasco JA, Sanchez Montesinos I, Espin Ferra J, Rodriguez Vazquez JF, Merida Velasco JR, Jimenez Collado J (1997) Development of the human knee joint. *Anat Rec* 248: 269–278
13 Morrison EH, Ferguson MW, Bayliss MT, Archer CW (1996) The development of articular cartilage: I. The spatial and temporal patterns of collagen types. *J Anat* 189: 9–22

14 Archer CW, Morrison H, Pitsillides AA (1994) Cellular aspects of the development of diarthrodial joints and articular cartilage. *J Anat* 184: 447–456
15 Gardner E, O'Rahilly R (1962) The development of the elbow joint of the chick and its correlation with the embryonic staging. *J Anat Entwicklungsgesch* 123: 174–179
16 Henrikson RC, Cohen AS (1965) Light and electron microscopic observation of the developing chick interphalangeal joints. *J Ultrastruct Res* 13: 129–162
17 O'Rahilly R, Gardner E (1956) The development of the knee joint of the chick and its correlation with embryonic staging. *J Morphol* 98: 49–88
18 Mitrovic DR (1977) Development of the metatarsophalangeal joint of the chick embryo: morphological, ultrastructural and histochemical studies. *Am J Anat* 150: 333–347
19 Ginsburg GT, Royster D, Kassabian G, Shuler CF, Dougherty WR, Sank AC (1995) Mesenchymal commitment to digital joint formation. *Ann Plast Surg* 35: 95–104
20 Takabatake K, Yamamoto T (1991) Morphology of the synovium during its differentiation and development in the mouse knee joint. A histochemical, SEM and TEM study. *Anat Embryol Berl* 183: 537–544
21 Mitrovic D (1978) Development of the diarthrodial joints in the rat embryo. *Am J Anat* 151: 475–485
22 Chevallier A, Kieny M, Mauger A (1977) Limb-somite relationship: origin of the limb musculature. *J Embryol Exp Morphol* 41: 245–258
23 Christ B, Jacob HJ, Jacob M (1977) Experimental analysis of the origin of the wing musculature in avian embryos. *Anat Embryol Berl* 150: 171–186
24 Kenny Mobbs T (1985) Myogenic differentiation in early chick wing mesenchyme in the absence of the brachial somites. *J Embryol Exp Morphol* 90: 415–436
25 Craig FM, Bentley G, Archer CW (1987) The spatial and temporal pattern of collagens I and II and keratan sulphate in the developing chick metatarsophalangeal joint. *Development* 99: 383–391
26 Thorogood PV, Hinchliffe JR (1975) An analysis of the condensation process during chondrogenesis in the embryonic chick hind limb. *J Embryol Exp Morphol* 33: 581–606
27 Chang SC, Hoang B, Thomas JT, Vukicevic S, Luyten FP, Ryba NJ, Kozak CA, Reddi AH, Moos M Jr (1994) Cartilage-derived morphogenetic proteins. New members of the transforming growth factor-beta superfamily predominantly expressed in long bones during human embryonic development. *J Biol Chem* 269: 28227–28234
28 Storm EE, Kingsley DM (1996) Joint patterning defects caused by single and double mutations in members of the bone morphogenetic protein (BMP) family. *Development* 122: 3969–3979
29 Storm EE, Huynh TV, Copeland NG, Jenkins NA, Kingsley DM, Lee SJ (1994) Limb alterations in brachypodism mice due to mutations in a new member of the TGF beta-superfamily. *Nature* 368: 639–643
30 Luyten FP (1997) Cartilage-derived Morphogenetic Protein-1. *Int J Biochem Cell Biol* 29: 1241–1244
31 Thomas JT, Lin K, Nandedkar M, Camargo M, Cervenka J, Luyten FP (1996) A human

chondrodysplasia due to a mutation in a TGF-beta superfamily member. *Nat Genet* 12: 315–317
32 Grebe H (1952) Die Achondrogenesis: ein einfach rezessives. *Erbmerkmal Folia Hered Path* 2: 23–28
33 Thomas JT, Kilpatrick MW, Lin K, Erlacher L, Lembessis P, Costa T, Tsipouras P, Luyten FP (1997) Disruption of human limb morphogenesis by a dominant negative mutation in CDMP1. *Nat Genet* 17: 58–64
34 Hamburger V, Hamilton HL (1951) A series of normal stages in the development of the chick embryo. *J Morphol* 88: 49–92
35 Macias D, Ganan Y, Sampath TK, Piedra ME, Ros MA, Hurle JM (1997) Role of BMP-2 and OP-1 (BMP-7) in programmed cell death and skeletogenesis during chick limb development. *Development* 124: 1109–1117
36 Rosen V, Thies RS, Lyons K (1996) Signaling pathways in skeletal formation: a role for BMP receptors. *Ann NY Acad Sci* 785: 59–69
37 Duprez D, Bell EJ, Richardson MK, Archer CW, Wolpert L, Brickell PM, Francis West PH (1996) Overexpression of BMP-2 and BMP-4 alters the size and shape of developing skeletal elements in the chick limb. *Mech Dev* 57: 145–157
38 Massague J, Chen YG (2000) Controlling TGF-beta signaling. Genes Dev 14: 627–644
39 Massague J,Wotton D (2000) Transcriptional control by the TGF-beta/Smad signaling system. *EMBO J* 19: 1745–1754
40 Merino R, Rodriguez-Leon J, Macias D, Ganan Y, Economides AN, Hurle JM (1999) The BMP antagonist Gremlin regulates outgrowth, chondrogenesis and programmed cell death in the developing limb. *Development* 126: 5515–5522
41 Hogan BL (1996) Bone morphogenetic proteins: multifunctional regulators of vertebrate development. *Genes Dev* 10: 1580–1594
42 Piccolo S, Sasai Y, Lu B, De Robertis EM (1996) Dorsoventral patterning in *Xenopus*: inhibition of ventral signals by direct binding of chordin to BMP-4. *Cell* 86: 589–598
43 Brunet LJ, McMahon JA, McMahon AP, Harland RM (1998) Noggin, cartilage morphogenesis, and joint formation in the mammalian skeleton. *Science* 280: 1455–1457
44 Francis West PH, Richardson MK, Bell E, Chen P, Luyten F, Adelfattah A, Barlow AJ, Brickell PM, Wolpert L, Archer CW (1996) The effect of overexpression of BMPs and GDF-5 on the development of chick limb skeletal elements. *Ann NY Acad Sci* 785: 254–255
45 Francis-West PH, Abdelfattah A, Chen P, Allen C, Parish J, Ladher R, Allen S, MacPherson S, Luyten FP, Archer CW (1999) Mechanisms of GDF-5 action during skeletal development. *Development* 126: 1305–1315
46 Tsumaki N, Tanaka K, Arikawa-Hirasawa E, Nakase T, Kimura T, Thomas JT, Ochi T, Luyten FP, Yamada Y (1999) Role of CDMP-1 in skeletal morphogenesis: promotion of mesenchymal cell recruitment and chondrocyte differentiation. *J Cell Biol* 144: 161–173
47 Gong Y, Krakow D, Marcelino C, Wilkin D, Chitayat D, Babul-Hirji R, Hudgins L, Cremers CW, Cremers FPM, Brunner HG et al (1999) Heterozygous mutations in the gene encoding noggin affect joint morphogenesis. *Nat Genet* 21: 302–304

48 Polinkovsky A, Robin NH, Thomas JT, Irons M, Lynn A, Goodman FR, Reardon W, Kant SG, Brunner HG, van der Burgt I et al (1997) Mutations in CDMP1 cause autosomal dominant brachydactyly type C. *Nat Genet* 17: 18–19
49 Baur ST, Mai JJ, Dymecki SM (2000) Combinatorial signaling through BMP receptor IB and GDF5: shaping of the distal mouse limb and the genetics of distal limb diversity. *Development* 127: 605–619
50 Yi SE, Daluiski A, Pederson R, Rosen V, Lyons KM (2000) The type I BMP receptor BMPRIB is required for chondrogenesis in the mouse limb. *Development* 127: 621–630
51 Nishitoh H, Ichijo H, Kimura M, Matsumoto T, Makishima F, Yamaguchi A, Yamashita H, Enomoto S, Miyazono K (1996) Identification of type I and type II serine/threonine kinase receptors for growth/differentiation factor-5. *J Biol Chem* 271: 21345–21352
52 Erlacher L, McCartney J, Piek E, Ten Dijke P, Yanagishita M, Oppermann H, Luyten FP (1998) Cartilage-derived morphogenetic proteins and osteogenic protein-1 differentially regulate osteogenesis. *J Bone Miner Res* 13: 383–392
53 Erlebacher A, Filvaroff EH, Gitelman SE, Derynck R (1995) Toward a molecular understanding of skeletal development. *Cell* 80: 371–378
54 Davis AP, Capecchi MR (1994) Axial homeosis and appendicular skeleton defects in mice with a targeted disruption of hoxd-11. *Development* 120: 2187–2198
55 Mortlock DP, Post LC, Innis JW (1996) The molecular basis of hypodactily (Hd): a deletion in Hoxa 13 leads to arrest of digital arch formation. *Nat Genet* 13: 284–289
56 Favier B, Rijli FM, Fromental-Romain C, Fraulob V, Chambon P, Pascal D (1996) Functional cooperation between the non-paralogous genes Hoxa-10 and Hoxd-11 in the developing forelimb and axial skeleton. *Development* 122: 449–460
57 Rountree R, Schoor M, Kingsley D (2000) Using Gdf5 control sequences to test the role of genes in joint development. *International Conference Bone Morphogenetic Proteins 2000* June 7–11, 2000, Granlibakken, Lake Tahoe, California
58 Sugiura T, Hötten G, Kawai S (1999) Minimal promoter components of the human growth/differentiation factor-5 gene. *Biochem Biophys Res Commun* 263: 707–713
59 Persson M (1983) The role of movements in the development of sutural and diarthrodial joints tested by long-term paralysis of chick embryos. *J Anat* 137: 591–599
60 Mitrovic D (1971) [Effect of pharmacological paralysis on the formation and evolution of articular fissures of the digital joints in chick embryo fleet] Effet de la paralysie pharmacologique sur la formation et l'évolution des fentes articulaires des articulations digitales des pattes chez l'embryon de poulet. *CR Acad Sci Hebd Seances Acad Sci D* 273: 1748–1751
61 Mitrovic D (1982) Development of the articular cavity in paralysed chick embryos and in chick embryo limb buds cultured in chorioallantoic membranes. *Acta Anatomica* 113: 313–324
62 Drachman DB, Sokoloff L (1966) The role of movement in embryonic joint development. *Dev Biol* 14: 401–420

63 Lelkes G (1958) Experiments *in vitro* on the role of movement in the development of joints. *J Embryol Exp Morphol* 6: 183–186
64 Thorogood P (1983) Morphogenesis of cartilage. In: BK Hall (ed): *Cartilage*. Academic Press, New York, 223–254
65 Mitrovic D (1971) [Physiological necrosis in the articular mesenchyma of rat and chick embryos] La nécrose physiologique dans le mesenchyme articulaire des embryons de rat et de poulet. *CR Acad Sci Hebd Seances Acad Sci D* 273: 642–645
66 Mitrovic D (1972) [Presence of degenerated cells in the developing articular cavity of the chick embryo] Présence de cellules dégénerées dans la cavité articulaire en développement chez l'embryon de poulet. *CR Acad Sci Hebd Seances Acad Sci D* 275: 2941–2944
67 Mori C, Nakamura N, Kimura S, Irie H, Takigawa T, Shiota K (1995) Programmed cell death in the interdigital tissue of the fetal mouse limb is apoptosis with DNA fragmentation. *Anat Rec* 242: 103–110
68 Craig FM, Bayliss MT, Bentley G, Archer CW (1990) A role for hyaluronan in joint development. *J Anat* 171: 17–23
69 Hasty P, Bradley A, Morris JH, Edmonson DG, Venuti JM, Olson EN (1993) Muscle deficiency and neonatal death in mice with a targeted mutation in the myogenin gene. *Nature* 364: 501–506
70 Pitsillides AA, Archer CW, Prehm P, Bayliss MT, Edwards JC (1995) Alterations in hyaluronan synthesis during developing joint cavitation. *J Histochem Cytochem* 43: 263–273
71 Nalin AM, Greenlee TK Jr, Sandell LJ (1995) Collagen gene expression during development of avian synovial joints: transient expression of types II and XI collagen genes in the joint capsule. *Dev Dyn* 203: 352–362
72 Ganan Y, Macias D, Duterque Coquillaud M, Ros MA, Hurle JM (1996) Role of TGF beta s and BMPs as signals controlling the position of the digits and the areas of interdigital cell death in the developing chick limb autopod. *Development* 122: 2349–2357
73 Zou H, Niswander L (1996) Requirement for BMP signaling in interdigital apoptosis and scale formation. *Science* 272: 738–741
74 Merino R, Macias D, Ganan Y, Economides AN, Wang X, Wu Q, Stahl N, Sampath KT, Varona P, Hurle JM (1999) Expression and function of Gdf-5 during digit skeletogenesis in the embryonic chick leg bud. *Dev Biol* 206: 33–45
75 Vortkamp A, Pathi S, Peretti GM, Caruso EM, Zaleske DJ, Tabin CJ (1998) Recapitulation of signals regulating embryonic bone formation during postnatal growth and in fracture repair. *Mech Dev* 71: 65–76
76 Liu Z, Luyten FP, Lammens J, Dequeker J (1999) Molecular signaling in bone fracture healing and distraction osteogenesis. *Histol Histopathol* 14: 587–595
77 Erlacher L, Ng CK, Ullrich R, Krieger S, Luyten FP (1998) Presence of cartilage-derived morphogenetic proteins in articular cartilage and enhancement of matrix replacement *in vitro*. *Arthritis Rheum* 41: 263–273
78 Chubinskaya S, Merrihew C, Cs-Szabo G, Mollenhauer J, McCartney J, Rueger DC,

Kuettner KE (2000) Human articular chondrocytes express osteogenic protein-1. *J Histochem Cytochem* 48: 239–250
79 Marinova-Mutafchieva L, Taylor P, Funa K, Maini RN, Zvaifler NJ (2000) Mesenchymal cells expressing bone morphogenetic protein receptors are present in the rheumatoid arthritis joint. *Arthritis Rheum* 43: 2046–2055
80 Yazaki Y, Matsunaga S, Onishi T, Nagamine T, Origuchi N, Yamamoto T, Ishidou Y, Imamura T, Sakou T (1998) Immunohistochemical localization of bone morphogenetic proteins and the receptors in epiphyseal growth plate. *Anticancer Res* 18: 2339–2344
81 Sakou T, Onishi T, Yamamoto T, Nagamine T, Sampath T, Ten Dijke P (1999) Localization of Smads, the TGF-beta family intracellular signaling components during endochondral ossification. *J Bone Miner Res* 14: 1145–1152
82 Serra R, Johnson M, Filvaroff EH, LaBorde J, Sheehan DM, Derynck R, Moses HL (1997) Expression of a truncated, kinase-defective TGF-beta type II receptor in mouse skeletal tissue promotes terminal chondrocyte differentiation and osteoarthritis. *J Cell Biol* 541 52 139: 451–452
83 Benya PD, Shaffer JD (1982) Dedifferentiated chondrocytes reexpress the differentiated collagen phenotype when cultured in agarose gels. *Cell* 30: 215–224
84 Dell'Accio F, De Bari C, Luyten FP (2001) Molecular markers predictive of the capacity of expanded human articular chondrocytes to form stable cartilage *in vivo. Arthritis Rheum* 44: 1608–1619
85 Vukicevic S, Luyten FP, Reddi AH (1989) Stimulation of the expression of osteogenic and chondrogenic phenotypes *in vitro* by osteogenin. *Proc Natl Acad Sci USA* 86: 8793–8797
86 Carrington JL, Chen P, Yanagishita M, Reddi AH (1991) Osteogenin (bone morphogenetic protein-3) stimulates cartilage formation by chick limb bud cells *in vitro. Dev Biol* 146: 406–415
87 Chen P, Carrington JL, Hammonds RG, Reddi AH (1991) Stimulation of chondrogenesis limb bud mesoderm cells by recombinant human bone morphogenetic protein 2B (BMP-2B) and modulation by transforming growth factor beta 1 and beta 2. *Exp Cell Res* 195: 509–515
88 Flechtenmacher J, Huch K, Thonar EJ, Mollenhauer JA, Davies SR, Schmid TM, Puhl W, Sampath TK, Aydelotte MB, Kuettner KE (1996) Recombinant human osteogenic protein 1 is a potent stimulator of the synthesis of cartilage proteoglycans and collagens by human articular chondrocytes. *Arthritis Rheum* 39: 1896–1904
89 Luyten FP, Yu YM, Yanagishita M, Vukicevic S, Hammonds RG, Reddi AH (1992) Natural bovine osteogenin and recombinant human bone morphogenetic protein-2B are equipotent in the maintenance of proteoglycans in bovine articular cartilage explant cultures. *J Biol Chem* 267: 3691–3695
90 Luyten FP, Chen P, Paralkar V, Reddi AH (1994) Recombinant bone morphogenetic protein-4, transforming growth factor-beta 1, and activin A enhance the cartilage phenotype of articular chondrocytes *in vitro. Exp Cell Res* 210: 224–229
91 Harrison ET Jr, Luyten FP, Reddi AH (1991) Osteogenin promotes reexpression of car-

tilage phenotype by dedifferentiated articular chondrocytes in serum-free medium. *Exp Cell Res* 192: 340–345

92 Harrison ET Jr, Luyten FP, Reddi AH (1992) Transforming growth factor-beta: its effect on phenotype reexpression by dedifferentiated chondrocytes in the presence and absence of osteogenin. *In Vitro Cell Dev Biol* 28A: 445–448

93 Pittenger MF, Mackay AM, Beck SC, Jaiswal RK, Douglas R, Mosca JD, Moorman MA, Simonetti DW, Craig S, Marshak DR (1999) Multilineage potential of adult human mesenchymal stem cells. *Science* 284: 143–147

94 Nakahara H, Goldberg VM, Caplan AI (1991) Culture-expanded human periosteal-derived cells exhibit osteochondral potential *in vivo*. *J Orthop Res* 9: 465–476

95 De Bari C, Dell'Accio F, Luyten FP (2000) Human periosteum-derived cells maintain phenotypic stability and chondrogenic potential throughout expansion regardless of donor age. *Arthritis Rheum* 44 (1): 85–95

96 De Bari C, Dell'Accio F, Tylzanowski P, Luyten FP (2001) Multipotent mesenchymal stem cells from adult human synovial membrane. *Arthritis Rheum* 44: 1928–1942

97 Maurice H, Crone M, Watt I (1988) Synovial chondromatosis. *J Bone Joint Surg [Br]* 70: 807–811

98 Rosen PS, Pritzker PH, Greenbaum J, Holgate RC, Noyek AM (1977) Synovial chondromatosis affecting the temporomandibular joint. Case report and literature review. *Arthritis Rheum* 20: 736–740

99 Fujita S, Iizuka T, Yoshida H, Segami N (1997) Transforming growth factor and tenascin in synovial chondromatosis of the temporomandibular joint. Report of a case. *Int J Oral Maxillofac Surg* 26: 258–259

100 van Beuningen HM, van der Kraan PM, Arntz OJ, van den Berg WB (1994) Transforming growth factor-beta 1 stimulates articular chondrocyte proteoglycan synthesis and induces osteophyte formation in the murine knee joint. Lab Invest 71: 279–290

101 van Beuningen HM, Glansbeek HL, van der Kraan PM, van den Berg WB (1998) Differential effects of local application of BMP-2 or TGF-beta 1 on both articular cartilage composition and osteophyte formation. *Osteoarthritis Cartilage* 6: 306–317

102 van Beuningen HM, Glansbeek HL, van der Kraan PM, van den Berg WB (2000) Osteoarthritis-like changes in the murine knee joint resulting from intra-articular transforming growth factor-beta injections. *Osteoarthritis Cartilage* 8: 25–33

103 O'Driscoll SW, Recklies AD, Poole AR (1994) Chondrogenesis in periosteal explants. An organ culture model for *in vitro* study. *J Bone Joint Surg Am* 76: 1042–1051

104 Wolfman NM, Hattersley G, Cox K, Celeste AJ, Nelson R, Yamaji N, Dube JL, DiBlasio-Smith E, Nove J, Song JJ et al (1997) Ectopic induction of tendon and ligament in rats by growth and differentiation factors 5, 6, and 7, members of the TGF-beta gene family. *J Clin Invest* 100: 321–330

105 Aspenberg P, Forslund C (1999) Enhanced tendon healing with GDF 5 and 6. *Acta Orthop Scand* 70: 51–54

106 Adams ME, Brandt KD (1991) Hypertrophic repair of canine articular cartilage in osteoarthritis after anterior cruciate ligament transection. *J Rheumatol* 18: 428–435

107 van der Kraan PM, Vitters EL, van Beuningen HM, van den Berg WB (1992) Proteoglycan synthesis and osteophyte formation in 'metabolically' and 'mechanically' induced murine degenerative joint disease: an *in-vivo* autoradiographic study. *Int J Exp Pathol* 73: 335–350

108 Serra R, Johnson M, Filvaroff EH, LaBorde J, Sheehan DM, Derynck R, Moses HL (1997) Expression of a truncated, kinase-defective TGF-beta type II receptor in mouse skeletal tissue promotes terminal chondrocyte differentiation and osteoarthritis. *J Cell Biol* 139: 541–552

109 von der MK, Kirsch T, Nerlich A, Kuss A, Weseloh G, Gluckert K, Stoss H (1992) Type X collagen synthesis in human osteoarthritic cartilage. Indication of chondrocyte hypertrophy. *Arthritis Rheum* 35: 806–811

110 Deng C-X, Weinstein M, Yang X (2000) Functions of mammalian Smad genes revealed by gene targeting in mice (abstract). International Conference Bone Morphogenetic Proteins

111 van Beuningen HM, Glansbeek HL, van der Kraan PM, van den Berg WB (1998) Differential effects of local application of BMP-2 or TGF-beta 1 on both articular cartilage composition and osteophyte formation. *Osteoarthritis Cartilage* 6: 306–317

112 Glansbeek HL, van Beuningen HM, Vitters EL, Morris EA, van der Kraan PM, van den Berg WB (1997) Bone morphogenetic protein 2 stimulates articular cartilage proteoglycan synthesis *in vivo* but does not counteract interleukin-1alpha effects on proteoglycan synthesis and content. *Arthritis Rheum* 40: 1020–1028

113 van Beuningen HM, van der Kraan PM, Arntz OJ, van den Berg WB (1994) *in vivo* protection against interleukin-1-induced articular cartilage damage by transforming growth factor-beta 1: age-related differences. *Ann Rheum Dis* 53: 593–600

114 Tomita T, Nakase T, Kaneko M, Tsuboi H, Takahi K, Hashimoto J, Takano H, Myoui A, Shi K, Yoshikawa H, Ochi T (2000) Distributions of BMP-2 and BMP receptors in the osteophyte of patients with osteoarthritis. *Arthritis Rheum* 43: S350

115 Erlacher L, Ng CK, Ullrich R, Krieger S, Luyten FP (1998) Presence of cartilage-derived morphogenetic proteins in articular cartilage and enhancement of matrix replacement *in vitro*. *Arthritis Rheum* 41: 263–273

116 Chubinskaya S, Merrihew C, Cs-Szabo G, Mollenhauer J, McCartney J, Rueger DC, Kuettner KE (2000) Human articular chondrocytes express osteogenic protein-1. *J Histochem Cytochem* 48: 239–250

117 Braun J, Bollow M, Neure L, Seipelt E, Seyrekbasan F, Herbst H, Eggens U, Distler A, Sieper J (1995) Use of immunohistologic and *in situ* hybridization techniques in the examination of sacroiliac joint biopsy specimens from patients with ankylosing spondylitis. *Arthritis Rheum* 38: 499–505

118 Cunningham NS, Paralkar V, Reddi AH (1992) Osteogenin and recombinant bone morphogenetic protein 2B are chemotactic for human monocytes and stimulate transforming growth factor beta 1 mRNA expression. *Proc Natl Acad Sci USA* 89: 11740–11744

119 Bhatia M, Bonnet D, Wu D, Murdoch B, Wrana J, Gallacher L, Dick JE (1999) Bone

morphogenetic proteins regulate the developmental program of human hematopoietic stem cells. *J Exp Med* 189: 1139–1148

120 Yoshikawa H, Takaoka K, Hamada H, Ono K (1985) Clinical significance of bone morphogenetic activity in osteosarcoma. A study of 20 cases. *Cancer* 56: 1682–1687

121 Yoshikawa H, Rettig WJ, Takaoka K, Alderman E, Rup B, Rosen V, Wozney JM, Lane JM, Huvos AG, Garin-Chesa P (1994) Expression of bone morphogenetic proteins in human osteosarcoma. Immunohistochemical detection with monoclonal antibody. *Cancer* 73: 85–91

122 Yoshikawa H, Takaoka K, Masuhara K, Ono K, Sakamoto Y (1988) Prognostic significance of bone morphogenetic activity in osteosarcoma tissue. *Cancer* 61: 569–573

123 Guo W, Gorlick R, Ladanyi M, Meyers PA, Huvos AG, Bertino JR, Healey JH (1999) Expression of bone morphogenetic proteins and receptors in sarcomas. *Clin Orthop* 175–183

124 Mehdi R, Shimizu T, Yoshimura Y, Gomyo H, Takaoka K (2000) Expression of bone morphogenetic protein and its receptors in osteosarcoma and malignant fibrous histiocytoma. *Jpn J Clin Oncol* 30: 272–275

# BMPs in articular cartilage repair

*Mislav Jelic[1,2], Marko Pecina[1], Miroslav Haspl[1], Anton Brkic[3] and Slobodan Vukicevic[2]*

[1]Departments of Orthopaedic Surgery and [2]Anatomy, School of Medicine, [3]Clinics of Surgery, Orthopaedic Surgery and Ophthalmology, Veterinary Faculty, University of Zagreb, Salata 11, 10000 Zagreb, Croatia

## Introduction

Over the past several decades, in clinical orthopedic work, from open Magnusson "housecleaning" arthroplasty to the autologous chondrocyte implantation, much has been learned about articular cartilage and its physiological capacity to restore itself. To date, no technique has been completely successful in restoring normal regenerative articular cartilage. Techniques to treat chondral defects include abrasion, drilling, microfracture technique, tissue autografts, allografts, and cell transplantation [1–12]. Bone marrow stimulation techniques such as abrasion, drilling, and microfractures produce only fibrocartilage and therefore do not offer a long-term cure. Subchondral bone plate microfracture (abrasion or drilling) has shown to enhance chondral resurfacing by providing a suitable environment for tissue regeneration and taking advantage of the body's own healing potential. The formation of a fibrin clot ("super clot") containing desired pluripotential stem cells is stimulated [10]. This clot then differentiates and remodels, resulting in a durable fibrocartilage repair tissue [1]. Perichondral and periostal interposition grafts produce repair tissue that is similar to hyaline cartilage but also lack the mechanical durability. Like bone marrow stimulation techniques, interposition grafts introduce precursor cells, which have a tendency to differentiate along lines other than cartilage [7]. Autologous osteochondral transplant systems have shown encouraging results, but graft matching and contouring to the recipient articular surface proved to be difficult. Moreover, the donor sites can be a limiting factor, and the fibrocartilaginous interface between the donor and recipient site may contribute to breakdown in the long run. Autologous chondrocyte implantation is a biological repair process with encouraging results. The procedure is expensive and so far it has not been demonstrated that autologous chondrocyte implantation can prevent degenerative cartilage changes [7]. In recent years, much has been learned about the various

Bone Morphogenetic Proteins, edited by Slobodan Vukicevic and Kuber T. Sampath

growth factors that stimulate chondrocyte differentiation and extracellular matrix production, but to date, a clinical technique has not been developed.

## Articular cartilage regeneration

Joint surface repair is still a major challenge in modern medicine, because the factors initiating cartilage formation, maturation, and repair are poorly understood. Specific biological challenges include the variable quality and quantity of the cartilage that is produced, decreasing responsiveness with age, bonding to the adjacent cartilage, and restoration of the subchondral bone [13]. Injury to cartilage initiates a specific reparative response. In lesions of the articular cartilage with no collagen damage, a loss of non-collagenous matrix occurs, leading eventually to complete repair of the damaged matrix [14]. In more severe cases, where there is a damage of the fibrillar network and cell death, the articular cartilage does not heal [15, 16].

Cartilage is a specialized connective tissue with a biomechanical function meant to bear compressive load. Over time, cartilage has been classified as hyaline, elastic and fibrous, based on histological and morphological appearance and developmental history. Articular cartilage is built only of hyaline cartilage and it does not contain nerves or blood vessels. It is made of extracellular matrix that is laid down and maintained by chondrocytes. A chondrocyte is a cell embedded in a dense cartilage matrix synthesized by chondrocytes themselves. Their differentiation is regulated by a number of humoral hormones and factors, and by locally produced cytokines.

Structurally, different layers formed by cells and matrix build mature cartilage. The superficially positioned tangential layer is made of horizontally directed chondrocytes. Upper radial and lower radial layers are made of hypertrophic chondrocytes which form columns and, in the bottom, a narrow calcified cartilage zone is interposited between the hyaline cartilage tissue and subchondral bone plate. This zone has a special meaning in the distinction of osteochondral (full thickness) and chondral (partial) defects in animal models of cartilage regeneration studies.

Two constituents, proteoglycans and collagens are responsible for cartilage behavior and metabolism. Collagens are the major component of cartilage extracellular matrix. They are specific products of phenotypic expression by differentiated cells. The collagen gene family consists of at least 30 genes making up a minimum of 18 different collagen types. Four of these collagen types, collagen II, IX, X and XI have been considered specific for cartilage. The collagen, principally type II, but also type IX and XI, forms a dense fibrillar network that is embedded in a high concentration of proteoglycans which creates a large osmotic pressure that draws water into the tissue and expands the collagen network. The most abundant proteoglycan in cartilage is aggrecan. Compressive properties of cartilage result from the balance between the osmotic swelling pressure of the proteoglycans and the tension in the collagen fibers [17].

## Bone morphogenetic proteins stimulate articular chondrocyte metabolism

So far there has not been shown any evidence that there is more than a little, if any, cell division in healthy adult articular cartilage. However, chondrocytes cultured in medium proliferate in response to serum growth factors. The time needed for the doubling of chondrocytes depends on the articular cartilage layer the cells were cultured from and the cell density. Chondrocyte proliferation is more rapid in low density than in high density cultures. Chondrocytes cultured from the deeper layers of tissue double more rapidly than those from the middle and superficial cartilage zones [18]. Subpopulations of human articular chondrocytes maintained in medium containing human adult serum, which has lower concentrations of growth factors than fetal serum, show little change in cell number during the culture period, and no difference in proliferation between cells from the superficial and deep zones [19].

*In vitro* studies performed through years by investigators in the field have identified bone morphogenetic proteins as modulators of articular cartilage chondrocyte metabolism, which is also seen through the fact that structural macromolecules of extracellular matrix bind BMPs. It is well known that chondrocytes in tissue culture progressively lose their phenotype in monolayer cultures. Dedifferentiation of chondrocytes is minimized in explant cultures of articular cartilage in which chondrocytes are encased in their own extracellular matrix [20].

In short-term cartilage explant cultures, BMP-4 stimulates dose-dependently both the proteoglycan synthesis [21] and the decrease in proteoglycancatabolism. BMP-4 also increases the levels of expression of type II collagen and proteoglycan aggrecan in short term cultures. This enhancement of cartilage phenotype by BMP-4 is largely independent of culture conditions. Moreover, BMP-4, besides promoting the chondrocyte phenotype, has also a weak mitogenic effect in monolayer and micromass cultures [22]. In studies on long-term monolayer articular chondrocyte cell cultures up to 28 days, BMP-2 was also found to stimulate proteoglycan synthesis [23], while not affecting cell proliferation and expression of type X collagen and osteocalcin synthesis. It also enhanced the expression of type II collagen and increased the expression of aggrecan [23].

When bovine articular chondrocytes are grown up to 5 weeks in the presence of 0.5% or 10% serum in combination with another BMP, BMP-7, they do not undergo hypertrophy, as determined by cell size, the absence of both type X collagen expression and synthesis, and of alkaline phosphatase activity. The presence of BMP-7 resulted in increased matrix synthesis. This data suggest that primary mammalian articular chondrocytes will not undergo hypertrophy in conditions previously shown to be permissive for hypertrophy of both chick sternal and chick articular chondrocytes. BMP-7 is crucial for maintanence of articular chondrocytes phenotype by preserving collagen II synthesis [24].

When extending these studies to chick sternal chondrocytes growth and maturation in high-density monolayers, suspension and agarose cultures up to 5 weeks,

BMP-7 dose dependently promoted chondrocyte maturation associated with enhanced alkaline phosphatase activity and increased mRNA levels and protein synthesis of type X collagen in both the presence and absence of serum [25]. The pivotal role of BMPs in the development and regeneration process of the skeleton suggests their role in articular cartilage defect repair.

In creating chondral defects, an investigation must not damage the calcified cartilage zone and the underlying subchondral bone. The borderline between hyaline articular cartilage and the zone of calcified cartilage is called the "tidemark" and represents the mineralization front [26].

Studying the healing phenomena of articular cartilage lesions led to a conclusion that it is essential to expand the existing cell population in order to increase the total pool of healthy cells contributing to the matrix repair. This might be obtained through increased cell proliferation and/or chemotaxis of cells from neighboring tissues such as the underlying bone and/or synovium [27]. Growth and differentiation factors can be used in this regard [28] with bone morphogenetic proteins (BMPs) being good candidates [29, 30]. Apart from BMPs, good candidates would also be recently discovered cartilage-derived morphogenetic proteins (CDMPs), novel TGF-beta superfamily members, with their cartilage-specific localization pattern that suggests their potential role in chondrocyte differentiation ([31, 32]; see the chapter by Luyten et al.).

## Cartilage regeneration in models using osteochondral defects

Regeneration of full-thickness cartilage defects which involves both cartilage and subchondral bone and bone marrow was studied by drilling holes in the articular cartilage of animal knee joints [27]. These defects undergo repair and a new layer of bone and cartilage is formed, but the macromolecular organization and the biochemical characteristics of the matrix are imperfect. The persistence of high levels of type I collagen and the substitution of the cartilage specific proteoglycans by other types, such as dermatan sulphate containing proteoglycans illustrate such imperfect healing [16, 33]. This culminates in a repair tissue with fibrillations and extensive degenerative changes after about 3 months, and finally a complete loss of tissue integrity occurs [34, 35]. Most investigations on articular cartilage healing *in vivo* have been performed on animal models using osteochondral, or full-thickness cartilage defects. Different BMPs have been tested in osteochondral defect models.

It has been demonstrated that recombinant human BMP-2 (rhBMP-2) with a collagen carrier significantly improves new tissue formation in osteochondral defects in NZW rabbits 6 months and 1 year following surgical procedure [36–38]. BMP-2 treated defects had a significantly better histological appearance than the untreated defect (those left empty or filled with a collagen sponge). The histological features that showed improvement were integration at the margin, cellular morphology, architecture within the defect and reformation of the tidemark. The total score was also better for the defects treated with rhBMP-2 than for the untreated defects [36, 37]. However, even though integration of new and old cartilage in treated animals was better in comparison to controls, it is still considered the weakest point of that study.

In another model, BMP-3 (osteogenin) combined with a porous HA in dog cartilage, full thickness defects significantly enhanced transformation of ingrowing fibrous tissue into the hyaline cartilage [39]. However, the integration at the margin of newly formed and old tissue was again incomplete.

Another BMP, BMP-7 can improve regeneration of full-thickness cartilage defects in rabbits 3 months following implantation. Histological examination of 20 osteochondral rabbit knee defects showed significant difference in healing of the defects treated with BMP-7 compared to those left empty or treated with a collagen gel only. Defects that were not treated with BMP-7 were filled with several tissue types 8 weeks following the procedure (data not shown). However, osteochondral defects treated with BMP-7 were completely bridged with abundant tissue resembling immature cartilage (Figs. 1A and B). New tissue consists of small rounded cells organized in columns (Fig. 1C) and embedded in compact extracellular matrix. Rebridgement was complete in superficial layers which protruded above the surface of intact chondrocytes (Figs. 1A and B). In some defects, deeper areas were still unfused with surrounding cartilage [40]. These results suggest the potential role of BMP-7 as an articular cartilage repair inducer, but 8 weeks is too early for conclusions on tissue integration and the architecture of newly formed cartilage.

BMP-7 was also evaluated in another study with NZW rabbits where osteochondral defects were made in the femoral patellar grove. Grossly, after 12 weeks it has been shown that BMP-7 treated defects showed repair that was continuous with the adjacent intact cartilage and was translucent. Maturing cartilage was present and it looked similar and was similarly thick when compared to the intact surrounding articular cartilage. In comparison, the repair tissue at control sites, that were treated either with no implant or matrix only, was filled primarily with fibrous tissue or fibrocartilage. That newly formed tissue was discontinuous with the surrounding cartilage and was opaque and inhomogenous. Histologically, moderate degeneration of the cartilage at the defect interfaces was noted, large clusters of chondrocytes were observed at the interface, and fissures were seen separating the intact cartilage from the repair tissue ([41]; see the chapter by Cook et al.). The integration of newly formed cartilage with old, intact cartilage was reported to be satisfactory. However, the observation time period of 12 weeks postoperatively was insufficient to evaluate the quality of integration and duration of the newly formed cartilage [41].

When osteochondral defects in goat knee joints were treated with rhBMP-7 implanted on a collagen carrier and studied 4 months after treatment partial or

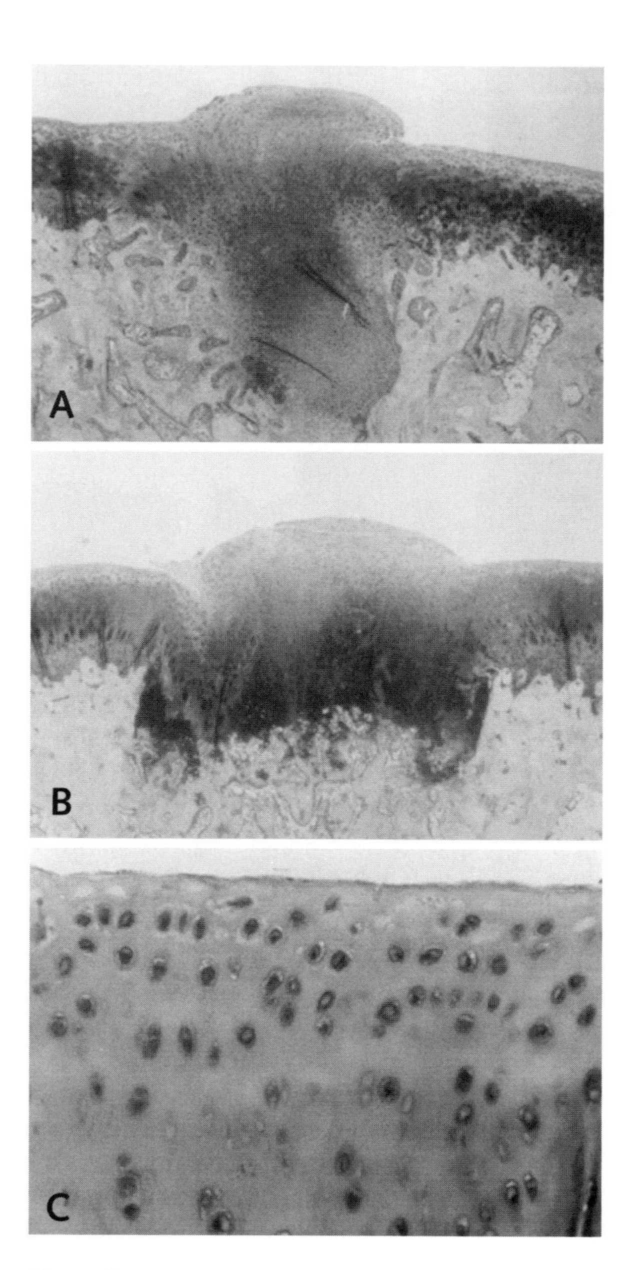


*Figure 1*
*Healing of osteochondral defects treated with BMP-7 in NZW rabbit knees. 8 weeks following surgery the defects are completely filled with tissue resembling immature cartilage, which is protruding above the surface of intact cartilage (A and B). Bonding of old and new cartilage is observed (A–C). On higher magnification small rounded cartilage cells have columnar orientation and are embedded in compact extracellular matrix (C).*

complete healing was observed in treated goats while only one of three untreated animals showed some cartilage formation [42].

Studies on articular cartilage healing using periosteum transplants in rabbits show that the periosteum, when transplanted into osteochondral defects, induces new cartilage-like tissue formation which contains 90% collagen II and is replaced by bone in the subchondral regions [43]. It is hypothesized that periosteum has an articular cartilage healing potential because of factors including orientation of the cambium layer and postoperative factors such as application of continuous passive motion and the maturity of the experimental animals [44, 45]. Even though the underlying molecular mechanism leading to periosteal articular cartilage healing in osteochondral defects is not understood, it has been shown by different investigators that periosteum contains chondrocyte precursor cells that form cartilage during limb development expressing various BMPs during fracture healing [44, 46].

## Cartilage regeneration in models using chondral defects

Regeneration of articular cartilage chondral defects was studied in sheep through damaging a complete chondral layer with a specially designed instrument (Fig. 2A), without damaging the subchondral bone, using a continuous application of BMP-7 that was delivered *via* an extraarticulary positioned mini-osmotic pump (Fig. 2B) [47]. Two 10 mm chondral defects were created in each knee; one on the medial condyle and the other on the trochlea of the femur, and randomly treated by either BMP-7 or by acetate buffer *via* an extraarticularly positioned mini-osmotic pump connected to a joint by a polyethylene tubing (Figs. 2B and C).

Commercially available mini-osmotic pumps (Alza Pharmaceuticals, Palo Alto, USA) were pretested *in vitro* and proved to be reliable in slow releasing of the protein which was biologically active in a cell-based assay that measures the alkaline phosphatase activity in an osteosarcoma cell line (ROS) *in vitro* [47].

In this study, for the first time, the termination time points of 3 and 6 months were determined by arthroscopy [48]. At 3 months following surgery defects treated with both low and high doses of BMP-7 were filled with newly formed cartilage, precartilagineous tissue and connective tissue at the top of the defect (Figs. 3A and B). The cartilage formation initially took place at the bottom progressing towards the surface of the defect (data not shown). In control knees there was no sign of cell ingrowth into the defect area (Fig. 3C). Defects treated with BMP-7 were filled with new cartilage except for areas filled with connective tissue and the new cartilage was well fused to the old cartilage (Fig. 3D). None of the control defects showed healing at six months following surgery. In BMP-7 treated knees newly formed cartilage was still well fused to the pre-existing one and stained positive for type II collagen (data not shown) [47].

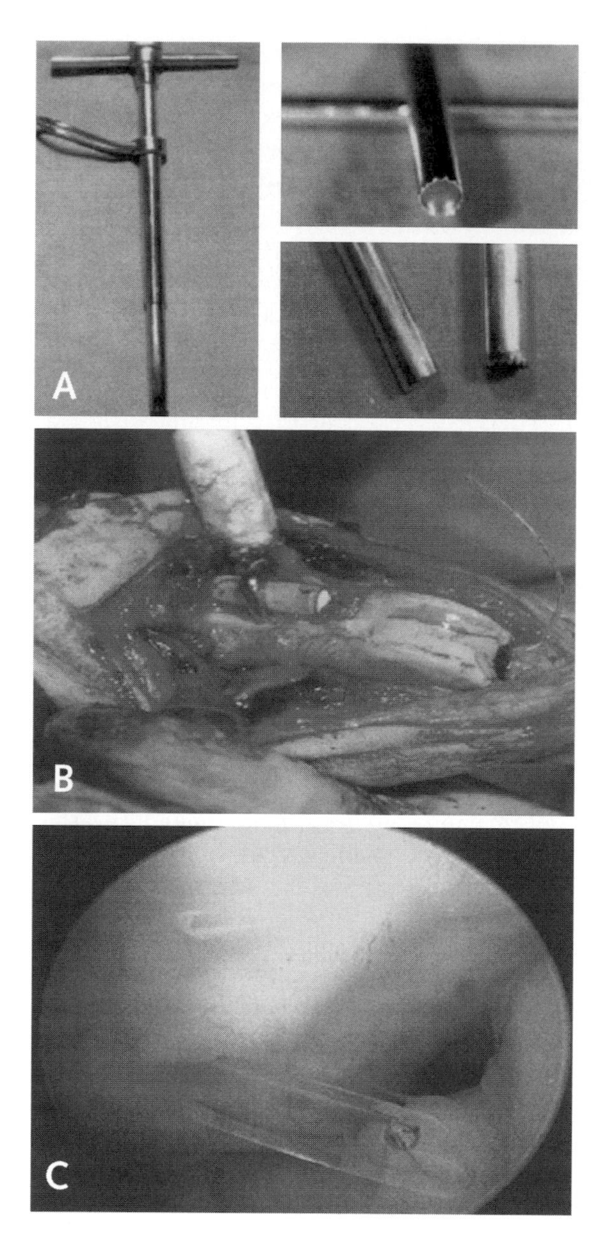


*Figure 2*
*Defects were created by an instrument consisting of an outside positioning ring and an inner rotating tube with a locking insert which allowed penetration up to 2 mm deep (A). The pump was stapled to the bone above the joint and connected with a catheter to the joint adjacent to chondral defects created in the sheep knee (B). Arthroscopic imaging of the tubing connecting a mini-osmotic pump with the joint space (C).*

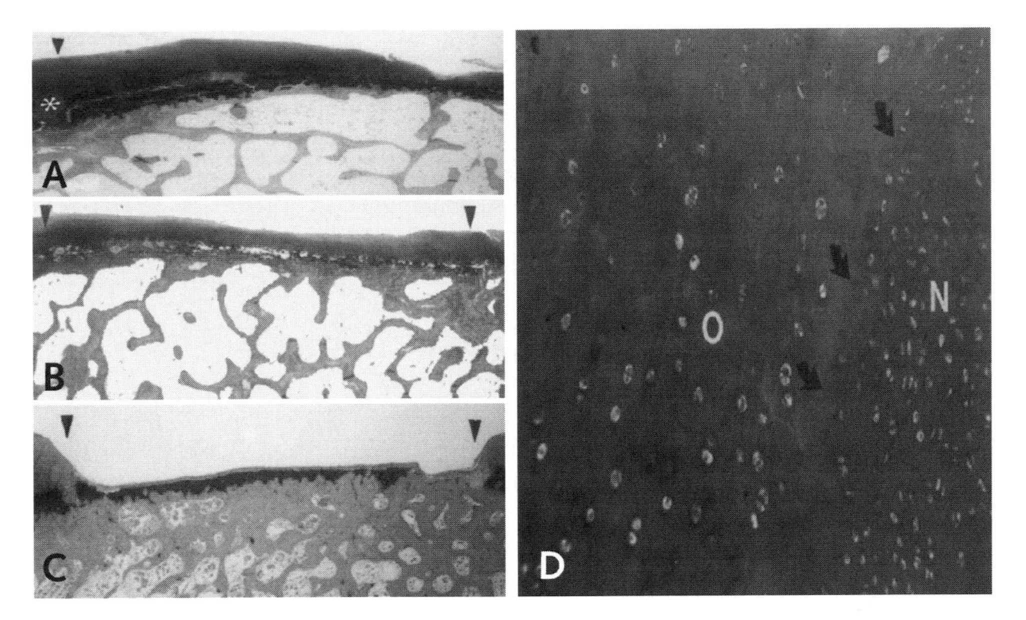


*Figure 3*
*(A) Regenerated joint cartilage filled the chondral defect area (indicated by two arrowheads) of a joint treated with BMP-7, 3 months post surgery (× 5, toluidine blue staining). (B) Regeneration of articular cartilage at 6 months following surgery and treatment with BMP-7. A condylar defect (arrows) treated with a low dose of BMP-7 was filled with newly regenerated cartilage (× 5, toluidine blue staining). (C) An empty defect treated with an acetate buffer vehicle 3 months following surgery (D). The bonding between old (O) and newly formed cartilage (N) in a chondral defect treated with BMP-7 is indicated by arrows (× 200, toluidine blue staining).*

Continuous presence of BMP-7 throughout a period of 2 to 4 weeks following surgery seems to have attracted the surrounding mesenchymal-like cells eventually originating from the synovium into the defect area, which then transformed into chondrocytes. BMP may, thus, be delivered to a joint space without a carrier in concentrations under the threshold for initiating ossification of surrounding soft tissues.

In studies using an osteochondral defect model in rabbits and the recombinant human BMP-2 [36–38] or BMP-7 [40] the repair tissue does not fuse well with the pre-existing adjacent cartilage neither in treated nor in untreated defects. The reason for a different ability of newly synthesized cartilage to fuse in osteochondral *versus* chondral defects could be based on the fact that in chondral defects the underlying bone supports the reparative process and that the ingrowing cells come from the synovium [49] and not from the bone marrow. Additional evidence supporting this concept came from the study of Sellers et al. [36, 37] demonstrating that BMP-

2 accelerated the rate of repair of subchondral bone with a subsequent improvement in the morphological features of cartilage in rabbits with osteochondral defects. Although it seems that the tissue integration in adult animals is unrelated to the method of treatment or the size of the defect, the majority of studies have used osteochondral defects which are lacking the support of the underlying bone resulting in the biomechanical instability of the regenerative tissue. It is of interest that articular cartilage defects undergo spontaneous repair in a fetal lamb joint repair model suggesting a different interaction between fetal chondrocytes and extracellular matrices [50].

A cytokine-based therapy for damaged cartilage would be clinically more useful and efficient than cell-based therapies, which involve removal of autologous cells derived from marrow [51] or from cartilage [52], followed by expansion in culture and then by a second operation for implantation into the defect. A single operation in which a cytokine is used to elicit repair of cartilage would substantially expedite the treatment process as well as reduce the costs. It has been recently reported that the expression of BMP-7 mRNA in human cartilage samples did not decrease with aging and was two-fold upregulated in OA cartilage, suggesting a role for BMPs in OA [53]. Apart from BMPs, good candidates in this regard would also be recently discovered cartilage-derived morphogenetic proteins (CDMPs), with their cartilage-specific localization pattern that suggests their potential role in chondrocyte differentiation ([31]; see the chapter by Luyten). The ability of BMP-7 to accelerate and improve cartilage repair in chondral defects emphasizes its importance as a candidate for cartilage repair in human osteoarthritis.

## Conclusion

BMPs have an important role in articular cartilage chondrocyte differentiation and production, as well as maintenance of the matrix. Animal experiments on articular cartilage defect healing have shown that BMPs act as differentiation factors depending on environmental conditions, suggesting that cartilage repair using BMPs may become an alternative and/or additive procedure for present clinical indications.

## References

1 Pecina M, Brezovecki-Bidin D (1985) Clinical, radiological and hystological investigations of degenerative changes in the articular cartilage of the knee joint. *Acta Orthop Iugosl* 16: 65–72

2 Johnson LL (1986) Arthroscopic abrasion arthroplasty historical and pathologic perspective: present status. *Arthroscopy* 2: 54–69

3 Bert JM (1993) Role of abrasion arthroplasty and debridgement in the management of osteoarthritis of the knee. *Rheum Dis Clin North Am* 19: 725–739
4 Haspl M, Pecina M (1995) Treatment of gonarthrosis with arthroscopic abrasion. *Lijec Vjesn* 117: 236–240
5 Menche DS, Frenkel SR, Blair B, Watnik NF, Toolan BC, Yaghoubain RS, Pitman MI (1996) A comparison of abrasion burr arthroplasty and subchondral drilling in the treatment of full-thickness cartilage lesions in the rabbit. *Arthroscopy* 12: 280–286
6 Akizuki S, Yasukawa Y, Takizawa T (1997) Does arthroscopic abrasion arthroplasty promote cartilage regeneration in osteoarthritic knees with eburnation? A prospective study of high tibial osteotomy with abrasion arthroplasty versus high tibial osteotomy alone. *Arthroscopy* 13: 9–17
7 Gilbert JE (1998) Current treatment options for the restoration of articulare cartilage. *Am J Knee Surg* 11: 42–46
8 Goymann V (1999) Abrasionsarthroplastik. *Orthopade* 28: 11–18
9 Steadman JR, Rodkey WG, Briggs KK, Rodrigo JJ (1999) The microfracture technic in the management of complete cartilage defects in the knee joint. *Orthopade* 28: 26–32
10 Passler HH (2000) Die Mikrofrakturierung zur Behandlung von Knorpeldefekten. *Zentralbl Chir* 125: 500–504
11 Kruger T, Wohlrab D, Reichel H, Hein W (2000) Der Effekt des arthroscopischen Gelenkdebridements bei fortgeschrittener Arthrose des Kniegelenkes. *Zentralbl Chir* 125: 490–493
12 Lahm A, Ergellet C, Steinwachs M, Reichelt A (2000) Arthroscopic management of osteochondral lesions of the talus: results of drilling and usefulness of magnetic resonance imaging before and after treatment. *Arthroscopy* 16: 299–304
13 O'Driscoll SW (1998) Current concepts review. The healing and regeneration of articular cartilage. *J Bone Joint Surg* 80: 1759–1812
14 Caterson B, Buckwalter JA (1990) Articular cartilage repair and remodeling. In: A Maroudas, KE Kuettner (eds): *Methods in cartilage research*. London: Academic Press, 313–319
15 Mankin HJ (1974) The reaction of the articular cartilage to injury and osteoarthritis. *New Engl J Med* 291: 1285–1292
16 Rosenberg L, Hunziker EB (1994) Cartilage repair in osteoarthrosis. The role of the dermatan sulfate proteoglycans. In: KE Kuettner, V Goldberg (eds): *Osteoarthrosis disorder*. Park Ridge, Illinois, The American Academy of Orthopaedic Surgeons, 341–356
17 Hardingham TE, Fosang AJ, Dudhia J (1992) In: KE Kuettner, R Schleyerbach, JG Peyron, VC Hascall (eds): *Articular cartilage and osteoarthritis*. Raven Press Ltd., New York, 5–21
18 Siczkowski M, Watt FM (1990) Subpopulations of chondrocytes from different zones of pig articular cartilage. Isolation, growth and proteoglycan synthesis in culture. *J Cell Sci* 97: 361–367
19 Archer CW, McDowell J, Bayliss MT, Stephens MD, Bentley (1990) G Phenotypic modulation in sub-populations of human articular chondrocytes *in vitro*. *J Cell Sci* 361–371

20 Reddi AH (1994) Bone and cartilage differentiation. *Curr Opin Gen Dev* 4: 737–744
21 Luyten FP, Yu YM, Yanagashita M, Vukicevic S, Hammonds RG, Reddi AH (1992) Natural bovine osteogenin and recombinant human bone morphogenetic protein-2B are equipotent in the maintenance of proteoglycans in bovine articular cartilage explant cultures. *J Biol Chem* 267: 3691–3685
22 Luyten FP, Chen P, Paralkar V, Reddi AH (1994) Recombinant bone morphogenetic protein-4, transforming growth factor beta1 and activin A enhance the cartilage phenotype of articular chondrocytes *in vitro*. *Exp Cell Res* 210: 224–229
23 Sailor LZ, Hewick RM, Morris EA (1996) Recombinant human bone morphogenetic protein-2 maintains the articular chondrocyte phenotype in long term culture. *J Orthop Res* 14: 937–945
24 Chen P, Vukicevic S, Sampath TK, Luyten FP (1995) Osteogenic protein-1 promotes growth and maturation of chick sternal chondrocytes in serum-free cultures. *J Cell Sci* 108: 105–114
25 Chen P, Vukicevic S, Sampath TK, Luyten FP (1993) Bovine articular chondrocytes do not undergo hypertrophy when cultured in the presence of serum and osteogenic protein-1. *Biochem Biophys Res Commun* 197: 1253–1259
26 Hunziker EB (1992) Articular cartilage structure in humans and experimental animals. In: KE Kuettner, R Schleyerbach, JG Peyron, VC Hascall (eds): Articular cartilage and osteoarthritis. Raven Press Ltd, New York, 183–201
27 Shapiro F, Koide S, Glimcher MJ (1993) Cell origin and differentiation in the repair of full-thickness defects of articular cartilage. *J Bone Joint Surg* 75: 532–553
28 Haaijman A, DeSouza RN, Bronckers AL, Goei SW, Burger EH (1997) OP-1 (BMP-7) affects mRNA expression of type I, II, X collagen, and matrix Gla protein in ossifying long bones *in vitro*. *J Bone Miner Res* 12: 1815–1823
29 Reddi AH (1998) Role of morphogenetic proteins in skeletal tissue engineering and regeneration. *Nat Biotechnol* 16: 247–252
30 Vukicevic S, Martinovic S, Basic M, Jelic M (1999) Bone morphogenetic proteins: First European Conference. *Bone* 24: 395–397
31 Chang SC, Hoang B, Thomas JT, Vukicevic S, Luyten FP, Ryba NJ, Kozak CA, Reddi AH, Moos M (1994) Cartilage derived morphogenetic proteins. *J Biol Chem* 269: 28227–28234
32 Vukicevic S, Stavljenic A, Pecina M (1995) Discovery and clinical applications of bone morphogenetic proteins. *Eur J Clin Chem Clin Biochem* 33: 661–671
33 Furukawa T, Eyre DR, Koide S, Glimcher MJ (1980) Biochemical studies on repair cartilage resurfacing experimental defects in the rabbit knee. *J Bone Joint Surg* 62: 79–89
34 Buckwalter JA (1990) Building on our strengths. *J Orthop Res* 8: 917–920
35 Metsaranta M, Kujala UM, Pelliniemi L, Osterman H, Aho H, Vuorio E (1996) Evidence of insufficient chondrocytic differentiation during repair of full thickness defects of articular cartilage. *Matrix Biol* 15: 39–47
36 Sellers RS, Peluso D, Morris EA (1997) The effect of recombinant human bone mor-

phogenetic protein-2 (rhBMP-2) on the healing of full-thickness defects of articular cartilage. *J Bone Joint Surg* 79: 1452–1463

37 Sellers RS, Zhang R, Glasson SS, Kim HD, Peluso D, D'Augusta DA, Beckwith K, Morris EA (2000) Repair of articular cartilage defects one year after treatment with recombinant human bone morphogenetic protein-2 (rhBMP-2). *J Bone Joint Surg* 82: 151–160

38 Frenkel SR, Saadeh PB, Mehrara BJ, Chin GS, Steinbrech DS, Brent B, Gittes GK, Longaker MT (2000) Transforming growth factor beta superfamily members: role in cartilage modeling. *Plast Reconstr Surg* 105: 980–990

39 Nimni M (2000) Osteogenic and chondrogenic effects of a recombinant BMP-3 with a collagen binding domain. *International Conference Bone Morphogenetic Proteins June 7-11, 2000*, Lake Tahoe, USA, abstract book, 49

40 Grgic M, Jelic M, Basic V, Basic N, Pecina M, Vukicevic S (1997) Regeneration of articular cartilage defects in rabbits by osteogenic protein-1 (bone morphogenetic protein-7). *Acta Med Croatica* 51: 23–27

41 Cook S, Rueger DC (1996) Osteogenic protein-1. Biology and applications. *Clin Orthop Rel Res* 324: 29–38

42 Louwerse RT, Heyligers IC, Klein-Nulend J, Sugihara S, van Kampen GP, Semeins CM, Goei SW, de Koning MH, Wuisman PI, Burger EH. (2000) Use of recombinant human osteogenic protein-1 for the repair of subchondral defects in articular cartilage in goats. *J Biomed Res* 49: 506–516

43 O'Driscoll SW, Keeley FW, Salter RB (1986) The chondrogenic potential of free autogenous periosteal grafts for biological resurfacing of major full-thickness defects in joint surfaces under the influence of continuous passive motion. An experimental investigation in rabbit. *J Bone Joint Surg* 68: 1017–1035

44 O'Driscoll SW (1999) Articular cartilage regeneration using periosteum. *Clin Orth Rel Res* 367: 186–203

45 Sanyal A, Sarkar G, Saris DB, Fitzsimmons J S, Bolander ME, O'Driscoll SW (1999) Initial evidence for the involvement of bone morphogenetic protein-2 early during periosteal chondrogenesis. *J Orthop Res* 17: 926–934

46 Hanada K, Solchaga LA, Caplan AI, Hering TM, Goldberg VM, Yoo JU, Johnstone B (2001) BMP-2 induction and TGF-beta1 modulation of rat periosteal cell chondrogenesis. *J Cell Biochem* 81: 284–294

47 Jelic M, Pecina M, Haspl M, Kos M, Taylor K, Maticic D, McCartney J, Yin S, Rueger D, Vukicevic S (2001) Regeneration of articular cartilage chondral defects by osteogenic protein-1 (bone morphogenetic protein-7) in sheep. *Growth Factors* 19: 101–113

48 Haspl M, Jelic M, Kos J, Vukicevic S, Pecina M (1999) Follow up arthroscopy in sheep knee chondral defect regeneration. First European Conference on Bone Morphogenetic Proteins, October 7–11, 1998, Zagreb, Croatia. *Bone* 24: A 418

49 Hunziker EB, Rosenberg LC (1996) Repair of partial-thickness in articular cartilage: cell recruitment from the synovial membrane. *J Bone Joint Surg* 78: 721–733

50 Namba RS, Meuli M, Sullivan KM, Le AX, Adzick NS (1998) Spontaneous repair of superficial defects in articular cartilage in a fetal lamb model. *J Bone Joint Surg* 80: 4–10

51 Wakitani S, Goto T, Pineda SJ, Young RG, Mansour JM, Caplan AI, and Goldberg VM (1994) Mesenchymal cell-based repair of large, full-thickness defects of articular cartilage. *J Bone Joint Surg* 76: 579–592
52 Brittberg M, Nilsson A, Lindahl A, Ohlsson C, and Peterson L (1996) Rabbit articular cartilage defects treated with autologous cultured chondrocytes. *J Clin Orthop* 326: 270–283
53 Chubinskaya S, Merrihew C, Szabo G, Mollenhauer J, McCartney J, Rueger DC, Kuettner KE (2000) Human articular chondrocytes express osteogenic protein-1. *J Histochem Cytochem* 48 (2): 239–250

# The role of bone morphogenetic proteins in kidney development and repair

*Fran Borovecki[1], Nikolina Basic[3], Mislav Jelic[1], Dunja Rogic[3], Haimanti Dorai[4], Ana Stavljenic-Rukavina[3], Kuber T. Sampath[2] and Slobodan Vukicevic[1]*

[1]Department of Anatomy, School of Medicine, University of Zagreb, Zagreb 10 000, Croatia;
[2]Genzyme Corporation, One Mountain Road, Framingham, MA 01701-9322, USA;
[3]Zagreb University Clinical Center, University of Zagreb, Zagreb 10 000, Croatia;
[4]Centocor, Inc., 200 Great Valley Parkway, Malvern, PA 19355, USA

## Introduction

Members of TGF-β superfamily are secreted glycoproteins and have been shown to regulate biological processes as diverse as migration, proliferation and differentiation of pluripotent progenitor cells involved in the development of several organ systems during embryogenesis and in adult tissue repair [1, 2]. The kidney has been identified as a major site of bone morphogenetic protein-7 (BMP-7) synthesis during embryonal and post-natal development [1, 3, 4]. Gene knock-out [5, 6] and *in vitro* experiments [4, 7] demonstrated the importance of BMP-7 in kidney development. Many developmental features are recapitulated during renal injury, and BMPs may be important in both preservation of function and resistance to injury [8, 9]. BMP-7 has a cytoprotective and anti-inflammatory effect in models of acute and chronic renal failure [8, 9].

## Bone morphogenetic proteins in kidney development

Mice lacking the BMP-7 gene died of uremia within 24 h following birth. One group reported the absence of tubules and immature glomeruli apparatus (S- and comma-shaped bodies) following the ingrowth of the ureteric bud into the metanephric mesenchyme in E-11 mice, suggesting that BMP-7 is necessary for the induction of the E-11 mesenchyme [6]. Another BMP-7 knock-out phenotype suggested that unaltered kidney development progressed up to E-14 in BMP-7 null mice, which was, however, followed by a rapid disappearance of the metanephric mesenchyme resulting in loss of kidney mass upon birth [5]. While this apparent discrepancy can be attributed to variance observed in mouse genetics, the precise role of BMP-7 in metanephric differentiation remains unknown.

Bone Morphogenetic Proteins, edited by Slobodan Vukicevic and Kuber T. Sampath

The permanent kidney of mammals, the metanephros, starts to develop when the ureteric buds emerge from the Wolffian ducts and enter the metanephric mesenchyme. The ureteric bud induces condensation of the surrounding metanephric mesenchyme, and reciprocally, the metanephric mesenchyme causes elongation and branching or the ureteric bud. At the tip of these branches, the ureteric bud induces aggregation of the mesenchymal cells. Each aggregate invaginates once to form a comma-shaped body and once again to form an S-shaped body. The blood vessels invaginate into one of the curves of the S-shaped bodies forming the future glomeruli. The epithelial cells begin to differentiate into the specific cell types such as podocytes, capsule cells, and proximal and distal tubule cells. The most distal part of the nephron and the newly formed tube connect, thereby, enabling passage of the materials [10] (Fig. 1).

The reciprocal induction was documented by *in vitro* experiments when the ureteric bud and the metanephric mesenchyme were cultured separately [11, 12]. The ureteric bud does not branch in the absence of the mesenchyme, and the mesenchyme dies without the ureteric bud. Although certain tissues (such as neural tube, spinal cord and salivary glands) enable the metanephric mesenchyme to form kidney tubules, the ureteric bud branches only under instructions from the metanephric mesenchyme [10]. However, the extrinsic influences, namely growth factors and protooncogenes, control the proliferation and differentiation of the metanephric cells. They act *via* the intracellular signalling pathways leading to activation of genes involved in the regulation of the growth processes. Current results propose existence of "cascade of events" with "checkpoints" at the beginning of each cascade. The cascade of development could not proceed after the checkpoint if a critical signal is missing [13].

Many genes are proposed to be essential for kidney development. However, a candidate gene should fulfill several criteria in order to be explicitly involved in the development. It must be expressed in appropriate time and space relative to the developing organ, and in the absence of the gene normal organ development should fail. So far, several genes satisfy these criteria. Gene knock-out studies enable identification of BMP-7, WT-1, Pax-2, c-ret, foxc1, foxc2, GDNF, BF-2, Eya1, Wnt-4, Emx2, PDGF B, PDGFRb, $\alpha8\beta1$ and $\alpha3\beta1$ as molecules that are required for kidney growth and development [14]. Recently, it has been shown that leukemia inhibitory factor (LIF) and members of the IL-6 family, including cardiotrophin, oncostatin and CNTF are expressed in the ureter and can induce nephrogenesis in culture. This possibly explains why the LIF knock-out has no obvious kidney phenotype [15].

The GDNF/GDNFR$\alpha$/ret receptor-ligand complex is necessary for growth and branching of the ureteric bud in the process of reciprocal inductive interaction between the epithelium of the Wolffian duct and the adjacent mesenchyme [16–20]. Inductive interaction in nephrogenesis is accompanied with elevation in the expression pattern of several factors. The Wilms tumor suppressor gene (WT-1) is already

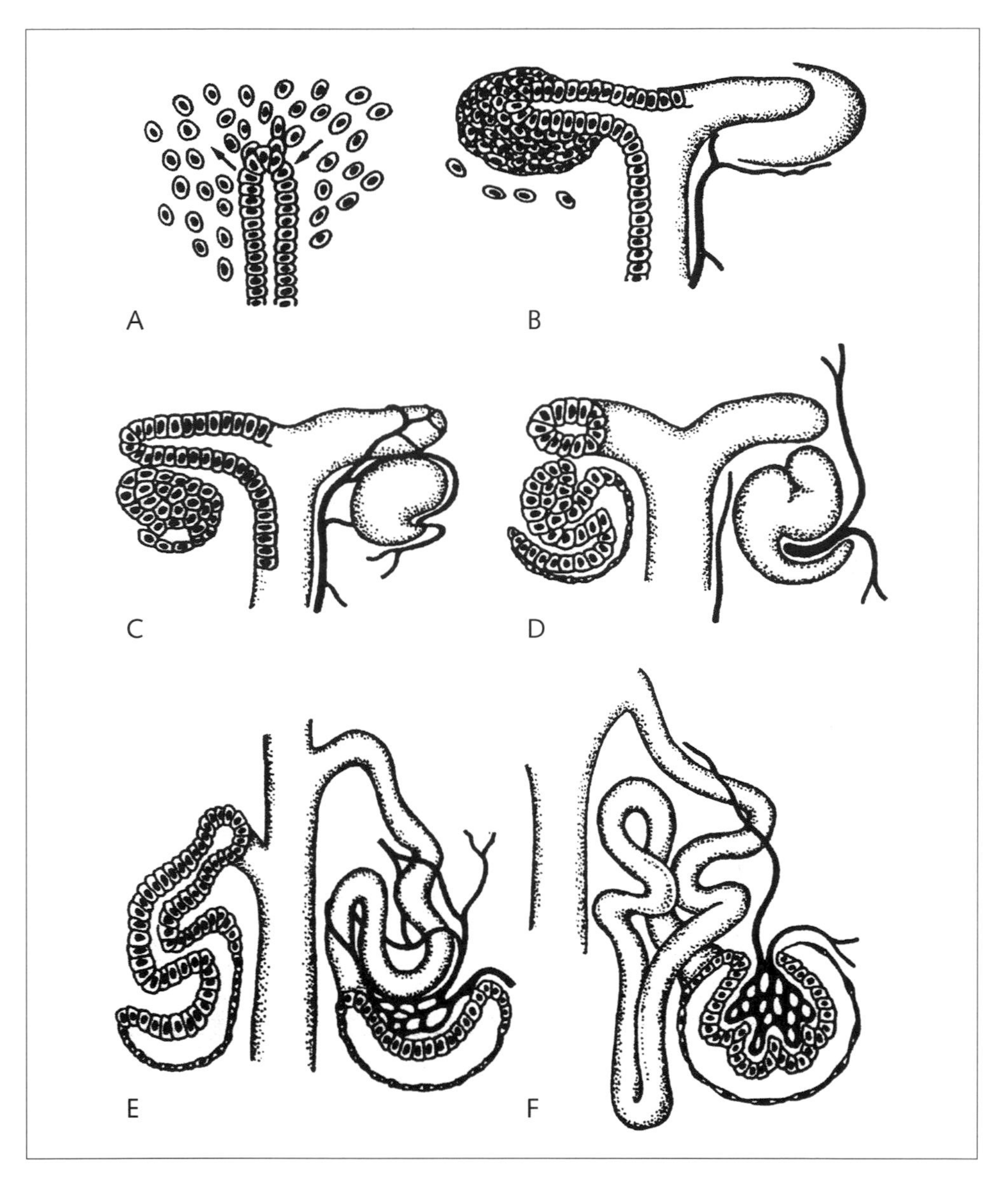


*Figure 1*
*Schematic drawing of various stages in the development of the nephron. As a result of inductive interaction between the ureteric bud and the metanephric mesenchyme (A), a condensate is formed (B). It goes through the comma-shape (C) and S-shape body stages (D). This is followed by tubule elongation and its connection with the nephric duct (E). When the blood vessels invade the distal curve of the S-shaped body, the future mature glomeruli begin to form (D, E and F).*

expressed in the uninduced mesenchyme but its expression is highly upregulated on induction. WT-1 null-mutant mice failed to develop kidneys because the metanephric mesenchyme cannot respond to inductive signals [21]. Using a microarray amphiregulin, a member of the epidermal growth factor (EGF) family has been characterized as a physiological target of WT-1. It stimulates ureteric branching in kidney organ cultures, but amphiregulin knock-out showed no renal phenotype [22]. Pax-2 is necessary for the mesenchymal aggregation and mesenchyme-to-epithelial transition during nephrogenesis, and it disappears after terminal differentiation of nephrons [23, 24]. WT-1 is a negative regulator of Pax-2 during kidney development [25]. Its expression is elevated in a variety of renal tumors [26].

After the initial induction, BMP-7 and Wnt-4 are required for subsequent mesenchymal differentiation by maintaining the inductive response. Wnt-4 is a cysteine-rich signaling molecule expressed in pretubular cells of the metanephric mesenchyme at the base of the ureteric bud. Its expression is absolutely necessary for kidney development and is lost upon fusion of nephron with the collecting duct [27]. As cell proliferation and differentiation proceed, more and more molecules are involved in the regulation. BF-2 is the "winged helix" transcription factor expressed in stromal cells. It is necessary for regulation of the nephrogenesis in the induced cell population that is destined to make epithelium [28]. In mice lacking PDGF B or its receptor PDGFRβ mesangial cells are absent thus disabling formation of the glomeruli [29].

BMP-7 is expressed in several tissues associated with inductive interactions and is required for proper nephrogenesis using gene targeting in mice [5, 6]. BMP-7 mRNA expression is the highest on day 13 of kidney development (Fig. 2) what corresponds with its proposed role in nephrogenesis. In the normal kidney, the highest expression of BMP-7 mRNA could be seen in tubules of the outer medulla, in cells at the periphery of the glomerular tuft, adventitia of renal arteries and epithelial cells of the renal pelvis and the ureter [7]. During development, BMP-7 transcripts are most abundantly present, first, in the epithelium of the branching ureteric buds, and later in the glomeruli (Fig. 3) [1]. Most of the homozygous animals die the first postnatal day from acute renal failure. Their kidneys failed to develop normally, and they also have microopthalmia and various degrees of skeletal deformities. The kidney starts to develop, reciprocal interactions occur, but further development ceases by approximately 14 days postcoitum accompanied with extensive apoptosis. Glomeruli and proximal convoluted tubules are well developed, so it seems that BMP-7 is absolutely necessary for the development of distal convoluted tubules and maintenance of the kidney structure. Multiple cysts are observed in the kidneys of animals that survived for a few days [30]. In the CNS and heart of the mutant animals, expression domains of the BMP family members completely overlap with that of BMP-7. It seems that at such places other BMP family members can substitute for BMP-7 [31].

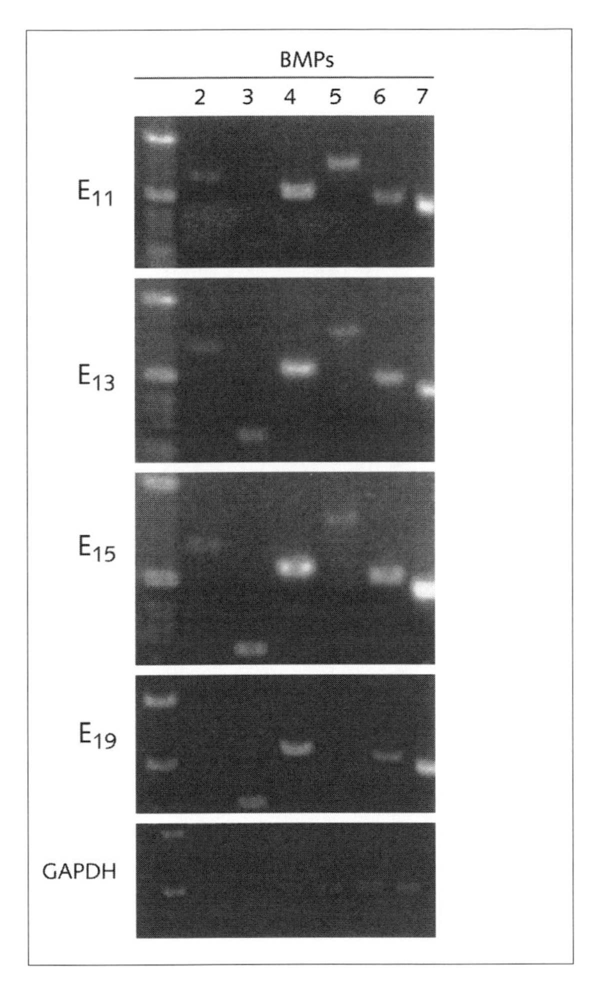


*Figure 2*
*Expression of bone morphogenetic proteins in developing mouse kidneys. Whole kidney RNA was isolated, cDNA was synthesized and analyzed by RT-PCR. GAPDH was used to normalize the reaction. At E11 of mouse development, BMPs 2–7 are expressed, with BMP-4 and BMP-7 being most abundant. BMP-3 and BMP-6 are gradually upregulated, while BMP-5 expression declines from E11 towards E19.*

It has been demonstrated that during kidney development, high doses of BMP-7 inhibit branching morphogenesis, whereas low doses are stimulatory [32]. Another study [62] showed that BMP-7 suppresses tubulogenesis and, in synergy with FGF–2, increases the cell population of stromal precursor cells in the developing kidney (Fig. 4). These results indicate an important function for BMP-7 in the main-

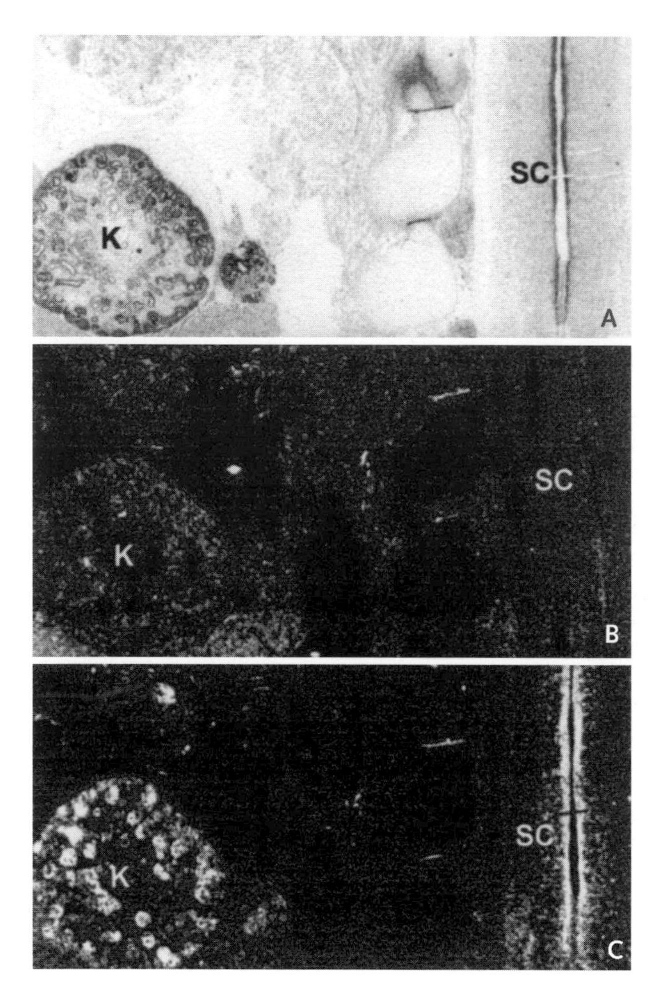


*Figure 3*
*BMP-7 mRNA expression in the kidney of a human embryo (9 weeks of gestation). (A) Toluidin blue-stained bright field image of a section through the kidney (K) and spinal cord (SC). Dark field images of sense (B) and antisense (C) mRNA probes [1] indicate synthesis in kidney glomeruli and the spinal cord.*

tenance of blastemal tissue and hence the continuous growth of the kidney during development. In cultured embryonic kidneys, BMP-7 mRNA expression was demonstrated in several glomerular cell types, such as mesangial, epithelial and endothelial cells. Distal tubule MDCK cells also expressed BMP-7 mRNA, but human proximal tubule HK-2 cells did not. Treatment with BMP-7 increased cellular proliferation of HK-2 cells, but not of the mesangial cells. These results suggest that BMP-7 is pro-

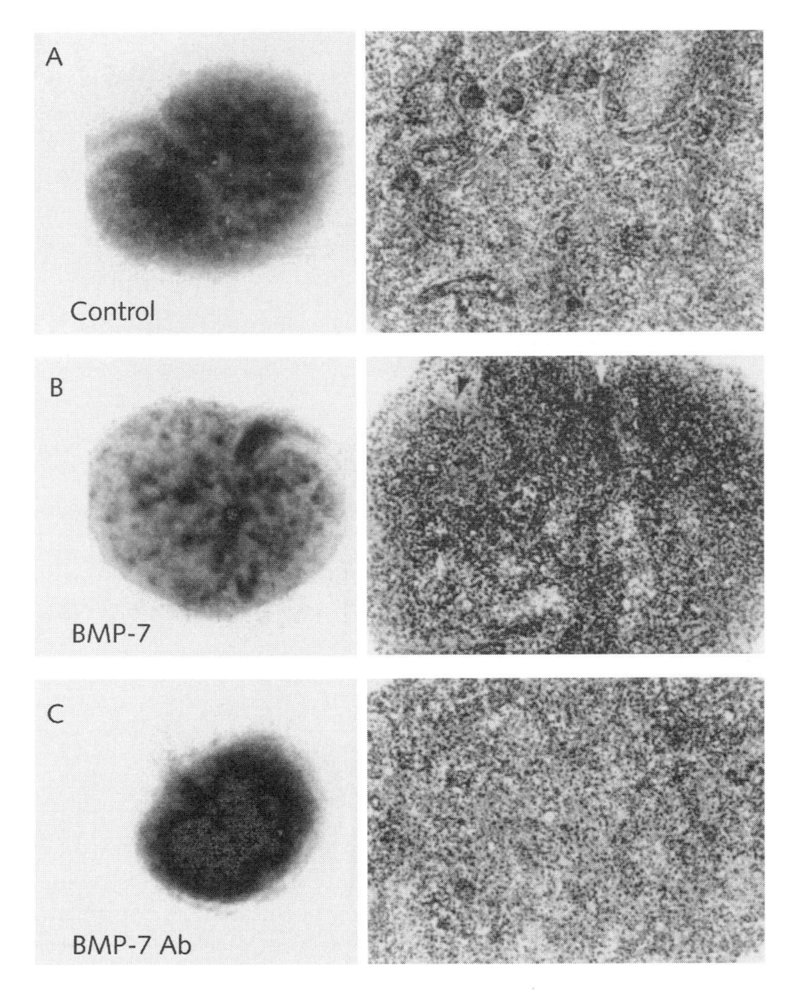


*Figure 4*
*Effect of BMP-7 on whole mouse embryo kidney explant cultures. E13 mouse kidneys were isolated and cultured [4] for 5 days in the presence of BMP-7 protein (B; 100 ng/ml daily) and BMP-7 antibody (C; 10 μg/ml daily). Addition of 100 ng/ml of BMP-7 induced more translucent kidney explants (B left) as a consequence of more pronounced stromal cell proliferation (B right). BMP-7 antibody induced atrophy of the mesenchyme and reduced tubulogenesis mimicking the phenotype of BMP-7 gene knock-out kidneys.*

duced in the renal glomerulus and then travels to the proximal tubule to regulate the proliferation of the cells in this region of the nephron [33]. BMP-7 expression in the epithelial components of the kidney is not dependent on cell-cell or cell-BMP-7 interactions with the metanephric mesenchyme. Disruption of proteoglycan synthesis results in the loss of BMP-7 expression in the mesenchyme. It seems that BMP-7 expression in the metanephric mesenchyme is dependent on proteoglycans and proper protein glycosylation [34]. The current data support a model in which signaling from the ureter induces metanephric expression of Pax2 and WT-1. They subsequently activate the signaling molecules BMP-7 and Wnt-4, which promote tubulogenesis and expression of stromal precursor cells. Several other BMPs are expressed during kidney development and in the postnatal life (Fig. 5)

BMP-4 is expressed in mesenchymal cells surrounding the Wolffian duct and the ureter stalk. It is important in the early morphogenesis of the kidney and urinary tract. It inhibits ectopic budding from the Wolffian duct or the ureter stalk by antagonizing inductive signals from the metanephric mesenchyme to the illegitimate sites on the Wolffian duct. Another function is to promote the elongation of the branching ureter within the metanephros. BMP-4 signaling can substitute for the surface ectoderm in supporting nephric duct morphogenesis [35]. BMP-4 null-mutant mice display abnormalities of the genitourinary tract including hypoplastic kidneys, hydroureter, ectopic ureterovesical junction and double collecting system ([36]; see the chapter by Martinovic). In the organ culture of the developing kidney, human recombinant BMP-4 diminishes the number of ureteric branches and changes the branching pattern *via* interfering with the differentiation of the metanephric mesenchyme [37]. In BMP-7 null-mutant mice, BMP-4 is expressed in the mesenchyme surrounding the ureteric bud in the early stages of development, then in the area of nephron development, and finally its expression is limited to the Bowmann capsule [30]. Its expression reaches maximal value from day 15 to 17 of embryonal development suggesting its role in tubulogenesis (Fig. 2).

BMP-2 and HGF function to control parallel pathways downstream of their respective cell surface receptors regulating the collecting duct morphogenesis [38]. In mesangial cells, BMP-2 inhibits PDGF-induced DNA synthesis and c-fos gene transcription [39]. BMP-2 expression is persistent during intrauterine and postnatal kidney development (Fig. 2), while its expression is downregulated in adult kidneys (Fig. 2).

Osteogenin (BMP-3) is mainly synthesized in the developing lung and kidney [40]. In normal rat kidneys, BMP-3 mRNA expression is limited to areas of tubule development, and is not found in the glomeruli [41]. On the contrary, Dudley and Robertson have found BMP-3 mRNA in the glomerular area of the future nephron in BMP-7 null-mutant mice [30]. Gradually, BMP-3 mRNA expression is upregulated from day 13 to 17 of embryonal development, and then decreases (Fig. 2). BMP-3 knock-out mice do not have kidney abnormalities (see the chapter by Martinovic).

BMP-5 expression is demonstrated in the cell layer adjacent to epithelial cells of the ureteric bud and in renal calices of the more mature kidneys in BMP-7 null-mutant mice [30]. In normal mouse embryos, BMP-5 expression is found in mesenchymal cells surrounding the ureter, but also in the renal calices at later stages of

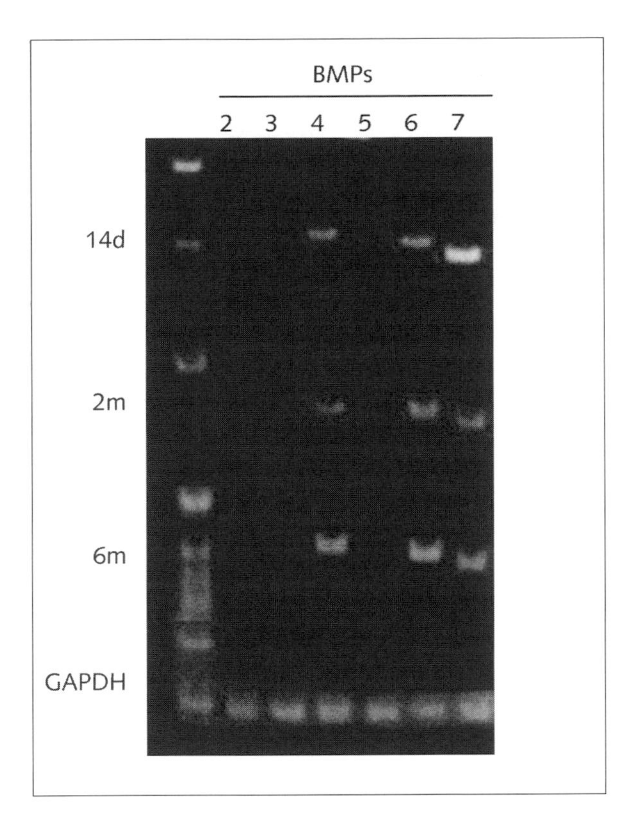


*Figure 5*
*Expression of bone morphogenetic proteins 2–7 in adult mouse kidneys. Whole kidney RNA was isolated, cDNA was synthesized and analyzed by RT-PCR. Reactions without cDNA were used as a negative control. GAPDH was used to normalize reactions. After two weeks, 2 and 6 months following delivery, BMP-7 is strongly expressed, while BMP-2 and BMP-5 appear low.*

development. BMP-5 mRNA is expressed in mice embryonal kidneys from day 12 to day 17 kidney during the postnatal life (Fig. 5). From the beginning of kidney development BMP-6 expression is upregulated, and the highest level is found in mature, adult kidneys (Figs. 2, 5).

## BMP-7 crosses the placental barrier during development

It is believed that knock-out studies of genes that transcribe circulating glycoproteins might give unreliable information as to their developmental function, due to

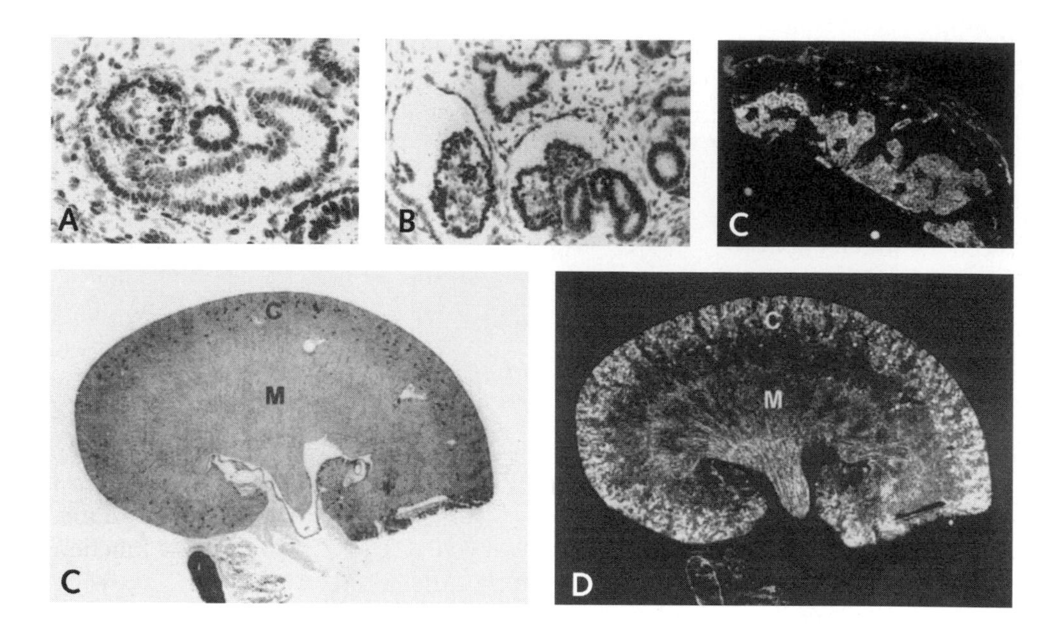


*Figure 6*
*Autoradiographs of systemically administered $^{125}$I-BMP-7 in pregnant rats. $^{125}$I-BMP-7 (0,237 mCi/mg) was administered intravenously to pregnant mice at different stages of the fetus development. Iodinated BMP-7 passed across the placenta and localized in developing fetal organs, the kidneys in particular, up to day 14 of gestation. Panels A and B show accumulation of radioactive grains in the fetal kidney tubules and mesenchyme in E12.5 kidneys. On E14, the grains accumulate in the fetal part of the placenta (C; arrowheads), but do not enter fetal blood vessels. $^{125}$I-BMP-7 accumulates in the kidney cortex (c) and the medulla (m) of the pregnant mice (D = toluidine blue stained bright field image; E = dark field image). Magnification $\times$5 in C, D, and E, and x 250 in A and B.*

their potential cross-over through the placental barrier, as it has been shown for TGF-β1 [42]. Among the BMP family of proteins, BMP-7 circulates in the bloodstream of mice and rats [8]. Whether intravenously administered BMP-7 in pregnant mice is made available to fetuses and thus masks the "true" developmental role of BMP-7 in gene-knock-out mice, was tested by analyzing the distribution of $^{125}$I-BMP-7 in fetal and maternal organs [43] (Fig. 6).

$^{125}$I-BMP-7 accumulates in fetuses during early pregnancy (days 8–12), while no $^{125}$I-BMP-7 is found after day 14 of pregnancy (Fig. 6).

On day 13 of gestation $^{125}$I-BMP-7 grains were detected in the developing kidney structures, localizing mainly above cells belonging to the kidney mesenchyme (Fig. 6). At later stages of pregnancy BMP-7 accumulated largely in the blood vessels of the mother and in the labyrinth (Fig. 6), which prevented the transport of

$^{125}$I-BMP-7 into the fetal capillaries. No trace of $^{125}$I-BMP-7 was found on day 18 of pregnancy in the blood vessels, which suggests that the placental membrane prevents transport of injected BMP-7 into the fetal bloodstream. Accordingly, no specific accumulation of $^{125}$I-BMP-7 is detected in any fetal tissue or in blood vessels of the umbilical cord [43].

The results suggest that BMP-7 from heterozygous mothers might have influenced the differentiation of the kidney during the early development of BMP-7 null-mutant fetuses [5, 6].

## The role of BMPs in acute and chronic kidney failure models

### Acute kidney failure

The finding that BMP-7 expression remains high in both the fetal and postnatal life, and is available in the circulation suggests that BMP-7 may have a systemic function and a role in the repair and regeneration of the adult kidney [3, 8].

Acute renal failure represents a clinical condition with persistently high mortality (40-80%), despite technical advances in both critical care medicine and dialysis. The successful treatment of patients with acute renal failure who require dialysis remains one of the greatest challenges facing nephrology today [44]. This condition can be fully understood and optimal treatment measures defined, only with knowledge of the underlying molecular and structural changes and events.

The damaged kidney is capable of complete repair and regeneration after acute injury and the process recapitulates features that occur during the development. It is assumed that regenerating cells take a step back, towards an earlier ontogenic stage, which makes the cells sensitive to embryonic stimuli [45, 46]. BMP-7 may be important in both preservation of function and resistance to injury [8].

The mechanisms controlling the cascade of cellular migration, growth and proliferation following acute renal failure undoubtedly comprise a number of autocrine and paracrine growth factors [47, 48], such as insulin-like growth factors (IGFs), epidermal growth factor (EGF), fibroblasts growth factor (FGF), transforming growth factors (TGF-$\alpha$, TGF-$\beta$), and hepatocytes growth factor (HGF) [49–52]. Animal studies dealing with acute renal failure due to ischemic-reperfusion insult have indeed proven that administration of BMP-7 has, for a period of 4 days following ischemia, a beneficial effect on the extent of injury and the regeneration of kidney function [8]. Bioavailability studies have shown that human BMP-7 has a serum half-life of about 30 min, and that significant amounts of $^{125}$I-BMP-7 can be found in both the kidney cortex and medulla shortly after iv administration [8].

Apart from being protective in ischemic acute renal failure, BMP-7 also influences the course of toxic kidney injury *in vitro*, as well as in acute nephrotoxic animal models utilizing administration of mercuric chloride and cisplatinum [53]. Both prophylactic and therapeutic systemic administration of BMP-7 to rats given mercuric chloride protected the kidney function and significantly extended the survival rate (Fig. 7). Similarly, BMP-7 protected the kidney function in rats treated with a high dose of cisplatinum (Fig. 9).

Mercuric chloride exerts its toxic effects on kidney cells through a variety of mechanisms, the principal target being S3 segment of proximal tubules. Intracellular pathways contributing to cell damage by mercury are primarly the consequence of its high affinity for sulfhydril groups. These protein groups are of utmost importance for cell function, since they are both located within active centers of various vital enzymes and they represent one of the main defense mechanisms against oxidative damage [54, 55]. Indeed, increased $H_2O_2$ production in mitochondria and heme oxygenase induction have been demonstrated both *in vitro* and in tubular cells isolated from rats treated with $HgCl_2$ [56]. Apart from interfering with respiratory chain and oxidative phosphorilation enzymes, mercury was shown in numerous studies to cause oxidative injury with subsequent lipid peroxidation, DNA damage and protein oxidation [57]. Thus, in terms of cytoprotection, since this toxicant may activate multiple pathways, multiple pathways may need to be blocked as well. *In vitro* studies show that BMP-7 significantly promotes cell survival and proliferation in human primary proximal tubule cells treated with mercury chloride, while it is ineffective in intact cells (Fig. 8). In rats with an ischemic-reperfusion kidney damage [8], BMP-7 was shown to ameliorate the course of injury through a variety of mechanisms, including inhibition of apoptosis, minimizing of infarction and cell necrosis and preventing intercellular adhesion molecule-1 (ICAM-1) expression, thus supressing the inflammatory response [8]. Whether the same mechanisms are responsible for its beneficial effects observed in nephrotoxic studies, remains to be elucidated. However, the oxidative damage is a principal cause of cell injury and death in both mercury-induced and ischemic-reperfusion insult to the kidney. Considering the fact that BMP-7 has a characteristic cystein-rich region in the carboxyterminal part of the polypeptide chain, it is conceivable that it might function as both mercury and/or free radical scavenger. On the other hand, the finding that BMP-7 is effective in promoting the proliferation and viability of renal tubular cells previously injured by mercury *in vitro*, while being ineffective in intact cells (Fig. 8), points to a difference in sensitivity to external stimuli between regenerating and intact cells. Indeed, the experiments dealing with liver regeneration [58] have shown that hepatocytes first need to be "primed" with either cytokines or reactive oxygen species in order to become fully competent to respond to growth factor stimuli. It is well established that kidney cells have a capacity for repair and function recovery after injury by recapitulating the molecular and cellular events that take place during nephrogenesis [50, 51] very similar to regenerating fractured bone [59, 60]. Since BMP-7 is a morphogenic protein involved in nephrogenesis during the embryogenesis, it may be postulated that injured cells exhibit *de novo* sensitivity to BMP-7 stimulation *in vitro*. During prenatal development of the mouse kidney,

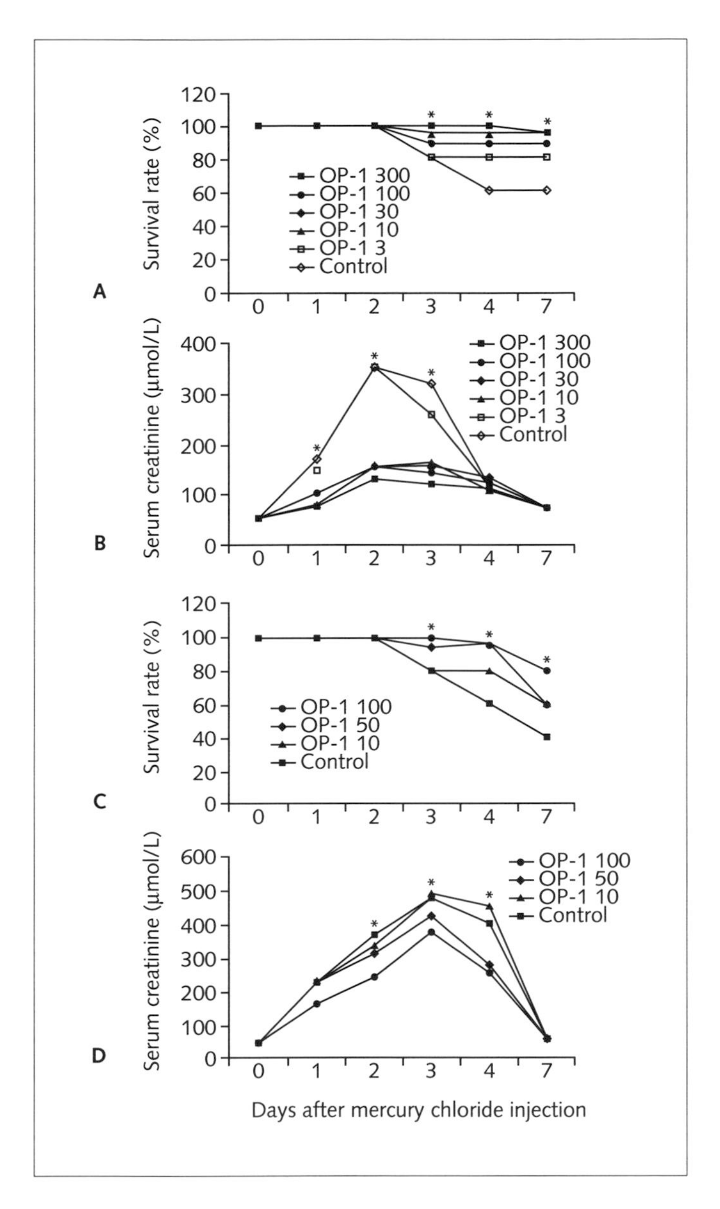


*Figure 7*
*Prophylactic (A and B) and therapeutic (C and D) effects of BMP-7 on the survival rate and serum creatinine values in rats subjected to acute toxic renal failure. Animals were given mercuric chloride (4 mg/kg) in a bolus at the beginning of the experiment. Vehicle (acetate buffer, pH 4,5) and BMP-7 were administered daily at 24-h intervals beginning on day 0, 10 min before the insult (data shown as mean ± SEM;* $p < 0.01$*, Student's* t*-test), or beginning 8 h following the insult. (Data shown as mean ± SEM;* $p < 0.01$*, Student's t-test.)*

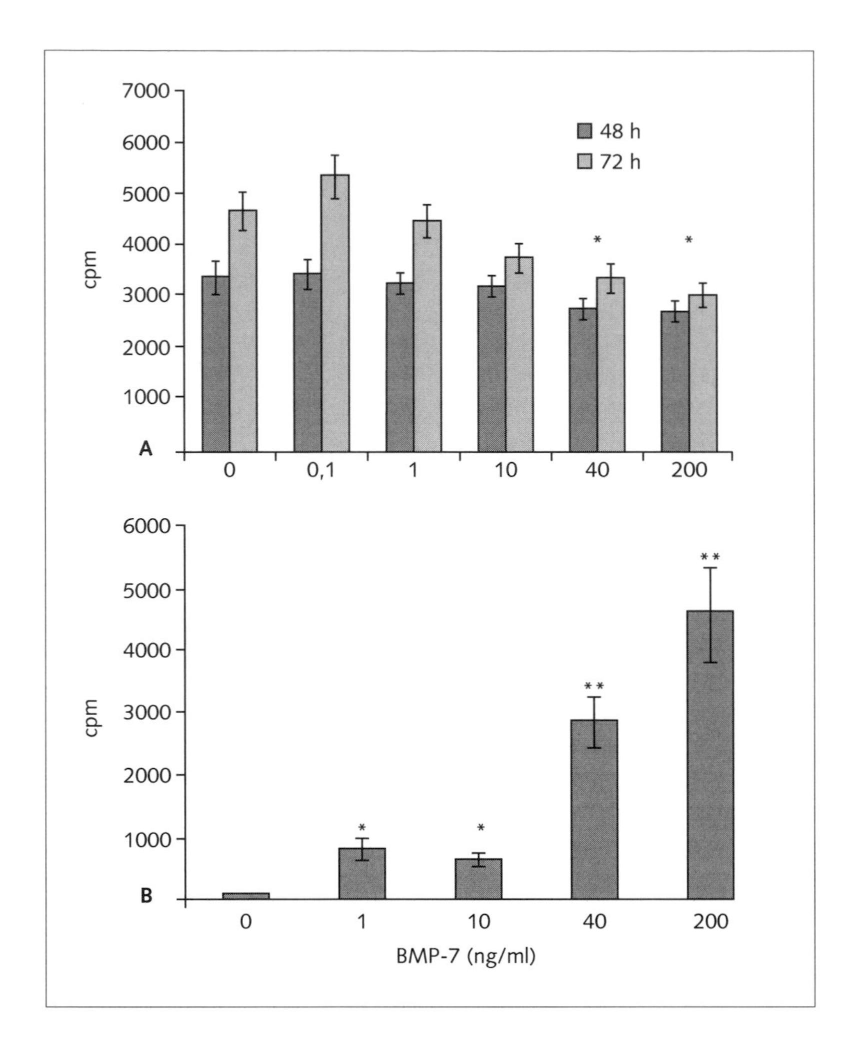


*Figure 8*

*Effect of BMP-7 on proliferation of intact or mercuric chloride ($HgCl_2$) exposed proximal kidney tubule cells (REPTEC).*

*(A) Human REPTEC cells were plated in serum-free medium for 24 h at a density of 4,000 cells per well in a 24-well plate. Cells were then incubated with different concentrations of BMP-7 for 48 (■) and 72 (□) h. The cells were pulsed with [$^3$H]-thymidine for the last 2 h of the culturing period. Data are shown as mean ± SEM. (*p < 0.05; Student's t-test)*

*(B) Human REPTEC cells were incubated with 30 μM of mercuric chloride for a period of 3 h. After exhaustive washing, the cells were incubated with BMP-7 for a period of 24 h and pulsed with [$^3$H]-thymidine for the last 2 h of the culturing period. BMP-7 dose-dependently protected against the toxic injury even when applied 3 h following $HgCl_2$ exposure. Data shown as mean ± SEM. (*p < 0.01; **p < 0.001; Student's t-test.)*

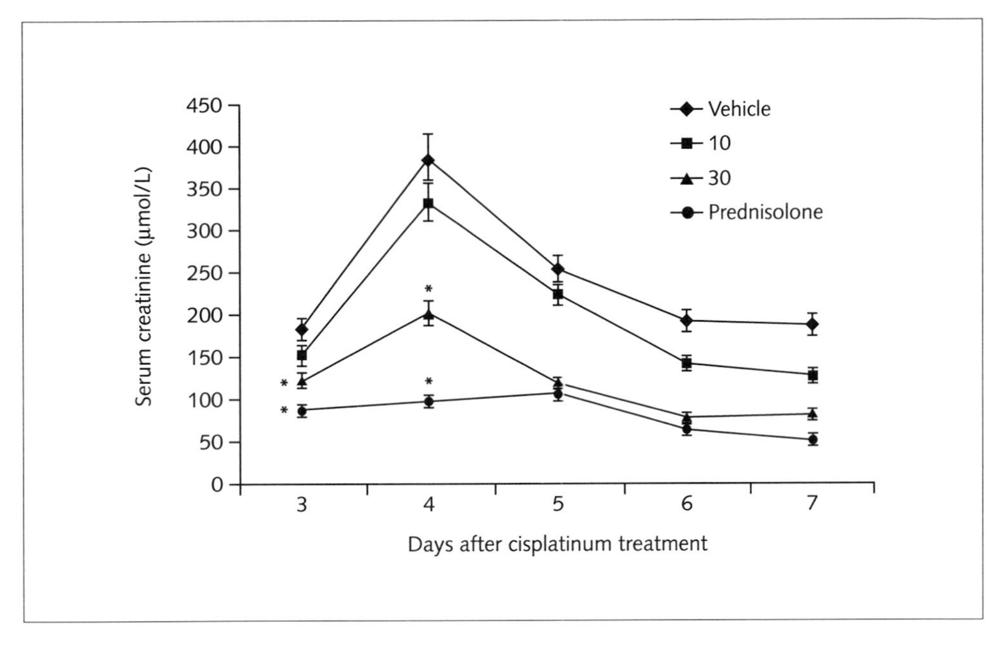


*Figure 9*
*Protection from kidney injury by BMP-7 in rats, following the application of 5 mg/kg of cisplatinum. Cisplatinum was administered intraperitoneally (5 mg/kg) to intact animals. Vehicle (◆, acetate buffer, pH 4,5), BMP-7 (10, ■, and 30, ▲, μg/kg) or prednisolone (●) as administered immediately before the application of cisplatinum and every 24 h thereafter for 4 days. Animals treated with 30 μg/kg of BMP-7 or with prednisolone showed significant reduction in serum creatinine values. Data shown as mean ± SEM. (*p < 0.001 on day 3 and p < 0.005 on day 4 following the application of cisplatinum, Student's t-test.)*

BMP-7 mRNA expression is most abundant on day 12, with a slow decline after day 15 (Fig. 2). It seems that there is a time frame during nephrogenesis in which the presence of BMP-7 is required for normal kidney development. In nephrogenic mesenchyme tissue explant cultures, BMP-7 was shown to prevent apoptosis [61] and the same effect was observed *in vivo* in ischemic-reperfusion injury [8]. Kidney BMP-7 mRNA and protein are selectively downregulated in the medulla after acute ischemic renal injury [8], thus BMP-7 modulation may be a key element for kidney repair [62]. Whether BMP-7 has a direct growth-promoting function either through early genes activation or apoptosis inhibition in damaged tubular cells, or it simply serves as a functional free radical scavenger, remains to be determined. Collectively, these data suggest that BMP-7 reduces the severity of renal damage associated with ischemia/reperfusion and nephrotoxic agents, and, as such, may provide a basis for the treatment of acute renal failure.

## BMP-7 kidney receptors are specific

Recently, membrane-bound, specific, high-affinity BMP-7 receptors in rat kidney tissues mediating BMP-7 actions have been characterized [63]. The major BMP-7-binding component of the kidney may be a long form of BMP type II receptor with a $M_r$ of 100 kDa. *In vivo* evidence suggests that the cellular target for BMP-7 in the kidney are the convoluted tubule epithelium and glomeruli in the cortex, and the collecting ducts in the medulla region. Moreover, *in situ* hybridization and immunostaining methods have shown localization of mRNA transcripts and the protein for BMP type II receptor in similar areas of the cortex and medulla.

It is noteworthy that plasma membranes from both the kidney cortex and medulla show the presence of specific receptors for BMP-7. The relative abundance of BMP-7 binding sites in cortex membranes is much higher than in the medulla region. Moreover, Scatchard analysis indicates that the receptors in the kidney cortex contained a single class of high-affinity BMP-7 binding sites, with a $K_a$ of $2.26 \times 10^9$ mol/L [63]. The calculated binding capacity of receptors per mg membrane protein is 1.01 pmol BMP-7. Recently, the presence of both high- and low-affinity binding sites for TGF-β have been identified in the proximal tubules isolated from the rabbit renal cortex [64]. However, so far there is no evidence of low-affinity BMP-7 binding sites in kidney cortex plasma membranes [63]. It is important to note that the endogenous levels of TGF-β and other related growth factors are normally low, and high-affinity and low-capacity receptors for these factors are implicated to mediate their actions.

The relative uptake of radiolabeled BMP-7 at 10 and 180 min in the cortex is 270 ng and 80 ng/g tissue, respectively. These values of BMP-7 are not considered to be low, since studies with TGF-β and activin also showed low tissue distribution [65, 66]. It has been shown that BMP-7 at these concentrations is effective in cell cultures in maintaining the epithelial phenotype of human proximal epithelial cells. Interestingly, tissue autoradiography, *in situ* hybridization, and immunostaining with a site-directed receptor antibody all identified the convoluted tubule epithelium, glomeruli and the collecting ducts of the medulla as cellular targets for BMP-7 [8, 63]. Previous studies have shown that the rat kidney is the major source for BMP-7 [3, 4] and that the major site of BMP-7 production is the epithelium of the collecting ducts within the medulla [8]. Taken together, these results suggest that BMP-7 might have both paracrine and autocrine roles in the kidney. It is pertinent to mention that tissue autoradiography has shown localization of radiolabeled BMP-7 in the S3 segment. Moreover, by *in situ* hybridization, it has been found that epithelial cells in the S3 zone synthesize BMP-7 mRNA. Therefore, it is likely that in case of an ischemic injury within the S3 zone, exogenously administered BMP-7 binds to cell receptors and protects from necrosis and infarction, as has been previously demonstrated [8]. When systemically administered, BMP-7 binds to α2-macroglobulin, which is present at high concentrations in blood. It is important to note that upon activation by protease, α2-macroglobulin undergoes a conformational change that exposes a previously buried domain close to the carboxyl terminus. That domain is then recognized by a cell surface receptor system in the liver, which mediates binding and endocytosis of the complex. This is the mechanism by which TGF-β is targeted to the liver by binding to activated α2-macroglobulin [67].

The degree of specificity with which BMP-7 interacts with the kidney receptors is high [63]. Other growth factors such as PDGF, TGF-β, IGF and FGF, even at high concentrations, fail to inhibit the binding of $^{125}$I-labeled BMP-7 to kidney plasma membrane receptors. Similarly, other members of the BMP family such as BMP-2 and CDMP-1 also fail to affect BMP-7 interaction with kidney receptors. Thus, BMP-7 does not share receptor-binding properties with other growth factors, and its mode of action in the kidney appears to be specific [63]. It is important to note that BMP-2 and CDMP-l show only 60 and 51% homology, respectively, with the primary sequence of BMP-7, suggesting that BMP-7 interaction with kidney cortex receptors may involve regions in BMP-7 that are not well conserved among these growth factors.

Miyazono and his associates cloned type I and type II receptors for BMPs and expressed them in COS cells [68, 69]. BMP-7 was shown to bind to two recombinant type I receptors, ALK-2 and ALK-6, and to ALK-3 less efficiently. These ALK receptors had $M_r$ values in the range of 50 to 58 kDa (see also the chapter by ten Dijke). On the other hand, the recombinant type II receptor is much larger and it has two forms, a truncated form with no C-terminus extension [70] and a long form with a $M_r$ of approximately 100 kDa [68]. The type II receptor can effectively bind BMP-7 on its own, while type I receptors are required to be coexpressed with the type II receptor for efficient binding to BMP-7. When plasma membranes isolated from the kidney cortex or medulla were analyzed by ligand blotting, each showed the presence of a prominent band with an $M_r$ of 100 kDa [63]. Interestingly, the size of the BMP-7-binding component of the rat kidney appears to match with Mr of the cloned BMP type II receptor. Further analysis by Western blot method using a site-directed receptor antibody identified the 100 kDa component as a BMP type II receptor. Consistent with this observation, both *in situ* hybridization and immunostaining methods have shown that mRNA transcripts and the protein for the BMP type II receptor are localized in glomeruli and adjacent convoluted tubules of the cortex, and in the collecting ducts of the medulla. Garcia-Ocana et al. have shown that hypertrophy of the proximal tubule is associated with an increased production of both TGF-β and TGF-β receptors [64]. On the other hand, in experimental membranous nephropathy, injury to glomerular epithelial cells is associated with an upregulation of the TGF-β2 and TGF-β3 isoforms, and an increase in TGF-β3 type I and type II receptor expression. Studies by Flyvbjerg et al. have shown that an initial increase in renal size and function in the experimental diabetic kidney is always preceded by an increase in renal IGF-I, IGF-binding proteins, and IGF receptor concentrations [71]. Clearly, those and the present studies signify the importance of

BMP-7, BMP-7 receptor, TGF-β, TGF receptors, IGF and IGF receptors as major regulators in kidney physiology and renal repair. Whether BMP-7 receptors in renal proximal tubules and glomeruli show similar concentration changes to regulate tubular cell growth and differentiation after renal injury remains to be elucidated. Moreover, these findings provide a molecular basis for the interaction of BMP-7 with different kidney regions [63].

## Chronic renal failure

Progressive and permanent reduction in the glomerular filtration rate (GFR), which is associated with the loss of functional nephron units, leads to chronic renal failure (CRF).

The subject progresses to end-stage renal disease when the GFR continues to decline to less than 10% of normal values (5–10 ml/min). At this point, renal failure will rapidly progress to cause death unless the subject receives renal replacement therapy, i.e. chronic hemodialysis, continuous peritoneal dialysis or kidney transplantation, or therapy that delays the progression of chronic renal disease.

The effect of systemically administered BMP-7 to delay or halt progression of end stage renal failure in a remnant kidney (5/6 nephrectomy) rat model was investigated. Recombinant human BMP-7 at doses of 10 μg/kg was administered three times per week intravenously beginning 2 days following surgery and continuing for 11 weeks. The effect of BMP-7 was monitored by serum creatinine values (Cr), GFR, and the survival rate. The results indicate that 2 weeks after the beginning of treatment, BMP-7 considerably decreased serum Cr values as compared to control animals. Rats treated with BMP-7 had better GFR and prolonged survival rate (Fig. 10).

The higher GFR observed in BMP-7-treated rats and the hystomorphometric analysis suggest that BMP-7 is capable of preventing rapid deterioration of the glomerular function in this model. In 18 weeks following nephrectomy the survival rate was 88% in BMP-7-treated rats as compared to 32% in controls. The experiment was terminated 30 weeks following nephrectomy with 60% survivors in BMP-7-treated and 15% survivors in control rats, respectively (Fig. 10). This result suggests that BMP-7 can delay the progression of the terminal phase of chronic renal failure. Since the process of the chronic kidney failure in humans lasts over years, delaying the progression is critical for the treatment of chronic kidney diseases. BMP-7 might provide a potential therapeutic basis for the treatment of end-stage renal failure.

In another model mimicking chronic renal injury human recombinant BMP-7 was systemically administered to rats with unilateral ureteral obstruction (UUO) and produced nearly complete protection for 5 days against tubulointerstitial fibrosis [9]. Tubulointerstitial fibrosis is a common final pathway contributing to pro-

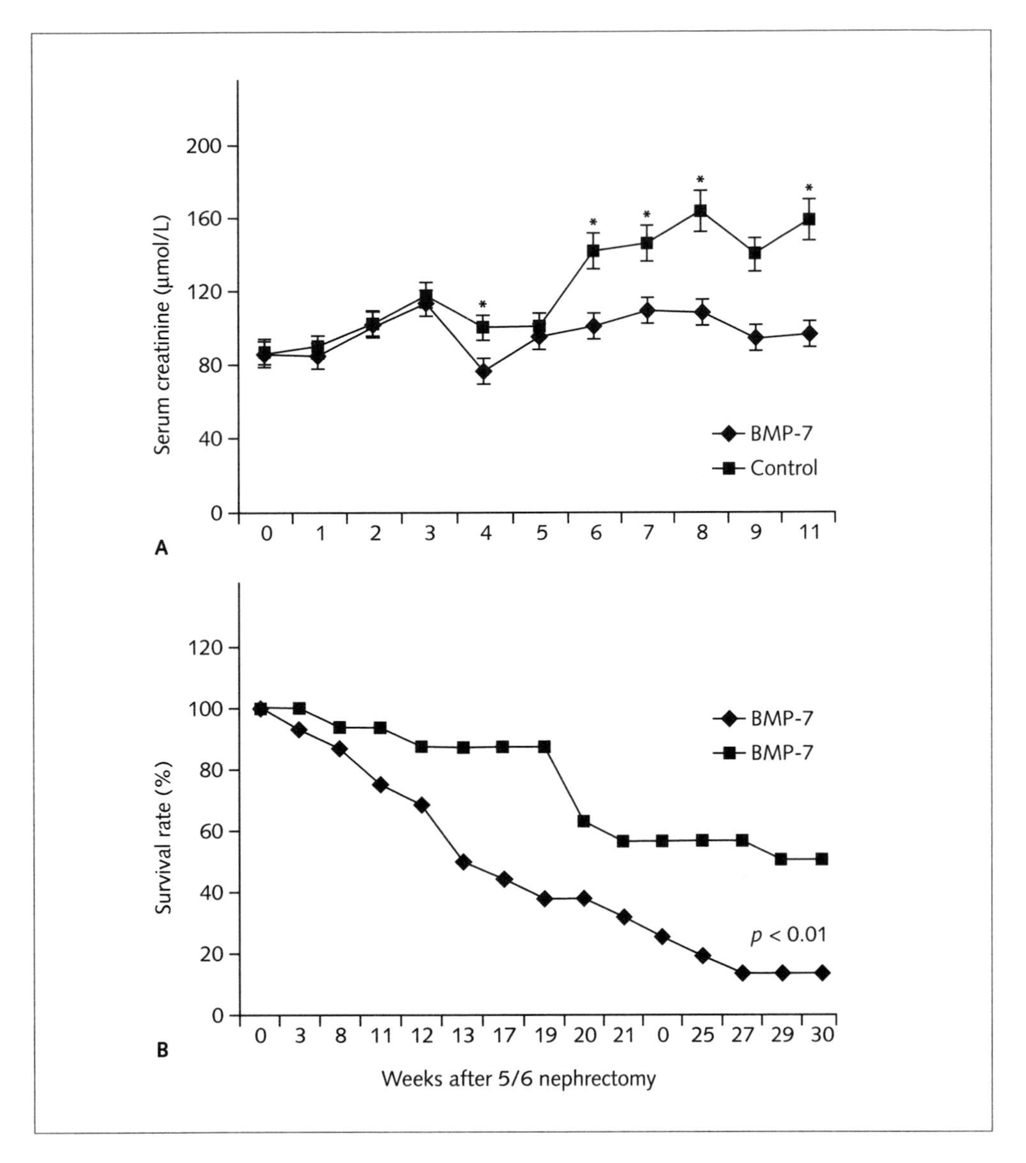


*Figure 10*

*Therapeutic effect of BMP-7 on serum creatinine values and the survival rate in rats following 5/6 nephrectomy. One week following removal of 5/6 kidney mass, rats were subjected to i.v. application of BMP-7 (10 µg/kg ) or a vehicle acetate buffer three times a week and serum creatinine values were measured throughout 11 weeks. Data are shown as average ± SEM;* $p \leq 0.01$ *for BMP-7 vs. vehicle treated rats, Student's* t*-test in (A) and Petö-Wilcoxon test (B) was used for the statistical analysis of the survival rate;* $p < 0.01$ *for BMP-7 treated vs. vehicle treated rats throughout 30 weeks of BMP-7 treatment.*

gression of many chronic kidney diseases [72–74]. UUO activates a cascade of events that produce tubulointerstitial fibrosis [75–77]. An early event in the damage cascade is angiotensin II upregulation, which stimulates tumor necrosis factor-α (TNF-α) production and TGF-β expression [78–82]. These cytokines activate nuclear factor κB (NF-κB), a crucial transcription factor in fibroblasts, macrophages and epithelial cells, involved in renal cellular transformation and apoptosis as well as interstitial inflammation and subsequent fibrosis. The damage cascade stimulated by UUO closely resembles that produced by several forms of renal injury [78, 83–85]. Suppression of this damage cascade might prevent fibrogenesis and preserve renal function. BMP-7 suppressed UUO-stimulated loss of the tubular epithelium due to apoptosis and prevented the transformation of renal cells into interstitial myofibroblasts [9]. This suggests that, whereas BMP-7 prevented tubular cell apoptosis as previously reported [8], it further appears to have maintained the phenotype of tubular cells and the interstitial fibroblasts. Both tubular cells and interstitial fibroblasts are subjected to phenotypic alterations as a result of UUO [74, 75, 86–88]. The preponderance of evidence is that phenotypic alteration of epithelial and fibroblastic cells to myofibroblasts is detrimental and leads to a progressive loss of renal function [75, 87–90]. BMP-7 administration was similar to but greater than enalapril in its protective action against tubulointerstitial fibrosis [9]. In addition, BMP-7 preserved the tubular epithelial structure and prevented tubular atrophy. In comparison, ACE inhibition decreases the activity of the damage cascade by suppressing UUO stimulation of TGF-β, TNF-α, and NF-κB, which are mediated by angiotensin II [76, 77]. Approximately 50% of the stimulation of this damage cascade, after UUO, is due to angiotensin II [75] and data suggest that more than 50% is suppressed by BMP-7. Thus BMP-7 may function as a renal homeostasis signal by providing a survival signal to epithelial cells, protecting the tubular epithelial cell phenotype, and suppressing gene activation associated with injury.

## Conclusion

BMPs may have important functions in kidney development and renal diseases. BMP-7 regulates kidney mesenchyme differentiation and preserves renal function by preventing inflammation and fibrosis following ischemia, nephrectomy and ureteral obstruction.

## References

1 Helder MN, Ozkaynak E, Sampath TK, Luyten FP, Latin V, Oppermann H, Vukicevic S (1995) Expression pattern of osteogenic protein-1 (bone morphogenetic protein-7) in human and mouse development. *J Histochem Cytochem* 43: 1035–1044

2 Vukicevic S, Stavljenic A, Pecina M (1995) Discovery and clinical applications of bone morphogenetic proteins. *Eur J Clin Chem Clin Biochem* 33: 661–671
3 Ozkaynak E, Schnegelsberg PN, Opperman H. (1991) Murine osteogenic protein-1 (OP-1): high levels of mRNA in kidney. *Biochem Biophys Res Commun* 179: 116–123
4 Vukicevic S, Kopp JB, Luyten FB, Sampath TK (1996) Induction of nephrogenic mesenchyme by osteogenic protein-1 (bone morphogenetic protein 7). *Proc Natl Acad Sci USA* 93: 9021–9026
5 Dudley AT, Lyons KM, Robertson EJ (1995) A requirement for bone morphogenetic protein-7 during development of the mammalian kidney and eye. *Genes Dev* 9: 2795–2807
6 Luo O, Hofmann A, Bronckers JJ, Sohocki M, Bradley A, Karsenty G (1995) BMP-7 is an inducer of nephrogenesis and is also required for eye development and skeletal patterning. *Genes Dev* 9: 2808–2820
7 Simon M, Maresh JG, Harris SE, Hernandez JD, Arar M, Olson MS, Abboud HE (1999) Expression of bone morphogenetic protein-7 mRNA in normal and ischemic adult rat kidney. *Am J Physiol* 276: 382–389
8 Vukicevic S, Basic V, Rogic D, Basic N, Shih M, Shepard A, Jin D, Dattatreyamurty B, Jones W, Dorai H et al (1998) Osteogenic protein-1 (bone morphogenetic protein-7) reduces severity of injury after ischemic acute renal failure in rat. *J Clin Invest* 102: 202–214
9 Hruska KA, Guo G, Wozniak M, Martin D, Miller S, Liapis H, Loveday K, Klahr S, Sampath TK, Morrissey J (2000) Osteogenic protein-1 prevents renal fibrogenesis associated with ureteral obstruction. *Am J Physiol Renal Physiol* 279: F130–F143
10 Saxen L (1987) *Organogenesis of the kidney*. Cambridge Univ. Press, Cambridge
11 Grobstein C (1953) Inductive epithelio-mesenchymal interactions in cultured organ rudiments of the mouse. *Science* 118: 52–55
12 Grobstein C (1956) Trans-filter induction of tubules in mouse metanephrogenic mesenchyme. *Exp Cell Res* 10: 434–440
13 Lipschutz JH (1998) Molecular development of the kidney: a review of the results of gene disruption studies. *Am J Kidney Dis* 31: 383–97
14 Schedl A, Hastie ND (2000) Cross-talk in kidney development. *Curr Opin Genet Dev* 10: 543–549
15 Barasch J, Yang J, Ware CB, Taga T, Yoshida K, Erdjument-Bromage H, Tempst P, Parravicini E, Malach S, Aranoff T, Oliver JA (1999) Mesenchymal to epithelial conversion in rat metanephros is induced by LIF. *Cell* 99: 377–386
16 Moore MW, Klein RD, Farinas I, Sauer H, Armanini M, Philips H, Reichardt LF, Ryan AM, Carver-Moore K, Rosenthal A (1996) Renal and neuronal abnormalities in mice lacking GDNF. *Nature* 382: 76–79
17 Suchardt A, D'Agati V, Larsson-Blomberg L, Constantini F, Pachinis V (1994) Defects in kidney and enteric nervous system of mice lacking the tyrosin kinase receptor Ret. *Nature* 367: 380–383
18 Vega QC, Worby CA, Lechner MS, Dixon JE, Dressler GR (1996) Glial cell line-derived

neurotrophic factor activates RET and promotes kidney morphogenesis. *Proc Natl Acad Sci USA* 93: 10657–10661
19 Jing S, Wen D, Yu Y, Holst PL, Luo Y, Fang M, Tamir R, Antonio L, Hu Z, Cupples R et al (1996) GDNF-induced activation of the Ret protein tyrosin kinase is mediated by GDNFR alpha, a novel receptor for GDNF. *Cell* 85: 1113–1124
20 Pichel JG, Shen L, Sheng HZ, Granholm AC, Drago J, Grinberg A, Lee EJ, Huang SP, Saarma M, Hoffer BJ et al (1996) Defects in enteric innervation and kidney development in mice lacking GDNF. *Nature* 382: 73–76
21 Kreidberg JA, Sariola H, Loring JM, Maeda M, Pelletier J, Housman D, Jaenisch R (1993) WT-1 is required for early kidney development. *Cell* 74: 679–691
22 Lee SB, Huang K, Palmer R, Truong VB, Herzlinger D, Kolquist KA, Wong J, Paulding C, Yoon SK, Gerald W et al (1999) The Wilms tumor suppressor WT1 encodes a transcriptional activator of amphiregulin. *Cell* 98: 663–673
23 Rothenpieler UW, Dressler GR (1993) Pax-2 is required for mesenchyme-to-epithelium conversion during kidney development. *Development* 119: 711–720
24 Torres M, Gomez Pardo E, Dressler GR, Gruss P (1995) Pax-2 controls multiple steps of urogenital development. *Development* 121: 4057–4065
25 Ryan G, Steele-Perkins V, Morris J, Rauscher FJ, Dressler GR (1995) Repression of Pax-2 by WT-1 during normal kidney development. *Development* 121: 867–875
26 Dressler GR, Wilkinson JE, Rothenpieler UW, Patterson LT, Williams-Simons L, Westphal H (1993) Deregulation of Pax-2 expression in transgenic mice generates severe kidney abnormalities. *Nature* 362: 65–67
27 Stark K, Vainio S, Vassileva G, McMahon AP (1994) Epithelial transformation of metanephric mesenchyme in the developing kidney regulated by Wnt-4. *Nature* 372: 679–683
28 Hatini V, Huh SO, Hertzlinger D, Soares VC, Lai E (1996) Essential role of stromal mesenchyme in kidney morphogenesis revealed by targeting disruption of winged helix transcription factor BF-2. *Genes Dev* 10: 1467–1478
29 Leveen P, Pekny M, Gebre-Medhin S, Swolin B, Larsson E, Betzholtz C (1994) Mice deficient for PDGF B show renal, cardiovascular and haematological abnormalities. *Genes Dev* 8: 1875–1887
30 Dudley AT, Robertson EJ (1997) Overlapping expression domains of bone morphogenetic protein family members potentially account for limited tissue defects in BMP-7 deficient embryos. *Dev Dyn* 208: 349–362
31 Karsenty G, Luo G, Hofmann C, Bradley A (1996) BMP-7 is required for nephrogenesis, eye development, and skeletal patterning. *Ann NY Acad Sci* 785: 98–107
32 Piscione TD, Yager TD, Gupta IR, Grinfeld B, Pei Y, Attisano L, Wrana J, Rosenblum ND (1997) BMP-2 and BMP-7 exert direct and opposite effects on renal branching morphogenesis. *Am J Physiol* 273:F961–F975
33 Kitten AM, Kreisberg JI, Olson MS (1999) Expression of osteogenic protein-1 mRNA in cultured kidney. *J Cell Physiol* 181: 410–415

34 Godin RE, Takaesu NT, Robertson EJ, Dudley AT (1998) Regulation of BMP-7 expression during kidney development. *Development* 125: 3473–3482
35 Obara-Ishihara T, Kuhlman J, Niswander L, Herzlinger D (1999) The surface ectoderm is essential for nephric duct formation in intermediate mesoderm. *Development* 126: 1103–1108
36 Miyazaki Y, Oshima K, Fogo A, Hogan BL, Ichikawa I (2000) Bone morphogenetic protein 4 regulates the budding site and elongation of the mouse ureter. *J Clin Invest* 105: 863–873
37 Raatikainen-Ahokas A, Hytonen M, Tehnunen A, Sainio K, Sariola H (2000) BMP-4 affects the differentiation of metanephric mesenchyme and reveals an early anterior-posterior axis of the embryonic kidney. *Dev Dyn* 217: 146–158
38 Gupta IR, Macias-Silva M, Kim S, Zhou X, Piscione TD, Whiteside C, Wrana JL, Rosenblum ND (2000) BMP-2/ALK3 and HGF signal in parallel to regulate renal collecting duct morphogenesis. *J Cell Sci* 113: 269–278
39 Ghosh Choundhury G, Kim YS, Simon M, Wozney J, Harris S, Ghosh Choundhury N, Abboud HE (1999) Bone morphogenetic protein 2 inhibits platelet-derived growth factor-induced c-fos gene transcription and DNA synthesis in mesangial cells. Involvement of mitogen-activated protein kinase. *J Biol Chem* 274: 10897–10902
40 Vukicevic S, Helder MN, Luyten FP (1994) Developing human lung and kidney are major sites for synthesis of bone morphogenetic protein-3 (osteogenin). *J Histochem Cytochem* 42: 869–875
41 Takahashi H, Ikeda T (1996) Transcripts for two members of the transforming growth factor-beta superfamily BMP-3 and BMP-7 are expressed in developing rat embryos. *Dev Dyn* 207: 439–449
42 Letterio JJ, Geiser AG, Kulkarni AB, Roche NS, Sporn MB, Roberts AB (1994) Maternal rescue of transforming growth factor-beta 1 null mice. *Science* 264: 1936–1938
43 Borovecki F, Jelic M, Sampath TK, Vukicevic S. Evidence that circulating bone morphogenetic protein-7 (osteogenic protein-1) in pregnant mice is available to the fetus through placental transfer during early stages of development. *Kidney Int; in press*
44 Thadhani R, Pascual M, Bonventre JV (1996) Acute renal failure. *N Engl J Med* 334: 1448–1460
45 Humes HD, MacKay SM, Funke AJ, Buffington DA (1997) Acute renal failure: growth factors, cell therapy and gene therapy. *Proc Assoc Am Physicians* 109: 547–557
46 Hirschberg R, Ding H (1998) Growth factors and acute renal failure. *Semin Nephrol* 18: 191–207
47 Humes DH, Liu S (1994) Cellular and molecular basis of renal repair in acute renal failure. *J Am Soc Nephrol* 5: 1–11
48 Witzgall R, Brown D, Schwarz C, Bonventre JV (1994) Localization of proliferating cell number antigen, vimentin, c-Fos and clusterin in the post-ischemic kidney: evidence for a heterogenous genetic response among nephron segments and a large pool of mitotically active and differentiated cells. *J Clin Invest* 93: 2175–2188

49 Hirschberg R, Kopple JD (1989) Evidence that insulin-like growth factor I increases renal plasma flow and glomerular filtration rate in fasted rats. *J Clin Invest* 83: 326–330
50 Andersson G, Jennische E (1988) IGF-I immunoreactivity is expressed by regenerating renal tubule cells after ischaemic injury in the rat. *Acta Physiol Scand* 132: 453–457.
51 Sugimura K, Goto T, Kasai S, Tsuchida K, Takemoto Y, Yamagami S (1998) The activation of serum hepatocyte growth factor in acute renal failure. *Nephron* 76: 364–365
52 Coimbra TM, Cieslinski DA, Humes HD (1990) Epidermal growth factor accelerates renal repair in mercuric chloride nephrotoxicity. *Am J Physiol* 259: 438–443
53 Weinberg JM (1993) The cellular basis of nephrotoxicity. In: RW Schrier, CW Gottschalk (eds): *Diseases of the kidney*. Little, Brown and Company, Boston, 1031–1098
54 Guillermina G, Adriana TM, Monica EM (1989) The implications of renal glutatione level in mercuric chloride nephrotoxicity. *Toxicology* 58: 187–195.
55 Houser MT, Milner LS, Kolbeck PC, Wei SH, Stohs SJ (1992) Glutathione monoethyl ester moderates mercuric chloride-induced acute renal failure. *Nephron* 61: 449–455
56 Nath KA, Croatt AJ, Likely S, Behrens TW, Warden D (1996) Renal oxidant injury and oxidant response induced by mercury. *Kidney Int* 50: 1032–1043
57 Southard J, Nitisewojo P, Green DE (1974) Mercurial toxicity and the perturbation of the mitochondrial control system. *Fed Proc* 33: 2147–2153
58 Fausto N (2000) Liver regeneration. *J Hepathol* 32: 19–31
59 Reddi AH, Huggins C (1972) Biochemical sequences in the transformation of normal fibroblasts in adolescent rats. *Proc Natl Acad Sci USA* 69: 1601–1605
60 Luyten FP, Cunningham NS, Vukicevic S, Paralkar V, Ripamonti U, Reddi AH (1992) Advances in osteogenin and related bone morphogenetic proteins in bone induction and repair. *Acta Orthop Belg* 58: 263–267
61 Dudley AT, Godin RE, Robertson EJ (1999) Interaction between FGF and BMP signaling pathways regulates development of metanephric mesenchyme. *Genes Dev* 15: 1601–1613
62 Almanzar MM, Kendall FS, Philip DH, Piqueras AI, Jones WK, Charette MF, Paredes AL (1998) Osteogenic protein-1 mRNA expression is selectively modulated after acute ischemic renal injury. *J Am Soc Nephrol* 9: 1456–1463
63 Bosukonda D, Shih MS, Sampath KT, Vukicevic S (2000) Characterization of receptors for osteogenic protein-1/bone morphogenetic protein-7 (OP-1/BMP-7) in rat kidneys. *Kidney Int* 58: 1902–1911
64 Garcia-Ocana A, Penaranda C, Esbrit P (1996) Transforming growth factor-beta and its receptors in rabbit renal proximal tubules after uninephrectomy. *Exp Nephrol* 4: 231–240
65 Dickson K, Philip A, Warshawsky H, O'Connor-McCourt M, Bergeron JJ (1995) Specific binding of endocrine transforming growth factor-beta 1 to vascular endothelium. *J Clin Invest* 95: 2539–2554
66 Niemuller CA, Randall KJ, Webb DJ, Gonias SL, LaMarre J (1995) Alpha 2-macroglob-

ulin conformation determines binding affinity for activin A and plasma clearance of activin A/alpha 2-macroglobulin complex. *Endocrinology* 136: 5343–5349

67 LaMarre J, Wollenberg GK, Gonias SL, Hayes MA (1991) Reaction of alpha 2-macroglobulin with plasmin increases binding of transforming growth factors-beta 1 and beta 2. *Biochim Biophys Acta* 1091: 197–204

68 Rosenzweig BL, Imamura T, Okadome T, Cox GN, Yamashita H, ten Dijke P, Heldin CH, Miyazono K (1995) Cloning and characterization of a human type II receptor for bone morphogenetic proteins. *Proc Natl Acad Sci USA* 92: 7632–7636

69 Yamashita H, ten Dijke P, Heldin CH, Miyazono K (1996) Bone morphogenetic protein receptors. *Bone* 19: 569–574

70 Liu F, Ventura F, Doddy J, Massague J (1995) Human type II receptor for bone morphogenic proteins (BMPs): Extension of the two-kinase receptor model to the BMPs. *Mol Cell Biol* 15: 3479–3486

71 Flyvbjerg A, Landau D, Domane H, Hernandez L, Gronback H, Leroith D (1995) The role of growth hormone, insulin-like growth factors (IGFs), and IGF-binding proteins in experimental diabetic kidney disease. *Metabolism* 44: 67–71

72 Couser WG (1993) Mediators of immune glomerular injury. *Clin Invest* 71: 8–11

73 Couser WG (1993) Pathogenesis of glomerulonephritis. *Kidney Int* 44: S519–S526

74 Johnson RJ, Hugo C, Hasley C, Pichler RH, Bassuk J, Thomas S, Suga S, Couser WG, Shankland SJ (1998) Mechanisms of progressive glomerulonecrosis and tubulointerstitial fibrosis. *Clin Exp Nephrol* 2: 307–312

75 Fern RJ, Yesko CM, Thornhill BA, Kim H-Y, Smithies O, Chevalier RL (1999) Reduced angiotensinogen expression attenuates renal interstitial fibrosis in obstructive nephropathy in mice. *J Clin Invest* 103: 39–46

76 Klahr SS (1998) Nephrology forum: obstructive nephropathy. *Kidney Int* 54: 286–300

77 Klahr S, Morrissey J (1998) Angiotensin II and gene expression in the kidney. *Am J Kidney Dis* 31: 171–176

78 Border WA, Noble NA (1998) Interactions of transforming growth factor-beta and angiotensin II in renal fibrosis. *Hypertension* 31: 181–188

79 Douglas JG, Romero M, Hopfer U (1990) Signaling mechanisms coupled to the angiotensin receptor of proximal tubular epithelium. *Kidney Int* 30: S43–S47

80 Kaneto H, Morrissey J, Klahr S (1993) Increased expression of TGF-β1 mRNA in the obstructed kidney of rats with unilateral ureteral ligation. *Kidney Int* 44: 313–321

81 Kaneto H, Morrissey J, McCracken R, Ishidoya S, Reyes A, Klahr S (1996) The expression of mRNA for tumor necrosis factor increases in the obstructed kidney of rats soon after unilateral ureteral ligation. *Nephrology* 2: 161–166

82 Kalahr S, Ishidoya S, Morrissey J (1995) Role of angiotensin II in the tubulointerstitial fibrosis of obstructive nephropathy. *Am J Kidney Dis* 26: 141–146

83 Johnson RJ, Alpers CE, Yoshimura A, Lombardi D, Pritzl P, Floege J, Schwartz SM (1992) Renal injury from angiotensin II-mediated hypertension. *Hypertension* 19: 464–474

84 Kagami S, Border WA, Miller DE, Noble NA (1994) Angiotensin II stimulates extracel-

lular matrix protein synthesis through induction of transforming growth factor-beta expression in rat glomerular mesangial cells. *J Clin Invest* 93: 2431–2437
85 Yoo KH, Thornhill BA, Wolstenholme JT, Chevalier RL (1998) Tissue-specific regulation of growth factors and clusterin by angiotensin II. *Am J Hypertens* 11: 715–722
86 Chevalier RL, Kim A, Thornhill BA, Wolstenholme JT (1999) Recovery following relief of unilateral ureteral obstruction in the neonatal rat. *Kidney Int* 55: 793–807
87 Nagle RB, Johnson ME, Jervis HR (1976) Proliferation of renal interstitial cells following injury induced by ureteral obstruction. *Lab Invest* 35: 18–22
88 Ng YY, Huang TP, Yang WC, Chen ZP, Yang AH, Mu W, Nikolic-Paterson DJ, Atkins RC, Lan HY (1998) Tubular epithelial-myofibroblast transdifferentiation in progressive tubulointerstitial fibrosis in 5/6 nephrectomized rats. *Kidney Int* 54: 864–876
89 Ishidoya S, Morrissey J, McCracken R, Reyes A, Klahr S (1995) Angiotensin II receptor antagonist ameliorates renal tubulointerstitial fibrosis caused by unilateral ureteral obstruction. *Kidney Int* 47: 1285–1294
90 Ishidoya S, Morrissey J, McCracken R, Klahr S (1996) Delayed treatment with enalapril halts tubulointerstitial fibrosis in rats with obstructive nephropathy. *Kidney Int* 49: 1110–1119

# Effects of bone morphogenetic proteins on neural tissues

*Pamela Lein*[1], *Karen M. Drahushuk*[2] *and Dennis Higgins*[2]

[1]Department of Environmental Health Sciences, School of Hygiene and Public Health, Johns Hopkins University, Baltimore, MD 21205, USA; [2]Department of Pharmacology and Toxicology, State University of New York, Buffalo, NY 14214, USA

## Introduction

Bone morphogenetic proteins (BMPs) were originally identified by their ability to induce ectopic bone formation [1]. However, it was subsequently found that BMPs are expressed in most, if not all, developing organs and that they profoundly alter the development of kidney [2], lung [3], blood [4], and heart [5], as well as cartilage, mesoderm and bone [1].

BMPs are also prominently expressed in the central and peripheral nervous systems and they have been implicated in the control of a host of critical developmental phenomena, including: neurulation, dorsal-ventral patterning, specification of neural and glial cell lineages, neural cell survival and proliferation, segmentation, axonal guidance, determination of neurotransmitter phenotype, regulation of dendritic growth and synapse formation. In addition, BMPs are neuroprotective in mature animals in models of ischemic and excitotoxin-induced injury. In this review, we try to summarize the major effects of BMPs, GDFs, and activins on neural development and function, with the greatest concentration being on the most recent literature. Due to space limitations, we did not consider neural BMP signaling mechanisms or the actions of TGF-β1, β2, and β3. However, these topics have been considered in recent comprehensive reviews [6, 7].

## Expression of BMPs, BMP receptors, and BMP antagonists

### Expression of BMPs

The cloning of BMP-7 (OP-1) from a human hippocampal cDNA library [8] provided the first indication that the nervous system expresses BMPs. This initial report was quickly followed by evidence of *BMP-7* transcriptional and translational products in the brain [9–12]. It is now clear that the nervous system expresses multiple BMPs from each of the known BMP subgroups. Specific BMPs identified in the nervous system thus far include: (1) BMP-5 [13, 14], BMP-6 (vgr) [15, 16] and BMP-7

Bone Morphogenetic Proteins, edited by Slobodan Vukicevic and Kuber T. Sampath

of the 60A subgroup; (2) BMP-2 [12] and BMP-4 [15] of the dpp subgroup; (3) dorsalin [17] and GDF-1 [18, 19] of the dorsalin subgroup; and (4) novel BMPs that have yet to be assigned to a subgroup such as BMP-9 [20], BMP-11 [21], GDF-10 [19, 22] and GDF-15 [23].

Spatiotemporal localization studies, which have primarily examined BMP expression at the mRNA level, indicate persistent and complex regional expression patterns at all stages of neural development into maturity. Downregulation of BMP expression appears critical to initial formation of the nervous system. Prior to gastrulation, BMPs -2, -4 and -7 are expressed throughout the blastula [24–27], but with the onset of gastrulation, BMP-2 expression is turned off everywhere [26, 28, 29] and BMP-4 is quickly downregulated in the organizer and the presumptive neural plate [29–32]. Following neurulation, BMP expression is upregulated in dorsal midline neural tube cells such that the developing roof plate expresses BMPs -4, -5 and -7 and dorsalin 1 in anterior regions [12, 17, 33–35] and BMP-6 in regions posterior of the telencephalon/diencephalon boundary [13]. BMP-11 is expressed in the dorsal-lateral edges of the neural tube adjacent to the roof plate [21].

As development continues, BMP expression continues to increase, generally reaching peak levels during the perinatal period. BMPs have been detected in every region of the developing nervous system including the forebrain [13, 36–38], midbrain and hindbrain [39–41], spinal cord (reviewed below, in the section "The role of BMPs in spinal cord patterning"), and ganglia of the peripheral nervous system [21, 36, 42–44]. Within any given brain structure, multiple BMPs are typically expressed in overlapping temporal and spatial patterns. For example, within the mouse hippocampus, BMP-2 transcripts reach peak levels at embryonic day 16 (E16), BMPs -6 and -7, at E18 [19, 36], BMP-5 at E18 and again in the adult, and BMP-4 at postnatal day 4 (PN4). Spatial patterns also vary between BMPs, e.g., in the PN6 hippocampus of mice, transcripts for GDF-1 are localized to CA1 through CA3 while GDF-10 mRNA is detected in CA3 and dentate gyrus [19]. BMP expression typically declines in the adult nervous system; however, strong signals are still detectable in discrete structures of the mature brain. For example, the adult hippocampus expresses relatively high levels of BMP-5, BMP-6, GDF-1 and GDF-10; the neocortex, BMP-5, BMP-6 and GDF-1; the cerebellum, BMP-5; the striatum, BMP-5 and BMP-7; and the brainstem, BMP-5 and BMP-6 [19, 45, 46].

At the cellular level, both neuronal and glial cells have been shown to express BMPs. *In vitro* studies indicate that transcripts for BMP-2 and BMP-7 are preferentially localized to microglia, astrocytes, and neurons of the forebrain while BMP-4 mRNA is associated primarily with bipotent oligodendroglial astroglial progenitor cells and oligodendrocytes [37, 47]. *In vivo*, BMP-4 protein is associated with type B astrocytes in the subventricular zone of adult mice [48], and BMP-6 protein has been localized to radial glia [49], neurons of the neocortex and hippocampus [45], and peripheral Schwann cells [50]. Similarly, transcriptional and translational products of BMP-6 and BMP-7 have been localized to neuronal and glial cells of perinatal sympathetic ganglia *in vivo* and *in vitro* (Chandrasekaran, Lein and Higgins, unpublished observations).

Although the expression of transcripts for BMPs within the nervous system has been well documented, there is a paucity of data concerning the localization of BMP translational products. However, recent immunohistochemical evidence of BMP expression suggests that BMP proteins are more widely distributed than would be predicted by *in situ* hybridization studies of BMP transcript expression [15, 37, 48, 51, 52]. Possibly, this is due to local diffusion from BMP-producing neural cells. Additional explanations may include delivery of BMPs *via* the fetal circulation or cerebrospinal fluid (CSF). It has been shown that placental tissue expresses BMPs -4 and -7 [53–55], but whether these are secreted into the fetal circulation and cross into the developing brain has yet to be determined. With respect to BMPs in the CSF, BMP-7 protein has been detected in bovine CSF [56] and transcripts for BMPs have been demonstrated in the choroid plexus of embryonic mice (BMPs -4, -5, -6 and –7; GDF-15) [13, 57] and adult rats (BMPs -6 and -7) [52]. These data suggest the potentially widespread distribution of BMPs in the developing and mature nervous system. In light of this, spatiotemporal expression patterns of BMP receptors and BMP antagonists may prove critical to regulation of BMP signaling in the nervous system

## Expression of BMP receptors

BMPs exert their biological effects by binding to type I and type II serine-threonine kinase receptors [58, 59]. Specific receptor subunits shown to bind BMPs include BMP receptor type IA (BMPR-IA), BMPR-IB, BMPR-II, activin receptor type I (ActR-I), and ActR-II [60–62]. BMP ligands can bind to either type I or type II receptor subunits independently, but both receptor types are required for high-affinity binding and signaling [59]. The combinatorial identity of BMP receptors that mediate BMP signaling in the nervous system is not known. Inferences can be made based on limited data regarding the ligand specificity of individual type I/type II heterodimers in non-neuronal cell systems [61–63], and the expression of BMP receptor subunits in neural tissue. RT-PCR and RNA blotting demonstrate that both the developing and adult nervous system express mRNA for BMPR-IA, BMPR-IB, ActR-I, BMPR-II, and ActR-II [60, 62, 64, 65] and localization of these transcripts by *in situ* hybridization studies suggests complex temporal and spatial regulation [6, 46]. The following is a summary of these findings; for a detailed description of BMP receptor expression in the nervous system, the reader is referred to the following references [42, 52, 66, 67].

BMPR-IA, BMPR-IB and BMPR-II are expressed within the CNS neuroepithelium as early as E11 in the rat and E12 in the mouse [42, 67]. During late embryonic and neonatal stages in the mouse and rat, transcript levels for BMPR-IA, BMPR-

IB, and BMPR-II increase significantly and ActR-I and ActR-II expression becomes apparent. BMPR-IA, ActR-I and BMPR-II mRNA exhibit widespread distribution in the brain with prominent expression in the subventricular zone, the hippocampus and the neocortex; BMPR-II is also strongly expressed in the substantia nigra and Purkinje cell layer of the cerebellum [42, 67]. During the early postnatal period, BMPR-IA, ActR-I and BMPR-II transcripts are maintained at high levels in these regions and BMPR-IA expression is upregulated in additional brain regions such as the thalamus, cerebellar Purkinje cell layer, and brain stem [42]. In contrast to the broad expression of BMPR-IA, ActR-I and BMPR-II in the brain, BMPR-IB and ActR-II exhibit more limited expression patterns. ActR-II is strongly expressed in the dorsal spinal cord and more diffusely in the developing hippocampus and olfactory cortex [66, 67]. BMPR-IB is strikingly restricted to the anterior olfactory nucleus and olfactory epithelium from late embryonic stages throughout the postnatal period into adulthood [42]. These patterns of expression are suggestive of roles for BMPR-IA, ActRI, BMPR-II and ActRII in multiple aspects of neural development including neurogenesis, neuronal and glial lineage determination, neuronal morphogenesis and synaptogenesis while suggesting a unique function for BMPR-IB in the development and function of the olfactory system.

BMP receptors are also detected in the developing PNS. Transcripts for BMPR-IA, BMPR-IB and BMPR-II are expressed in cranial ganglia, sympathetic ganglia, and DRG. High levels are evident from E15 throughout development into adulthood [52, 67, 68]. ActR-I and ActR-II are similarly localized to these ganglia but under different temporal regulation: expression is first evident at E21 and is strongly downregulated in the adult [67]. These expression patterns are consistent with proposed roles for BMPs in regulating cell fate and differentiation of neural crest-derived progenitor cells.

Transcripts for type I and type II receptors have also been detected in the CNS of adult animals, but generally at much lower levels than observed in late embryonic and early postnatal development [42, 67]. However, these levels may be upregulated in response to brain injury [52]. *In situ* hybridization patterns suggest unique patterns of regional expression for the different receptor subunits, although there are some significant discrepancies between studies. Zhang et al. [42] reported in a comparative analysis of BMPR-IA, BMPR-IB and BMPR-II expression that BMPR-IA was the most abundant with widespread distribution in gray matter and the choroid plexus and particularly robust expression within the neocortex, cerebellar Purkinje cell layer and brainstem nuclei. BMPR-IA was notably absent from white matter. BMPR-II transcripts were seen in the cerebellar Purkinje cell layer, the hippocampus and the choroid plexus while BMPR-IB mRNA was restricted to the olfactory nucleus. In contrast, Soderstrom et al. [67] and Charytoniuk et al. [52] observed mRNA for BMPR-II in the cortex, dentate gyrus, hippocampus, substantia nigra and ventral horn of the spinal cord. The former study also reported ActR-I and ActR-II mRNA in the dentate gyrus. However, in neither study were transcripts for BMPR-

IA and BMPR-IB detected in adult brain. The reason(s) for these discrepancies are not known but may be attributable to differences in the probes used for *in situ* hybridization since Zhang et al. [42] were able to corroborate their positive observations of BMPR-IA and IB using nuclease protection assays and Western blot analyses.

There is a paucity of data concerning BMP receptor expression at the protein level. Western blot analyses have detected BMPR-IA and BMPR-II in whole mouse brain with peak levels occurring from E13 through P7 followed by significant downregulation in the adult [42]. Immunocytochemical analyses revealed diffuse staining for BMPR-IA, BMPR-IB and BMPR-II in cultured O2A cells [47]. The latter observation raises questions of cellular and subcellular distribution *in vivo*. Adult brain slices double-labeled with antibodies selective for neurons, astrocytes or oligodendrocytes and antisense probe for BMPR mRNA suggest that BMP receptors are localized to neurons [42]. The exclusion of BMPR-IA from white matter in the adult brain would suggest subcellular localization to the somatodendritic domain of mature neurons; however, evidence that BMPs function in axon guidance [69] would argue that in some contexts, BMP receptors are also expressed in axons.

What conclusions can be made on the basis of the available data regarding expression patterns of BMP receptors? First, these data support a role for receptor-mediated BMP signaling in both the developing and adult nervous system. Second, significant overlap in the expression patterns of BMP receptors and BMP ligands raises the possibility of autocrine, paracrine and cooperative signaling loops. Third, the incomplete overlap between type I and type II BMP receptors, particularly in the adult brain, has interesting implications regarding the combinatorial identity of neural BMP receptors. If the current model for BMP signal transduction through type I/type II receptor heterodimers is correct, then it is likely that additional subtypes of type I and type II BMP receptors have yet to be identified in neural tissue. Relevant to this issue, a novel type I serine/threonine kinase named activin-receptor-like kinase-7 (ALK-7) that is preferentially expressed in the brain was recently cloned [70–72]. The ligand(s) and physiologically relevant type II BMP receptor(s) that interact with ALK7 have yet to be determined. Alternatively, there is evidence that homodimeric BMP receptors transduce BMP signals, albeit with lower efficiency than heterodimeric receptors [71]. If this observation is physiologically relevant to BMP signaling in the nervous system, then differential expression of heterodimeric and homodimeric receptors may represent yet another mechanism for regulating the efficacy of BMP signaling in the brain.

## Expression of BMP antagonists

BMP signaling is determined not only by the spatiotemporal expression of BMP ligands and receptors, but also by that of soluble BMP antagonists, which directly bind

BMPs and prevent functional receptor/ligand interaction [74–76]. It now appears that at least four distinct classes of inhibitory BMP binding proteins have evolved independently in vertebrates: follistatin, noggin, chordin and the DAN family of binding proteins, which includes DAN [77], cerberus [77–80] and DRM/gremlin [79]. These BMP antagonists bind to various BMPs and other TGF-β family members with differing degrees of specificity. For example, follistatin binds both activin and BMP-7 avidly, but does not compete with the type I receptor for BMP-4 binding [81], whereas noggin and chordin bind to BMPs -2 and -4 with greater affinity than BMP-7 [74, 75]. Profiling the BMP binding affinities as well as the expression patterns of these BMP antagonists will be critical to understanding their role in BMP signaling in the nervous system.

Consistent with their proposed role in neural induction [82, 83], the BMP antagonists noggin, chordin, follistatin and cerberus are expressed in the *Xenopus* organizer at the gastrula stage [78, 84–87]. Similarly, transcripts of these antagonists are expressed in the organizers of birds, fish and mammals with the following exceptions: neither noggin nor follistatin is expressed in the zebrafish organizer and follistatin is not present in the mouse organizer [88–90]. After neurulation, transcripts for follistatin [85] and noggin [91] are detected in the notochord; translation products of the former are known to be secreted by mesodermal cell types that flank the ventral neural tube [35]. Noggin is also expressed along the longitudinal extent of the dorsal neural tube in a gradient of expression that decreases caudorostrally [91-93]. Gremlin is expressed in the developing neural crest [79] and chordin mRNA is localized to the neuroepithelium of the neural tube, hind, mid and forebrain [94]. At later stages of development, follistatin is stably expressed in the hindbrain [95] and noggin is detected in cortical structures but its spatial pattern is dependent on developmental age. In E15 mice, noggin mRNA and protein is present at very low levels in the cortex [37, 48, 96]. In neonatal mice, noggin protein is abundant in the developing subcortical white matter and corpus callosum and present at much lower levels in the rest of the cortex [37].

BMP antagonists have also been found in the adult nervous system. *In situ* hybridization of neural tissue from adult mice revealed noggin transcripts in the tufted cells of the olfactory bulb, the piriform cortex, and cerebellar Purkinje cells [96]. Transcriptional and translational products of noggin were also detected in the subventricular zone of adult mice [48]. In mice, chordin mRNA is expressed in the granular layer of the cerebellum, the dentate gyrus, and subfields CA1, CA2 and CA3 of the hippocampus [94]; chordin has also been detected in RNA blots of adult human cerebellum [97]. *In situ* hybridization of adult rat brains using probes for DRM, the mammalian homologue of gremlin, demonstrated strong expression in neurons and glial cells of the cortex and in the molecular and granular layers of the cerebellum [98]. The expression of these antagonists at the protein level is largely unknown. However, based on patterns of transcript expression, it would appear that BMP antagonists are important in not only the developing but also the adult nervous system. Functional studies have provided considerable insight regarding the role of BMP antagonists during neural development, but their physiological significance in the mature brain is largely unknown.

## BMPs and neural induction

The formation of the vertebrate nervous system is initiated during gastrulation when the ectoderm gives rise to the neural plate. Grafting experiments in amphibians, fish, birds and mammals have shown that signals from a distinct cluster of mesodermal cells, known as the organizer, induce ectoderm to adopt a neural rather than an epidermal fate [99, 100]. Surgical or genetic ablation of the organizer does not necessarily preclude the formation of a neural plate [101–107], suggesting the existence of additional inductive interactions that promote neuralization. These observations have stimulated an intensive search for molecules with neural inducing activity. Thus far, several candidate molecules have been implicated in the mechanism of neural induction, and all share the property that they inhibit BMP signaling.

The first indication that blocking BMP signaling pathways might be important in neural induction came from observations that expression of dominant negative ActR-II in isolated animal caps causes the generation of neural tissue [108]. Similar results were obtained in animal caps treated with follistatin, a potent inhibitor of activin [109]. Since *Xenopus* animal caps (which are ectodermal explants from blastula stage embryos) typically form epidermis unless recombined with organizer grafts, these data suggested that neural differentiation is the default state of ectoderm, and signaling by activin or a related TGF-β ligand promotes epidermal rather than neural differentiation. To test this hypothesis directly, researchers exploited an earlier observation that dissociation of animal caps causes ectoderm to form neural tissue even in the absence of signals from the organizer, presumably because epidermis-inducing factors are diluted under these conditions [110]. Thus, adding back these epidermalizing factors to dissociated ectodermal cells should block neuralization and cause epidermis to form. Using this bioassay, activin was observed to induce mesoderm, not epidermis [111]; however, BMP-4 proved to be a potent epidermal inducer and its epidermalizing effects could be blocked by dominant negative ActR-II and by follistatin [112]. It was subsequently shown that BMP-2 and -7 also induce epidermis in this bioassay [113] and that BMP-2, -4 and -7 are expressed in *Xenopus* gastrula ectoderm [27, 29].

Further evidence that BMPs bias ectoderm towards an epidermal fate comes from observations that activation of BMP signaling components downstream of the ligand also induces epidermis in dissociated ectoderm. Thus, overexpression of constitutively active Type I receptors, [82, 113], Smad1 or Smad5 [114, 115], or Msx1, an immediate early response to BMP signaling [116], effectively inhibits neuralization and promotes epidermalization of ectoderm in dissociated cell cultures. Conversely, inhibition of BMP signaling promotes neuralization in intact animal caps. Injection of mutant BMP-4 or BMP-7 [27], antisense *BMP4* [86] or dominant-negative Type I BMP receptors [109, 117] promotes the generation of neural tissue. Neural induction also occurs when signaling elements downstream of the BMP receptor are blocked. Thus, overexpression of the inhibitory Smads, Smad6 or Smad7 [118–120] or dominant negative forms of the early response elements Vent-1, Vent-2 and Msx1 [82, 121] causes animal cap ectoderm to adopt a neural rather than epidermal fate.

Based on these observations, it has been proposed that the organizer and other regions of the embryo neuralize ectoderm through inhibition of BMP signaling. There are data to support this hypothesis. The BMP antagonists noggin, chordin, follistatin, cerberus and Xnr3 are expressed in the *Xenopus* organizer at the gastrula stage when neural induction is thought to occur, and ectopic expression of these antagonists causes neural development in blastula-stage animal caps [78, 84, 86, 87, 109]. These effects occur in the absence of mesoderm induction, providing critical evidence that BMP antagonists are direct neural inducers. Studies in *Drosophila* have shown that overexpression of noggin antagonizes the epidermalizing activity of the *Drosophila* BMP-4 homologue (dpp), but does not inhibit the epidermis-inducing effects of constitutively active BMP receptors [122], suggesting that these BMP antagonists target BMP signaling upstream of the receptor. Additional biochemical and genetic studies in *Xenopus* and *Drosophila* support the conclusion that noggin, chordin, follistatin, cerberus and Xnr3 induce neural fates by directly binding BMPs and preventing functional interaction with their receptors [82, 83]. Despite notable interspecies differences in gene expression patterns for BMP ligands and antagonists [88, 89], there is evidence suggesting that BMP antagonism is a conserved mechanism of neural induction across frogs, fish, birds and mammals, and that soluble BMP antagonists constitute an important component of this mechanism [82, 83, 90].

While BMP antagonists appear sufficient for neural induction, there is no definitive evidence yet that the BMP antagonists are necessary for neural induction. Mice with targeted deletion of noggin [91], follistatin [91], or cerberus [123] still develop a neural plate. Similarly, deletion of chordin in zebrafish does not block neural induction, although it does reduce the size of the neural plate [124, 125]. The minimal effect of single mutations on neural induction is perhaps not surprising, because of the overlapping expression and redundant activities of the BMP antagonists. However, a neural plate still forms in mice doubly mutant for noggin and chordin [126], and in mice with genetic deletion of the organizer (the node) that effectively eliminates noggin, chordin and other node-derived neural inducing signals [103, 104, 107]. These data suggest the existence of additional families of signaling molecules that are not derived from the organizer. Two candidates include FGF and Wnt/β-catenin signals [127–130].

There is evidence that both FGF and Wnt signals may induce neuralization by suppressing BMP transcription in the prospective neural plate. In explants of chick ectoderm, the FGF receptor inhibitor SU5402 inhibits BMP downregulation and under these conditions the explants differentiate into epidermis. Application of noggin or other BMP antagonists to explants treated with the FGF inhibitor restores neural fate [130]. Ectopic activation of Wnt signaling in *Xenopus* animal caps is sufficient to both suppress BMP-4 expression and induce neural differentiation [131]. Thus two different signaling pathways cause downregulation of BMP expression which is correlated with neural induction.

In summary, these data suggest that in the developing embryo, ectodermal cells exposed to BMPs are fated to become epidermis, while inhibition of BMP signaling drives ectoderm towards a neural fate. Prior to gastrulation, FGF and Wnt signaling promote neural differentiation by repressing the expression of BMP genes from prospective neural plate; during gastrulation the activity of BMP proteins is antagonized by soluble factors derived from the organizer region [83, 132].

## BMPs and the specification of neural/glial cell fate

Development of a functional nervous system requires precise regulation of neuronal and glial cell differentiation from a common neural progenitor cell [133–135]. BMP signaling influences progenitor cell fate decisions, but the precise effects of BMPs vary according to progenitor cell type and/or cell stage [37, 48, 136, 137]. Thus, BMPs inhibit neuronal lineage development in the olfactory epithelium [138], but promote neuronal cell fate specification in neural crest stem cells [139], in cerebellar granule cell precursors [140] and in spinal cord neural precursors [141]. BMPs have also been shown to selectively promote astroglial cell development in neural cultures from the embryonic midbrain [40] and hindbrain [41]. There is evidence that progenitor cell response to BMPs may be influenced by the cellular and cytokine context of the local environment [36, 137, 142, 143], and by the relative balance of BMPs and BMP antagonists [37, 48, 137, 144]. Much of what is known about BMPs in neuronal and glial lineage commitment has been derived from studies of cortical development and thus the remainder of this discussion will focus on lineage determination in the cortex.

During early embryogenesis, neurons and glia of the neocortex are generated from multipotent progenitors located within the neuroepithelium of the ventricular zone (VZ). During the later perinatal period, neurons, astrocytes and oligodendrocytes are generated from multipotent cells present within the subventricular zone (SVZ). *In vitro* studies of VZ progenitors indicate that BMPs decrease proliferation and trigger both neuronal and astroglial differentiation with concurrent suppression of oligodendroglial lineages [37, 137, 144]. Comparative analyses of other TGF-β family members suggest that this activity is unique to BMPs [46]. In the absence of exogenous BMPs, overexpression of dominant negative BMPRI inhibits neurite outgrowth and neuronal migration in VZ explants [51] and noggin blocks neuronal lineage elaboration in dissociated VZ cultures [144]. These data suggest that endogenous BMPs regulate VZ progenitor cell fate, a conclusion supported by spatiotemporal expression patterns of BMPs *in vivo* [37, 51, 143].

The response of cultured SVZ progenitors to BMPs varies in that BMPs promote astroglial differentiation while suppressing both neuronal and oligodendroglial differentiation [36, 37, 39, 137, 143, 145]. BMP-induced astroglial differentiation of SVZ progenitors appears to require concurrent signaling by LIF [38, 143], and BMP suppression of oligodendroglial differentiation appears to be mediated by active mechanisms. The latter is based on observations that sonic hedgehog (Shh) [147] and noggin [37, 137] promote the generation of oligodendroglial lineages from cultured SVZ progenitors, presumably *via* similar mechanisms since Shh has been shown to increase noggin expression [148]. The differential effects of BMPs on astroglial and oligodendroglial fate are maintained during later stages of lineage specification: exposure of postnatal subcortical bipotent oligodendroglial-astroglial (O-2A) progenitor cells to BMPs promotes dose-dependent elaboration of astrocytes and inhibition of oligodendroglial lineage expression [36, 47, 136]. *In vivo* studies are consistent with the proposal that BMPs actively suppress oligodendroglial lineages. Thus, BMPs are expressed primarily in the dorsal aspect of the neural tube, whereas oligodendroglia arise predominantly along the ventral neural axis [14, 136]. Noggin is predominantly expressed in the developing subcortical white matter, but not in the remainder of the cortex, corresponding to sites enriched with oligodendroglia or astroglia, respectively [37]. More convincingly, there is a paucity of oligodendroglia in the noggin knock-out mouse [91].

Evidence that neurogenesis persists in the mammalian CNS throughout adult life [149] raises the question of whether BMPs influence neuronal versus glial fate decisions in the adult nervous system. This question has been addressed by Alvarez-Buylla and colleagues [48] who found that adult SVZ cells express BMPs and their cognate receptors, whereas the ependymal cells adjacent to the SVZ express noggin. In SVZ cells cultured from adult brains, the addition of exogenous BMPs or overexpression of constitutively active type I BMPRs inhibits neurogenesis and promotes glial differentiation, similar to observations reported by others [150]. In contrast, exogenous noggin promotes neurogenesis and inhibits glial differentiation. *In vivo*, overexpression of BMP7 in the ependyma inhibits neurogenesis while stimulating generation of glial cell types and ectopic expression of noggin in the striatum promotes neuronal differentiation of SVZ cells grafted to the striatum. These data suggest that noggin production in the ependyma creates a neurogenic environment in the adjacent SVZ by blocking endogenous BMP signaling. There is evidence to suggest that interactions between noggin and BMP may similarly influence neuronal and glial lineage elaboration in the developing cortex [37, 144]. Observations that interplay between BMPs and BMP antagonists similarly regulate neuronal *versus* glial specification in the developing and adult cortex and neuronal *versus* epidermal specification during neural induction suggest a conserved paradigm of BMP signaling that is repeated in various contexts throughout the life of the organism.

## The role of BMPs in spinal cord patterning

BMPs play essential roles in the development of caudal spinal cord character and the differentiation of the dorsal-most ventral interneurons as well as dorsal commissural interneurons [34, 69, 151, 152]. Cooperative signaling from fibroblast growth factor, paraxial mesoderm caudalizing activity, retinoids, and BMPs guide the rostrocaudal development of the neural tube [152]. For example, BMP-7 enhances expression of rostral characteristics whereas fibroblast growth factor controls the acquisition of caudal traits [153]. The establishment of the rostrocaudal axis produces the embryonic midbrain, hindbrain, and spinal cord in the chick [154]. Subsequent to the foundation of spinal cord caudal characteristics, BMPs regulate specific aspects of the dorsal-ventral differentiation processes. Expresssion of BMP-2, -4, -5 and –7 in the dorsal cord forms a concentration gradient from the highly concentrated dorsal region to the less concentrated ventral; therefore, BMPs exert their greatest influence on the differentiation of dorsal commissural interneurons, as well as the dorsal-most ventral midline cells *via* inhibition of sonic hedgehog signaling from the notochord [152].

The epidermal ectoderm sets off a BMP-mediated differentiation cascade by transiently secreting BMP-4 and BMP-7 [34], promoting the development of the dorsal midline from multipotent neuroepithelial cells [155]. Progressive dorsalization activity is regulated by BMP-4, -5, -6, and -7 in murine and chick tissues, as well as BMP-2 in the mouse, which are secreted from the dorsal midline as it forms from the closing neural fold [46]. The BMP induction cascade continues with the development of the BMP-secreting primitive roof plate from the neural fold and dorsal midline [34, 69, 151]. GDF-7 is also expressed by the roof plate cells, and is required for the differentiation of D1A and D1B sensory interneurons [34, 151]. In addition, expression of BMP-7 from the roof plate acts as a chemorepellent to guide dorsal commissural axon projections toward the ventral cord [69]. Collectively, the roof plate expresses a variety of BMPs at different developmental stages, long after the cessation of BMP secretion from the epidermal ectoderm.

BMP-7 and sonic hedgehog are expressed from the prechordal mesoderm and together regulate the differentiation of ventral midline cells [153, 156], whereas sonic hedgehog expressed by the notochord independently controls the induction of floor plate cells [153]. BMPs diffuse from the dorsal cord and regulate the response of ventral cord precursors to sonic hedgehog. Notochord-derived sonic hedgehog and BMP-binding proteins, including the BMP antagonists noggin and follistatin, in turn attenuate BMP effects in this region [34, 83, 91, 152, 157, 158]. The role of BMPs in dorsoventral regulation in attenuation of ventral neuronal differentiation was confirmed by utilizing BMP mutants in zebrafish [159]. The resultant pattern of differentiation factors in the ventral and dorsal spinal cord can be described as a dual concentration gradient system, with BMPs -2, -4 and -7 concentrated in the dorsal region, diffusing out towards the ventral, and sonic hedgehog in the ventral diffusing dorsally, away from the notochord and floor plate [35, 152, 160].

At each level of the dorsal-ventral gradient, neural precursors at specific locales require a certain concentration of differentiation factor to develop into the correct neuronal type [46, 152]. For example, neural precursors in the dorsal cord need high concentrations of BMPs for maturation, whereas ventral neuronal precursors require high levels of sonic hedgehog. In addition, precursor cells require such signals to contact them during specific periods of development, as cells become competent to respond appropriately to precise developmental progression. In this way, the BMP influence on neuronal precursor cells is temporally and spatially modified, resulting in a specific patterning effect throughout the neural tube.

## Effects of BMPs on brain development

In many regions of the developing brain, BMPs regulate the generation and differentiation of neuronal cells during various stages of ontogeny [137]. For example, BMPs induce the differentiation of cerebellar granule neurons [140], and striatal GABAergic neurons [161]. In addition, BMPs-2, -4, -6, -7, -12 and –13 stimulate differentiation [51, 137], and signaling through the BMP receptor promotes migration [51] of cortical neurons from neocortical precursor cells within the VZ. BMP-2 regulates proliferation in the forebrain [13], and BMP-7 induces serotonergic characteristics during the development of hindbrain raphe neural precursors [41].

Not unlike the developing spinal cord, the activity of BMPs is counterbalanced by antagonistic factors. For instance, BMPs and the BMP antagonist chordin coordinately regulate rostrocaudal patterning of ventral midline cells, as chordin inhibits BMP support of rostral characteristics and promotes the enhancement of caudal properties [153]. BMP-2 and -4 control the number and the properties of developing cortical precursors in conjunction with the inhibitor noggin [37]. In addition to these agents, sonic hedgehog modifies the influence of BMPs on the proliferation and differentiation of CNS neural precursors, and also induces the expression of noggin to elicit a number of its effects [147]. Sonic hedgehog attenuates BMP signaling, promoting the acquisition of ventral properties and inhibiting the anti-proliferative effects of BMP-2 on neuronal precursor cells [147]. Lastly, during the development of the early forebrain, BMP-7, BMP-4 and sonic hedgehog are jointly secreted from the prechordal mesoderm to induce the expression of rostral diencephalon ventral midline cells [153, 156]. In this way, BMP-7 and BMP-4 alter the effect of sonic hedgehog to induce differentiation of rostral diencephalon cells rather than floor plate cells.

BMPs regulate the survival of developing neurons in both the CNS and the PNS, and they exert their effects by both independent means and in conjunction with other trophic factors. BMP-2 promotes the survival of striatal GABAergic neurons in a manner that does not require additional growth factor signaling [161], and GDF-15 also acts directly to increase survival of dopaminergic neurons [57]. In contrast, BMPs -2 and -6 promote the survival of dopaminergic neurons in an indirect manner, most likely through secretion of glial cell growth factors [40]. GDF-5 and BMP-2, -4, -7 and -12 have minor survival promoting effects on dorsal root sensory neurons, although they exhibit strong synergistic interactions with neurotrophin 3 and NGF [162, 163]. Synergistic interactions of BMPs with neurotrophins and glial cell line-derived neurotrophic factor have also been observed in sympathetic, nodose and ciliary ganglia [164, 165]. One of the mechanisms by which BMPs increase neuronal survival may be by the stimulation of expression of the neurotrophin receptor trkC [42, 166].

BMPs also refine brain development by inducing selective apoptotic events. While BMPs -2 and -4 have been demonstrated to inhibit cell death in an early cerebellar cell line [167], these BMPs promote cell death in the dorsal forebrain [13]. BMP-4 is expressed in the dorsal r3 and r5 rhombomeres, upregulating the expression of an apoptosis-associated gene, Msx2, and triggering cell death. BMP-4 subsequently initiates the formation of discrete paths of neural crest cells migrating out of the hindbrain [168]. Finally, BMPs -2 and -4 promote apoptosis in the absence of fibroblast growth factor and nerve growth factor in an early sympathoadrenal progenitor cell line [169].

## Effects of BMPs on the neural crest, development of the peripheral nervous system and specification of neurotransmitter phenotype in both central and peripheral neurons

The neural crest gives rise to most of the neurons and glia in the peripheral nervous system, including those populating sympathetic, parasympathetic, enteric, and dorsal root ganglia. BMPs critically influence the development of neural crest cells and thereby the development of the entire peripheral nervous system. In fact, BMPs regulate several distinct stages in the development of this cell population, and at each stage their effect is different. One of their most prominent effects is on the specification of the neurotransmitter phenotype and this has also been observed in neurons derived from the central nervous system.

In the caudal regions of the neural crest that give rise to the peripheral nervous system, BMP-4 and -7 are initially present in the epidermis, and subsequently BMP-4 and other TGF-β superfamily members appears in dorsal neural tube [33, 34,

151]. BMP-4 and -7 cause neural tube cells to begin expressing genes characteristic of neural crest cells [33, 34, 170], such as slug and HNK, and mutations in either BMP gene interfere with neural crest development in zebrafish [159]. Neural crest abnormalities were not, however, noted in mice lacking the BMP-7 gene [171–173], suggesting that other BMPs can substitute for it in some species.

After the neural crest has formed, it is necessary for the epithelial premigratory crest to convert into mesenchyme and begin its dispersal. BMP-4 regulates the initial stages of neural crest migration [93] and one of the ways it does this is by stimulating the expression of rhoB [174], a GTP-binding protein that is required for the delamination of neural crest cells. In addition, BMP regulates the expression of several cadherins which might be involved in cell-cell interactions [93, 175, 176]. In all of these interactions, the concentration of free BMP is critical, and inappropriate levels of BMPs can interfere with rather than promote neural crest development [177]. In the neural crest, the level of free BMP4 appears to be determined not only by the pattern of expression of its mRNA, but also by dynamic changes in the expression of its antagonist, noggin [93].

When migrating neural crest cells coalesce to form sympathetic ganglia, they are again exposed to BMPs [139, 178] and this interaction is required for normal ganglionic development [44]. Exposure to BMP-2 induces neurogenesis and suppresses gliogenesis in cultures of rat neural crest stem cells, and it acts in an instructive manner to induce expression of neural characteristics, rather than by supporting a subpopulation of previously committed precursors [139]. BMP-2 also induces expression of the MASH1, a transcription factor required for the development of autonomic neurons [139, 146]. Thus BMPs appear to commit neural crest cells to an autonomic motor rather than a sensory phenotype [139, 178]. Under most conditions, BMP-2, -4 and -7 also promote the initial expression of tyrosine hydroxylase and increase the synthesis of catecholamines and thereby determine the neurotransmitter phenotype of the sympathetic neuroblasts [44, 178-180]. However, under certain conditions *in vitro* [139, 181], the effects of BMPs on neurogenesis can be separated from effects on the adrenergic phenotype, suggesting that they may represent separate and dissociable phenomena.

The signaling cascade that mediates the effects of BMPs on sympathetic neuroblasts involves at least three classes of transcription factors: MASH1 [139, 182, 146], the Phox2 homeodomain proteins [182-184], and dHAND [185]. MASH1 and Phox2a have also been implicated in the generation of catecholaminergic neurons in the central nervous system [186-188], and so it might be expected that the differentiation of these neurons would also be affected by BMPs. Consistent with this possibility, it has been found that BMPs are required for the generation of noradrenergic neurons in zebrafish hindbrain [189] and dopaminergic neurons in C. elegans [190] and that BMP-2 and activin stimulate the differentiation of dopaminergic neurons in cultures derived from ventral mesencephalon [191] and basal forebrain ventricular zone [192], respectively.

However, the effects of BMPs are not restricted to catecholaminergic neurons, because BMPs also induce expression of GABAergic and cholinergic phenotypes in cultures derived from striatum [BMP-2, 161] and septum [BMP-9, 20], respectively. In addition, activin and/or BMP-2, -4 and -6 have been found to regulate neuropeptide expression in sensory [43], sympathetic [193] and parasympathetic [194] neurons. Thus, specification of the neurotransmitter phenotype represents one of the most pervasive actions of this family of proteins.

## Effects of BMPs on process growth

BMPs stimulate the differentiation of neocortical [51, 144], striatal [161], and mesencephalic dopaminergic [191] neurons and PC12 cells [195, 196] and this inductive activity is associated with increased growth of unspecified processes, i.e., neurites. In addition, BMPs act as roof-plate derived chemorepellents for commissural axons [69] and growth cones and stimulate the growth of long, axon-like processes from retinal ganglion cells [197]. However, some of the most striking responses to BMPs occur in dendrites.

Cultured sympathetic neurons extend only axons when grown in the presence of nerve growth factor. In contrast, subsequent exposure to BMP-2, -4, -6, or -7 causes these neurons to begin forming dendrites within 24 h [198]. This represents a specific morphogenic effect of BMPs, because it occurs without changes in either cell survival or axonal growth. Moreover, in the presence of BMP-7, sympathetic neurons eventually generate an arbor equivalent in size to that observed *in vivo*, suggesting that BMPs are a sufficient stimulus for normal morphological development. Stimulation of dendritic growth has also been observed in cultured hippocampal [199] and cortical neurons [200] and in spinal motor neurons developing in ocular implants [201]. BMP-induced dendritic growth requires Smad1 and activity of the proteasome [Guo and Higgins, unpublished observations] and is associated with expression of MAP-2, a dendrite specific cytoskeletal protein [202]. In addition, in hippocampal cultures increased dendritic growth results in an increase in the rate of synapse formation. It is not yet known whether BMPs also regulate these critical activities *in vivo*. However, relevant BMPs are expressed in hippocampus [19, 36, 45, 46], cortex [51], spinal cord and sympathetic ganglia ([139, 178]; Lein and Higgins, unpublished observations).

The dendrite-promoting activities of BMPs are antagonized by retinoic acid [203], a morphogen that is synthesized in sympathetic ganglia [203, 204] and that also interacts with BMPs in regulating the sensitivity of these cells to neurotrophins and GDNF [164, 166]. Leukemia inhibitory factor [LIF] and other members of the IL-6 cytokine family also block the dendrite-promoting effects of BMPs [205]. In addition, they cause retraction of existing dendrites [206]. These activities are of interest because axotomy is known to induce both dendritic retraction [207] and the synthesis of LIF [208, 209]. It is, therefore, likely that some of the regressive effects of axonal injury are mediated by cytokine-induced changes in the responsiveness to BMPs. The mechanism by which IL-6 related cytokines block BMP action in sympathetic ganglia is unclear. However, Nakashima et al. [143] reported that Stat3 and Smad1, which are the respective downstream signaling elements for LIF and BMP-2, bind to the p300 transcriptional activator and this tripartite complex was implicated in synergistic interactions between BMP-2 and LIF in neuroepithelial cultures.

## BMPs in adult brain

Although BMPs have been extensively studied in developing animals, their potential functions in the mature brain have received limited attention. In fact, endogenous BMPs have only one known role, the regulation of neurogenesis [48] in the adult rodent subventricular zone (reviewed in the section "Neural/glial cell fate").

The effects of exogenous BMPs have received more attention because of potential clinical applications. Pretreatment with BMP-7 reduces ischemia induced-injury and infarct size in the rat cerebral cortex [210, 211] and, under these conditions, expression of the BMPR-II is also increased [52]. BMP-7 also enhances functional motor recovery when given up to 3 days after occlusion of the middle cerebral artery and the fact that there is such a wide window of opportunity for drug administration has led to the suggestion that BMP-7 might be useful in the treatment of stroke [212, 213]. In this case, the BMP-7 effect seems to represent a stimulation of regeneration rather than a change in the size of infarct. Decreased neuronal death was also reported in hypoxic infant rats that were treated with activin, but it is not known whether this protein also protects mature neurons [214].

GDF-15 protects nigrostriatal neurons exposed to 6-hydroxydopamine [57], and GDF-5 reduces toxicity in dopaminergic neurons exposed to $MPP^+$ [215]. In addition, GDF-5 and BMP-2 enhance the survival of dopaminergic neurons that have been grafted into lesioned striatum [216, 217]. Thus, BMPs have neuroprotective effects in animal models of Parkinson's disease. Moreover, activin A has been found to protect striatal neurons in a quinolinic lesion model of Huntington's disease [218] and to reduce excitotoxin-induced cell death in the hippocampus [219].

## Concluding remarks

The first papers on neural effects of BMPs were published in the early 1990s. Since that time the field has expanded rapidly, with the greatest growth occurring in the last 3 years. There is now compelling evidence for the involvement of BMPs and BMP antagonists in many early developmental events, including neurulation, dorsal/ventral patterning, and neural crest development. In addition, there are strong indications that BMPs are involved in later aspects of neural development, such dendritic growth, synapse formation and specification of some glial cell lineages. However, progress in the latter areas of investigation has been hampered by the fact that deletion of many of the BMP genes leads to early embryonic lethality in transgenic knockout mice. In these areas, conditional BMP mutations may be helpful. Currently, there is also limited knowledge as to the role of BMPs in the mature nervous system or their therapeutic potential and this would seem to be an important area for future exploration.

*Acknowledgements*
This work was supported by grants from the National Science Foundation (DH, IBN 9808565) and Johns Hopkins University, School of Hygiene and Public Health (PL Faculty Innovation Award).

## References

1 Wozney JM (1992) The bone morphogenetic family and osteogenesis. *Mol Reprod Dev* 32: 160–167
2 Lipschutz JH (1998) Molecular development of the kidney – a review of the results of gene disruption studies. *Am J Kidney Dis* 31: 383–397
3 Warburton D, Schwarz M, Tefft D, Flores-Delgado G, Anderson KD, Cardoso WV (2000) The molecular basis of lung morphogenesis. *Mech Develop* 92: 55–81
4 Davidson AJ, Zon LI (2000) Turning mesoderm into blood. *Current Topics in Develop Biol* 50: 45–60
5 Nakajima Y, Yamagishi T, Hokari S, Nakamura H (2000) Mechanisms involved in valvulosepta endocardial cushion formation in early embryogenesis: Roles of TGG-beta and bone morphogenetic protein. *Anat Record* 258: 119–127
6 Ebendal T, Bengtsson H, Soderstrom S (1998) Bone morphogenetic proteins and their receptors: potential functions in the brain. *J Neurosci Res* 51: 139
7 Bottner M, Krieglstein K, Unsicker K (2000) The transforming growth factor-βs: Structure, signaling, and roles in nervous system development and functions. *J Neurochem* 75: 2227–2240
8 Ozkaynak E, Rueger DC, Drier EA, Corbett C, Ridge RJ, Sampath TK, Oppermann H (1990) OP-1 cDNA encodes an osteogenic protein in the TGF-beta family. *EMBO J* 9: 2085–2093
9 Ozkaynak E, Schnegelsberg PN, Oppermann H (1991) Murine osteogenic protein (OP-1): high levels of mRNA in kidney. *Biochem Biophys Res Commun* 179: 116–123
10 Vukicevic S, Latin V, Chen P, Batorsky R, Reddi AH, Sampath TK (1994) Localization of osteogenic protein-1 (bone morphogenetic protein-7) during human embryonic devel-

opment: high affinity binding to basement membranes. *Biochem Biophys Res Commun* 198: 693–700
11 Helder MN, Ozkaynak E, Sampath KT, Luyten FP, Latin V, Oppermann H, Vukicevic S (1995) Expression pattern of osteogenic protein-1 (bone morphogenetic protein-7) in human and mouse development. *J Histochem Cytochem* 43: 1035–1044
12 Lyons KM, Hogan BL, Robertson EJ (1995) Colocalization of BMP 7 and BMP 2 RNAs suggests that these factors cooperatively mediate tissue interactions during murine development. *Mech Dev* 50: 71–83
13 Furuta Y, Piston DW, Hogan BLM (1997) Bone morphogenetic proteins (BMPs) as regulators of dorsal forebrain development. *Development* 124: 2203–2212
14 Golden JA, Bracilovic A, McFadden KA, Beesley JS, JL RR, Grinspan JB (1999) Ectopic bone morphogenetic proteins 5 and 4 in the chicken forebrain lead to cyclopia and holoprosencephaly. *Proc Natl Acad Sci USA* 96: 2439–2444
15 Jones CM, Lyons KM, Hogan BL (1991) Involvement of bone morphogenetic protein-4 (BMP-4) and Vgr-1 in morphogenesis and neurogenesis in the mouse. *Development* 111: 531–542
16 Wall NA, Blessing M, Wright CV, Hogan BL (1993) Biosynthesis and *in vivo* localization of the decapentaplegic-Vg-related protein, DVR-6 (bone morphogenetic protein-6). *J Cell Biol* 120: 493–502
17 Basler K, Edlund T, Jessell TM, Yamada T (1993) Control of cell pattern in the neural tube: regulation of cell differentiation by dorsalin-1, a novel TGF beta family member. *Cell* 73: 687–702
18 Lee SJ (1991) Expression of growth/differentiation factor 1 in the nervous system: conservation of a bicistronic structure. *Proc Natl Acad Sci USA* 88: 4250–4254
19 Soderstrom S, Ebendal T (1999) Localized expression of BMP and GDF mRNA in the rodent brain. *J Neurosci Res* 56: 482–492
20 Lopez-Coviella I, Berse B, Krauss R, Thies RS, Blusztajn JK (2000) Induction and maintenance of the neuronal cholinergic phenotype in the central nervous system by BMP-9. *Science* 289: 313–316
21 Gamer LW, Wolfman NM, Celeste AJ, Hattersley G, Hewick R, Rosen V (1999) A novel BMP expressed in developing mouse limb, spinal cord, and tail bud is a potent mesoderm inducer in *Xenopus* embryos. *Dev Biol* 208: 222–232
22 Zhao R, Lawler AM, Lee SJ (1999) Characterization of GDF-10 expression patterns and null mice. *Dev Biol* 212: 68–79
23 Bottner M, Suter-Crazzolara C, Schober A, Unsicker K (1999) Expression of a novel member of the TGF-beta superfamily, growth/differentiation factor-15/macrophage-inhibiting cytokine-1 (GDF-15/MIC-1) in adult rat tissues. *Cell Tissue Res* 297: 103–110
24 Dale L, Howes G, Price BM, Smith JC (1992) Bone morphogenetic protein 4: a ventralizing factor in early *Xenopus* development. *Development* 115: 573–585
25 Nishimatsu S, Suzuki A, Shoda A, Murakami K, Ueno N (1992) Genes for bone mor-

phogenetic proteins are differentially transcribed in early amphibian embryos. *Biochem Biophys Res Commun* 186: 1487–1495

26 Clement JH, Fettes P, Knochel S, Lef J, Knochel W (1995) Bone morphogenetic protein 2 in the early development of *Xenopus* laevis. *Mech Dev* 52: 357–370

27 Hawley SH, Wunnenberg-Stapleton K, Hashimoto C, Laurent MN, Watabe T, Blumberg BW, Cho KW (1995) Disruption of BMP signals in embryonic *Xenopus* ectoderm leads to direct neural induction. *Genes Dev* 9: 2923–2935

28 Shoda A, Murakami K, Ueno N (1994) Biologically active BMP-2 in early *Xenopus* laevis embryos. *Biochem Biophys Res Commun* 198: 1267–1274

29 Hemmati-Brivanlou A, Thomsen GH (1995) Ventral mesodermal patterning in *Xenopus* embryos: expression patterns and activities of BMP-2 and BMP-4. *Dev Genet* 17: 78–89

30 Fainsod A, Steinbeisser H, De Robertis EM (1994) On the function of BMP-4 in patterning the marginal zone of the *Xenopus* embryo. *EMBO J* 13: 5015–5025

31 Re'em-Kalma Y, Lamb T, Frank D (1995) Competition between noggin and bone morphogenetic protein 4 activities may regulate dorsalization during *Xenopus* development. *Proc Natl Acad Sci USA* 92: 12141–12145

32 Schmidt JE, Suzuki A, Ueno N, Kimelman D (1995) Localized BMP-4 mediates dorsal/ventral patterning in the early *Xenopus* embryo. *Dev Biol* 169: 37–50

33 Liem KF, Tremml G, Roelink H, Jessell TM (1995) Dorsal differentiation of neural plate cells induced by BMP-mediated signals from epidermal ectoderm. *Cell* 82: 969–979

34 Liem KF, Tremml, G, Jessell TM (1997) A role for the roof plate and its resident TGFβ-related proteins in neuronal patterning in the dorsal spinal cord. *Cell* 91: 127–138

35 Liem KF, Jessell TM, Briscoe J (2000) Regulation of the neural patterning activity of sonic hedgehog by secreted BMP inhibitors expressed by notochord and somites. *Development* 127: 4855–4866

36 Mehler MF, Marmur R, Gross R, Mabie PC, Zang Z, Papavasiliou A, Kessler JA (1995) Cytokines regulate the cellular phenotype of developing neural lineage species. *Int J Dev Neurosci* 13: 213–240

37 Mabie PC, Mehler MF, Kessler JA (1999) Multiple roles of bone morphogenetic protein signaling in the regulation of cortical cell number and phenotype. *J Neurosci* 19: 7077–7088

38 Nakashima K, Yanagisawa M, Arakawa H, Taga T (1999) Astrocyte differentiation mediated by LIF in cooperation with BMP2. *FEBS Lett* 457: 43–46

39 Gross RE, Mehler MF, Mabie PC, Zang Z, Santschi L, Kessler JA (1996) Bone morphogenetic proteins promote astroglial lineage commitment by mammalian subventricular zone progenitor cells. *Neuron* 17: 595–606

40 Jordan J, Böttner M, Schluesener HJ, Unsicker K, Krieglstein K (1997) Bone morphogenetic proteins: neurotrophic roles for midbrain dopaminergic neurons and implications of astroglial cells. *Eur J Neurosci* 9: 1699–1710

41 Galter D, Böttner M, Krieglstein K, Schömig E, Unsicker K (1999) Differential regulation of distinct phenotypic features of serotonergic neurons by bone morphogenetic proteins. *Eur Neurosci Assoc* 11: 2444–2452

42 Zhang D, Mehler MF, Song Q, Kessler JA (1998) *Development* of bone morphogenetic protein receptors in the nervous system and possible roles in regulating trkC expression. *J Neurosci* 18: 3314–3326
43 Ai XB, Cappuzzello J, Hall AK (1999) Activin and bone morphogenetic proteins induce calcitonin gene-related peptide in embryonic sensory neurons *in vitro*. *Molec Cellular Neurosci* 14: 506–518
44 Schneider C, Wicht H, Enderich J, Wegner M, Rohrer H (1999) Bone morphogenetic proteins are required *in vivo* for the generation of sympathetic neurons. *Neuron* 24: 861–870
45 Tomizawa K, Matsui H, Kondo E, Miyamoto K, Tokuda M, Itano T, Nagahata S, Akagi T, Hatase O (1995) *Development*al alteration and neuron-specific expression of bone morphogenetic protein-6 (BMP-6) mRNA in rodent brain. *Brain Res Mol Brain Res* 28: 122–128
46 Mehler MF, Mabie PC, Zhang D, Kessler JA (1997) Bone morphogenetic proteins in the nervous system. *Trends Neurosci* 20: 309–317
47 Mabie PC, Mehler MF, Marmur R, Papavasiliou A, Song Q, Kessler JA (1997) Bone morphogenetic proteins induce astroglial differentiation of oligodendroglial-astroglial progenitor cells. *J Neurosci* 17: 4112–4120
48 Lim DA, Tramontin AD, Trevejo JM, Herrera DG, Garcia-Verdugo JM, Alvarez-Buylla A (2000) Noggin antagonizes BMP signaling to create a niche for adult neurogenesis. *Neuron* 28: 713–726
49 Schluesener HJ, Meyermann R (1994) Expression of BMP-6, a TGF-beta related morphogenetic cytokine, in rat radial glial cells. *Glia* 12: 161–164
50 Schluesener HJ, Meyermann R, Jung S (1995) Immunolocalization of vgr (BMP-6, DVR-6), a TGF-beta related cytokine, to Schwann cells of the rat peripheral nervous system: expression patterns are not modulated by autoimmune disease. *Glia* 13: 75–78
51 Li W, Cogswell CA, LoTurco JJ (1998) Neuronal differentiation of precursors in the neocortical ventricular zone is triggered by BMP. *J Neurosci* 18: 8853–8862
52 Charytoniuk DA, Traiffort E, Pinard E, Issertial O, Seylaz J, Ruat M (2000) Distribution of bone morphogenetic protein and bone morphogenetic protein receptor transcripts in the rodent nervous system and up-regulation of bone morphogenetic protein receptor type II in hippocampal dentate gyrus in a rat model of global cerebral ischemia. *Neurosci* 100: 33–43
53 Oida S, Iimura T, Maruoka Y, Takeda K, Sasaki S (1995) Cloning and sequence of bone morphogenetic protein 4 (BMP-4) from a human placental cDNA library. *DNA Seq* 5: 273–275
54 Martinovic S, Latin V, Suchanek E, Stavljenic-Rukavina A, Sampath KI, Vukicevic S (1996) Osteogenic protein-1 is produced by human fetal trophoblasts *in vivo* and regulates the synthesis of chorionic gonadotropin and progesterone by trophoblasts *in vitro*. *Eur J Clin Chem Clin Biochem* 34: 103–109
55 Ozkaynak E, Jin DF, Jelic M, Vukicevic S, Oppermann H (1997) Osteogenic protein-1 mRNA in the uterine endometrium. *Biochem Biophys Res Commun* 234: 242–246

56 Dattatreyamurty B, Roux E, Kaplan PL, Robak LA, Horbinski C, Lein P, Higgins D, Chandrasekaran V (2001) Cerebrospinal fluid contains biologically active bone morphogenetic protein-7. *Exp Neurol* 172: 273–281
57 Strelau J, Sullivan A, Bottner M, Lingor P, Falkenstein E, Suter-Crazzolara C, Galter D, Jaszain J, Krieglstein K, Unsicker K (2000) Growth/Differentiation Factor-15/Macrophage inhibitory cytokine-1 is a novel trophic factor for midbrain dopaminergic neurons *in vivo*. *J Neurosci* 20: 8597–8603
58 Massague J (1996) TGF-β signaling: receptors, transducers, and Mad proteins. *Cell* 85: 947–950
59 ten Dijke P, Miyazono K, C-H H (1996) Signaling via hetero-oligomeric complexes of type I and type II serine/threonine kinase receptors. *Curr Opin Cell Biol* 8: 139–145
60 ten Dijke P, Yamashita H, Sampath TK, Reddi AH, Estevez M, Riddle DL, Ichijo H, Heldin CH, Miyazono K (1994) Identification of type I receptors for osteogenic protein-1 and bone morphogenetic protein-4. *J Biol Chem* 269: 16985–16988
61 Liu F, Ventura F, Doody J, Massague J (1995) Human type II receptor for bone morphogenetic proteins (BMPs): extension of the two-kinase receptor model to the BMPs. *Mol Cell Biol* 15: 3479–3486
62 Rosenzweig BL, Imamura T, Okadome T, Cox GN, Yamashita H, ten Dijke P, Heldin CH, Miyazono K (1995) Cloning and characterization of a human type II receptor for bone morphogenetic proteins. *Proc Natl Acad Sci USA* 92: 7632–7636
63 Yamashita H, ten Dijke P, Huylebroeck D, Sampath TK, Andries M, Smith JC, Heldin C-H, Miyazono K (1995) Osteogenic protein-1 binds to activin type II receptors and induces certain activin-like effects. *J Cell Biol* 130: 217–226
64 Matsuo T (1993) The genes involved in the morphogenesis of the eye. *Jpn J Ophthalmol* 37: 215–251
65 Dewulf N, Verschueren K, Lonnoy O, Moren A, Grimsby S, Vande Spiegle K, Miyazono K, Huylebroeck D, Ten Dijke P (1995) Distinct spatial and temporal expression patterns of two type I receptors for bone morphogenetic proteins during mouse embryogenesis. *Endocrinology* 136: 2652–2663
66 Bengtsson H, Soderstrom S, Ebendal T (1995) Expression of activin receptors type I and II only partially overlaps in the nervous system. *Neuroreport* 7: 113–116
67 Soderstrom S, Bengtsson H, Ebendal T (1996) Expression of serine/threonine kinase receptors including the bone morphogenetic factor type II receptor in the developing and adult rat brain. *Cell Tissue Res* 286: 269–279
68 Zhang H, Bradley A (1996) Mice deficient for BMP2 are nonviable and have defects in amnion/chorion and cardiac development. *Development* 122: 2977–2986
69 Ausburger A, Schuchardt A, Hoskins S, Dodd J, Butler S (1999) BMPs as mediators of roof plate repulsion of commissural neurons. *Neuron* 24: 127–141
70 Lorentzon M, Hoffer B, Ebendal T, Olson L, Tomac A (1996) Habrec1, a novel serine/threonine kinase TGF-beta type I-like receptor, has a specific cellular expression suggesting function in the developing organism and adult brain. *Exp Neurol* 142: 351–360
71 Ryden M, Imamura T, Jornvall H, Belluardo N, Neveu I, Trupp M, Okadome T, ten

Dijke P, Ibanez CF (1996) A novel type I receptor serine-threonine kinase predominantly expressed in the adult central nervous system. *J Biol Chem* 271: 30603–30609
72 Tsuchida K, Sawchenko PE, Nishikawa S, Vale WW (1996) Molecular cloning of a novel type I receptor serine/threonine kinase for the TGF beta superfamily from rat brain. *Mol Cell Neurosci* 7: 467–478
73 Hogan BL (1996) Bone morphogenetic proteins: multifunctional regulators of vertebrate development. *Genes Dev* 10: 1580–1594
74 Piccolo S, Sasai Y, Lu B, De Robertis EM (1996) Dorsoventral patterning in *Xenopus*: inhibition of ventral signals by direct binding of chordin to BMP-4. *Cell* 86: 589–598
75 Zimmerman LB, De Jesus-Escobar JM, Harland RM (1995) The Spemann organizer signal noggin binds and inactivates bone morphogenetic protein 4. *Cell* 86: 599–606
76 Fainsod A, Deissler K, Yelin R, Marom K, Epstein M, Pillemer G, Steinbeisser H, Blum M (1997) The dorsalizing and neural inducing gene follistatin is an antagonist of BMP-4. *Mech Dev* 63: 39–50
77 Stanley E, Biben C, Kotecha S, Fabri L, Tajbakhsh S, Wang CC, Hatzistavrou T, Roberts B, Drinkwater C, Lah M et al (1998) DAN is a secreted glycoprotein related to *Xenopus* cerberus. *Mech Dev* 77: 173–184
78 Bouwmeester T, Kim S, Sasai Y, Lu B, De Robertis EM (1996) Cerberus is a head-inducing secreted factor expressed in the anterior endoderm of Spemann's organizer. *Nature* 382: 595–601
79 Hsu DR, Economides AN, Wang X, Eimon PM, Harland RM (1998) The *Xenopus* dorsalizing factor Gremlin identifies a novel family of secreted proteins that antagonize BMP activities. *Mol Cell* 1: 673–683
80 Piccolo S, Agius E, Leyns L, Bhattacharyya S, Grunz H, Bouwmeester T, De Robertis EM (1999) The head inducer Cerberus is a multifunctional antagonist of Nodal, BMP and Wnt signals. *Nature* 397: 707–710
81 Iemura S, Yamamoto TS, Takagi C, Uchiyama H, Natsume T, Shimasaki S, Sugino H, Ueno N (1998) Direct binding of follistatin to a complex of bone-morphogenetic protein and its receptor inhibits ventral and epidermal cell fates in early *Xenopus* embryo. *Proc Natl Acad Sci USA* 95: 9337–9342
82 Weinstein DC, Hemmati-Brivanlou A (1999) Neural induction. *Annu Rev Cell Dev Biol* 15: 411–433
83 Harland R (2000) Neural induction. *Curr Opin Genet Dev* 10: 357–362
84 Lamb TM, Knecht AK, Smith WC, Stachel SE, Economides AN, Stahl N, Yancopolous GD, Harland RM (1993) Neural induction by the secreted polypeptide noggin. *Science* 262: 713–718
85 Hemmati-Brivanlou A, Kelly OG, Melton DA (1994) Follistatin, an antagonist of activin, is expressed in the Spemann organizer and displays direct neuralizing activity. *Cell* 77: 283–295
86 Sasai Y, Lu B, Steinbeisser H, De Robertis EM (1995) Regulation of neural induction by the Chd and Bmp-4 antagonistic patterning signals in *Xenopus* [published errata appear

in *Nature* 1995 Oct 26; 377 (6551): 757 and 1995 Nov 23; 378 (6555): 419]. *Nature* 376: 333–336

87 Hansen CS, Marion CD, Steele K, George S, Smith WC (1997) Direct neural induction and selective inhibition of mesoderm and epidermis inducers by Xnr3. *Development* 124: 483–492

88 Albano RM, Smith JC (1994) Follistatin expression in ES and F9 cells and in preimplantation mouse embryos. *Int J Dev Biol* 38: 543–547

89 Bauer H, Meier A, Hild M, Stachel S, Economides A, Hazelett D, Harland RM, Hammerschmidt M (1998) Follistatin and noggin are excluded from the zebrafish organizer. *Dev Biol* 204: 488–507

90 Streit A, Stern CD (1999) Neural induction. A bird's eye view. *Trends Genet* 15: 20–24

91 McMahon JA, Takada S, Zimmerman LB, Fan CM, Harland RM, McMahon AP (1998) Noggin-mediated antagonism of BMP signaling is required for growth and patterning of the neural tube and somite. *Genes Dev* 12: 1438–1452

92 Shimamura K, Hartigan DJ, Martinez S, Puelles L, Rubenstein JL (1995) Longitudinal organization of the anterior neural plate and neural tube. *Development* 121: 3923–3933

93 Sela-Donenfeld D, Kalcheim C (1999) Regulation of the onset of neural crest migration by coordinated activity of BMP4 and noggin in the dorsal neural tube. *Development* 126: 4749–4762

94 Scott IC, Steiglitz BM, Clark TG, Pappano WN, Greenspan DS (2000) Spatiotemporal expression patterns of mammalian chordin during postgastrulation embryogenesis and in postnatal brain. *Dev Dyn* 217: 449–456

95 Feijen A, Goumans MJ, van den Eijnden-van Raaij AJ (1994) Expression of activin subunits, activin receptors and follistatin in postimplantation mouse embryos suggests specific developmental functions for different activins. *Development* 120: 3621–3637

96 Valenzuela DM, Economides AN, Rojas E, Lamb TM, Nunez L, Jones P, Lp NY, Espinosa R, Brannan CI, Gilbert DJ, et al. (1995) Identification of mammalian noggin and its expression in the adult nervous system. *J Neurosci* 15: 6077–6084

97 Pappano WN, Scott IC, Clark TG, Eddy RL, Shows TB, Greenspan DS (1998) Coding sequence and expression patterns of mouse chordin and mapping of the cognate mouse chrd and human CHRD genes. *Genomics* 52: 236–239

98 Topol LZ, Marx M, Laugier D, Bogdanova NN, Boubnov NV, Clausen PA, Calothy G, Blair DG (1997) Identification of drm, a novel gene whose expression is suppressed in transformed cells and which can inhibit growth of normal but not transformed cells in culture. *Mol Cell Biol* 17: 4801–4810

99 Gerhart J, Doniach T, Stewart R (1991) *Organizing the* Xenopus *organizer*. Plenum Press, New York

100 Doniach T (1993) Planar and vertical induction of anteroposterior pattern during the development of the amphibian central nervous system. *J Neurobiol* 24: 1256–1275

101 Smith JL, Schoenwolf GC (1989) Notochordal induction of cell wedging in the chick neural plate and its role in neural tube formation. *J Exp Zool* 250: 49–62

102 Sater AK, Jacobson AG (1990) The restriction of the heart morphogenetic field in *Xenopus* laevis. *Dev Biol* 140: 328–336
103 Ang SL, Rossant J (1994) HNF-3 beta is essential for node and notochord formation in mouse development. *Cell* 78: 561–574
104 Weinstein DC, Ruiz i Altaba A, Chen WS, Hoodless P, Prezioso VR, Jessell TM, Darnell JE (1994) The winged-helix transcription factor HNF-3 beta is required for notochord development in the mouse embryo. *Cell* 78: 575–588
105 Shih J, Fraser SE (1996) Characterizing the zebrafish organizer: microsurgical analysis at the early-shield stage. *Development* 122: 1313–1322
106 Davidson BP, Kinder SJ, Steiner K, Schoenwolf GC, Tam PP (1999) Impact of node ablation on the morphogenesis of the body axis and the lateral asymmetry of the mouse embryo during early organogenesis. *Dev Biol* 211: 11–26
107 Klingensmith J, Ang SL, Bachiller D, Rossant J (1999) Neural induction and patterning in themouse in the absence of the node and its derivatives. *Dev Biol* 216: 535–549
108 Hemmati-Brivanlou A, Melton DA (1992) A truncated activin receptor inhibits mesoderm induction and formation of axial structures in *Xenopus* embryos. *Nature* 359: 609–614
109 Hemmati-Brivanlou A, Melton DA (1994) Inhibition of activin receptor signaling promotes neuralization in *Xenopus*. *Cell* 77: 273–281
110 Weinstein DC, Hemmati-Brivanlou A (1997) Neural induction in *Xenopus* laevis: evidence for the default model. *Curr Opin Neurobiol* 7: 7–12
111 Smith JC, Price BM, Van Nimmen K, Huylebroeck D (1990) Identification of a potent *Xenopus* mesoderm-inducing factor as a homologue of activin A. *Nature* 345: 729–731
112 Wilson PA, Hemmati-Brivanlou A (1995) Induction of epidermis and inhibition of neural fate by Bmp-4. *Nature* 376: 331–333
113 Suzuki A, Kaneko E, Ueno N, Hemmati-Brivanlou A (1997) Regulation of epidermal induction by BMP2 and BMP7 signaling. *Dev Biol* 189: 112–122
114 Suzuki A, Chang C, Yingling JM, Wang XF, Hemmati-Brivanlou A (1997) Smad5 induces ventral fates in *Xenopus* embryo. *Dev Biol* 184: 402–405
115 Wilson PA, Lagna G, Suzuki A, Hemmati-Brivanlou A (1997) Concentration-dependent patterning of the *Xenopus* ectoderm by BMP4 and its signal transducer Smad1. *Development* 124: 3177–3184
116 Suzuki A, Ueno N, Hemmati-Brivanlou A (1997c) *Xenopus* msx1 mediates epidermal induction and neural inhibition by BMP4. *Development* 124: 3037–3044
117 Xu RH, Kim J, Taira M, Zhan S, Sredni D, Kung HF (1995) A dominant negative bone morphogenetic protein 4 receptor causes neuralization in *Xenopus* ectoderm. *Biochem Biophys Res Commun* 212: 212–219
118 Casellas R, Brivanlou AH (1998) *Xenopus* Smad7 inhibits both the activin and BMP pathways and acts as a neural inducer. *Dev Biol* 198: 1–12
119 Hata A, Lagna G, Massague J, Hemmati-Brivanlou A (1998) Smad6 inhibits BMP/Smad1 signaling by specifically competing with the Smad4 tumor suppressor. *Genes Dev* 12: 186–197

120 Nakayama T, Gardner H, Berg LK, Christian JL (1998) Smad6 functions as an intracellular antagonist of some TGF-beta family members during *Xenopus* embryogenesis. *Genes Cells* 3: 387–394

121 Onichtchouk D, Glinka A, Niehrs C (1998) Requirement for Xvent-1 and Xvent-2 gene function in dorsoventral patterning of *Xenopus* mesoderm. *Development* 125: 1447–1456

122 Holley SA, Neul JL, Attisano L, Wrana JL, Sasai Y, O'Connor MB, De Robertis EM, Ferguson EL (1996) The *Xenopus* dorsalizing factor noggin ventralizes *Drosophila* embryos by preventing DPP from activating its receptor. *Cell* 86: 607–617

123 Simpson EH, Johnson DK, Hunsicker P, Suffolk R, Jordan SA, Jackson IJ (1999) The mouse Cer1 (Cerberus related or homologue) gene is not required for anterior pattern formation. *Dev Biol* 213: 202–206

124 Hammerschmidt M, Pelegri F, Mullins MC, Kane DA, van Eeden FJ, Granato M, Brand M, Furutani-Seiki M, Haffter P, Heisenberg CP et al C (1996) dino and mercedes, two genes regulating dorsal development in the zebrafish embryo. *Development* 123: 95–102

125 Schulte-Merker S, Lee KJ, McMahon AP, Hammerschmidt M (1997) The zebrafish organizer requires chordino. *Nature* 387: 862–863

126 Bachiller D, Klingensmith J, Kemp C, Belo JA, Anderson RM, May SR, McMahon JA, McMahon AP, Harland RM, Rossant J, De Robertis EM (2000) The organizer factors Chordin and Noggin are required for mouse forebrain development. *Nature* 403: 658–661

127 Alvarez IS, Araujo M, Nieto MA (1998) Neural induction in whole chick embryo cultures by FGF. *Dev Biol* 199: 42–54

128 Hongo I, Kengaku M, Okamoto H (1999) FGF signaling and the anterior neural induction in *Xenopus*. *Dev Biol* 216: 561–581

129 Streit A, Berliner AJ, Papanayotou C, Sirulnik A, Stern CD (2000) Initiation of neural induction by FGF signalling before gastrulation. *Nature* 406: 74–78

130 Wilson SI, Graziano E, Harland R, Jessell TM, Edlund T (2000) An early requirement for FGF signalling in the acquisition of neural cell fate in the chick embryo. *Curr Biol* 10: 421–429

131 Baker JC, Beddington RSP, Harland RM (1999) Wnt signaling in *Xenopus* embryos inhibits BMP4 expression and activates neural development. *Genes Dev* 13: 3149–3159

132 Jessell TM, Sanes JR (2000) *Development*. The decade of the developing brain. *Curr Opin Neurobiol* 10: 599–611

133 Weiss S, Reynolds BA, Vescovi AL, Morshead C, Craig CG, van der Kooy D (1996) Is there a neural stem cell in the mammalian forebrain? *Trends Neurosci* 19: 387–393

134 McKay R (1997) Stem cells in the central nervous system. *Science* 276: 66–71

135 Edlund T, Jessell TM (1999) Progression from extrinsic to intrinsic signaling in cell fate specification: a view from the nervous system. *Cell* 96: 211–224

136 Grinspan JB, Edell E, Carpio DF, Beesley JS, Lavy L, Pleasure D, Golden JA (2000)

Stage-specific effects of bone morphogenetic proteins on the oligodendrocyte lineage. *J Neurobiol* 43: 1–17
137 Mehler MF, Mabie PC, Zhu G, Gokhan S, Kessler JA (2000) *Development*al changes in progenitor cell responsiveness to bone morphogenetic proteins differentially modulate progressive CNS lineage fate. *Dev Neurosci* 22: 74–85
138 Shou J, Rim PC, Calof AL (1999) BMPs inhibit neurogenesis by a mechanism involving degradation of a transcription factor. *Nat Neurosci* 2: 339–345
139 Shah NM, Groves AK, Anderson DJ (1996) Alternative neural crest fates are instructively promoted by TGFβ superfamily members. *Cell* 85: 331–343
140 Alder J, Lee KJ, Jessell TM, Hatten ME (1999) Generation of cerebellar granule neurons *in vivo* by transplantation of BMP-treated neural progenitor cells. *Nat Neurosci* 2: 535–540
141 Kalyani AJ, Piper D, Mujtaba T, Lucero MT, Rao MS (1998) Spinal cord neuronal precursors generate multiple neuronal phenotypes in culture. *J Neurosci* 18: 7856–7868
142 Bartlett PF, Brooker GJ, Faux CH, Dutton R, Murphy M, Turnley A, Kilpatrick TJ (1998) Regulation of neural stem cell differentiation in the forebrain. *Immunol Cell Biol* 76: 414–418
143 Nakashima K, Yanagisawa M, Arakawa H, Kimura N, Hisatsune T, Kawasbata M, Miyazono K, Taga T (1999) Synergistic signaling in fetal brain by STAT3-Smad1 complex bridged by p300. *Science* 284: 479–482
144 Li W, LoTurco JJ (2000) Noggin is a negative regulator of neuronal differentiation in developing neocortex. *Dev Neurosci* 22: 68–73
145 Zhu G, Mehler MF, Mabie PC, Kessler JA (2000) *Development*al changes in neural progenitor cell lineage commitment do not depend on epidermal growth factor receptor signaling. *J Neurosci Res* 59: 312–320
146 Ernsberger U, Patzke H, Tissier-Sat, Reh T, Goridis C, and Rohrer H (1995) The expression of tyrosine hydroxylase and transcription factors cPhox-2 and Cash1: Evidence for distinct inductive steps in the differentiation of chick sympathetic precursor cells. *Mech Dev* 52: 125–136
147 Zhu G, Mehler MF, Zhao J, Yu Yung S, Kessler JA (1999) Sonic hedgehog and BMP2 exert opposing actions on proliferation and differentiation of embryonic neural progenitor cells. *Dev Biol* 215: 118–129
148 Hirsinger E, Duprez D, Jouve C, Malapert P, Cooke J, Pourquie O (1997) Noggin acts downstream of Wnt and Sonic Hedgehog to antagonize BMP4 in avian somite patterning. *Development* 124: 4605–4614
149 Gage FH (2000) Mammalian neural stem cells. *Science* 287: 1433–1438
150 Zhu G, Mehler MF, Mabie PC, Kessler JA (1999) *Development*al changes in progenitor cell responsiveness to cytokines. *J Neurosci Res* 56: 131–145
151 Lee KJ, Mendelsohn M, Jessell TM (1998) Neuronal patterning by BMPs: a requirement for GDF7 in the generation of a discrete class of commissural interneurons in the mouse spinal cord. *Genes Dev* 12: 3394–3407

152 Jessell TM (2000) Neuronal specification in the spinal cord: inductive signals and transcriptional codes. *Nature Rev Genet* 1: 20–29
153 Dale JK, Vesque C, Lints TJ, Sampath TK, Furley A, Dodd J, Placzek M (1997) Cooperation of BMP7 and SHH in the induction of forebrain ventral midline cells by prechordal mesoderm. *Cell* 90: 257–269
154 Muhr J, Graziano E, Wilson S, Jessell TM, Edlund T (1999) Convergent inductive signals specify midbrain, hindbrain, and spinal cord identity in gastrula stage chick embryos. *Neuron* 23: 689–702
155 Kalyani AJ, Rao MS (1998) Cell lineage in the developing neural tube. *Biochem Cell Biol* 76: 1051–1068
156 Dale K, Sattar N, Heemskerk J, Clarke JD, Placzek M, Dodd J (1999) Differential patterning of ventral midline cells by axial mesoderm is regulated by BMP7 and chordin. *Development* 126: 397–408
157 Knecht AK, Harland RM (1997) Mechanisms of dorsal-ventral patterning in noggin-induced neural tissue. *Development* 124: 2477–2488
158 Chuang PT, McMahon AP (1999) Vertebrate hedgehog signaling modulated by induction of a hedgehog-binding protein. *Nature* 397: 617–621
159 Nguyen VH, Trout J, Connors SA, Anderman P, Weinberg E, Mullins MC (2000) Dorsal and intermediate neuronal cell types of the spinal cord are established by a BMP signaling pathway. *Development* 127: 1209–1220
160 Barth KA, Kishimoto Y, Rohr KB, Seydler C, Schulte-Merker S, Wilson SW (1999) Bmp activity establishes a gradient of positional information throughout the entire neural plate. *Development* 126: 4977–4987
161 Hattori A, Katayama M, Iwasaki S, Ishii K, Tsujimoto M, Kohno M (1999) Bone morphogenetic protein-2 promotes survival and differentiation of striatal GABAergic neurons in the absence of glial cell proliferation. *J Neurochem* 72: 2264–2271
162 Farkas LM, Scheuermann S, Pohl J, Unsicker K, Krieglstein K (1997) Characterization of growth/differentiation factor 5 (GDF-5) as a neurotrophic factor for cultured neurons from chicken dorsal root ganglia. *Neurosci Lett* 236: 120–122
163 Farkas LM, Jászai J, Unsicker K, Krieglstein K (1999) Characterization of bone morphogenetic protein family members as neurotrophic factors for cultured sensory neurons. *Neuroscience* 92: 227–235
164 Thang SH, Kobayashi M, Matsuoka I (2000) Regulation of glial cell line-derived neurotrophic factor responsiveness in developing rat sympathetic neurons by retinoic acid and bone morphogenetic protein-2. *J Neurosci* 20: 2917–2925
165 Bengtsson H, Söderström S, Kylberg A, Charette MF (1998) Potentiating interactions between morphogenetic protein and neurotrophic factors in developing neurons. *J Neurosci Res* 53: 559–568
166 Kobayashi M, Fujii M, Kurihara K, Matsuoka I (1998) Bone morphogenetic protein-2 and retinoic acid induce neurotrophin-3 responsiveness in developing rat sympathetic neurons. *Molec Brain Res* 53: 206–217
167 Iantosca MR, McPherson CE, Ho SY, Maxwell GD (1999) Bone morphogenetic pro-

teins-2 and -4 attenuate apoptosis in a cerebellar primitive neuroectodermal tumor cell line. *J Neurosci Res* 56: 248–258

168 Graham A, Francis-West P, Brickell P, Lumsden A (1994) The signalling molecule BMP4 mediates apoptosis in the rhombencephalic neural crest. *Nature* 372: 684–686

169 Song Q, Mehler MF, Kessler JA (1998) Bone morphogenetic proteins induce apoptosis and growth factor dependence of cultured sympathoadrenal progenitor cells. *Dev Biol* 196: 119–127

170 Selleck MAJ, Garcia-Castro MI, Artinger KB, Bronner-Fraser M (1998) Effects of Shh and noggin on neural crest formation demonstrate that BMP is required in the neural tube but not ectoderm. *Development* 125: 4919–4930

171 Luo G, Hofmann C, Bronckers AL, Sohocki M, Bradley A, Karsenty G (1995) BMP-7 is an inducer of nephrogenesis, and is also required for eye development and skeletal patterning. *Genes Dev* 9: 2808–2820

172 Dudley AT, Lyons KM, Roberston EJ (1995) A requirement for bone morphogenetic protein-7 during development of the mammalian kidney and eye. *Genes & Dev* 9: 2795–2807

173 Solursh, M, Langille RM, Wood J, Sampath TK (1996) Osteogenic protein-1 is required for mammalian eye development. *Biochem Biophys Res Comm* 218: 438–443

174 Liu J-P, Jessel TM (1998) A role for rhoB in the delamination of neural crest cells from the dorsal neural tube. *Development* 125: 5055–5067

175 Akitaya T, Bronner-Fraser M (1992) Expression of cell adhesion molecules during initiation and cessation of neural crest migration. *Dev Dyn* 194: 12–20

176 Nakagawa S, Takeichi M (1998) Neural crest emigration from the neural tube depends on regulated cadherin expression. *Development* 125: 2963–2971

177 Marchant L, Linker C, Ruiz P, Guerrero N, Mayor R (1998) The inductive properties of mesoderm suggest that the neural crest cells are specified by a BMP gradient. *Dev Biol* 198: 319–329

178 Reissman E, Ernsberger U, Francis-West PH, Rueger D, Brickell PM, Rohrer H (1996) Involvement of bone morphogenetic protein-4 and bone morphogenetic protein-7 in the differentiation of the adrenergic phenotype in developing sympathetic neurons. *Development* 122: 2079–2088

179 Varley JE, Wehby RG, Rueger DC, Maxwell GD (1995) Number of adrenergic cells and islet-1 immunoreactive cells is increased in avian trunk neural crest cultures in the presence of human recombinant osteogenic protein-1. *Dev Dyn* 203: 434–447

180 Varley JE, Maxwell GD (1996) BMP2 and BMP4, but not BMP6, increase the number of adrenergic cells which develop in quail neural crest cultures. *Exp Neurol* 140: 84–94

181 Morrison SJ, Csete M, Groves AK, Melega W, Wold B, and Anderson DJ (2000) Culture in reduced levels of oxygen promotes clonogenic sympathoadrenal differentiation by isolated neural crest stem cells. *J Neurosci* 20: 7370–7376

182 Lo L, Tiveron MC, Anderson DJ (1998) MASH1 activates expression of the paired homeodomain transcription factor Phox2a and couples pan-neuronal and subtype-specific components of autonomic neuronal identity. *Development* 125: 609–620

183 Lo L, Morin X, Brunet JF, Anderson DJ (1999) Specification of neurotransmitter identity by Phox2 proteins in neural crest stem cells. *Neuron* 22: 693–705
184 Ernsberger U, Reissmann E, Mason I, Rohrer H (2000) The expression of dopamine-β-hydroxylase, tyrosine hydroxylase, and Phox2 transcription factors in sympathetic neurons: evidence for common regulation during nor adrenergic induction and divergin regulation later in development. *Mech Dev* 92: 169–177
185 Howard MJ, Stanke M, Schneider C, Wu X, Rohrer H (2000) The transcription factor dHAND is a downstream effector of BMPs in sympathetic neurons specification. *Development* 127: 4073–4081
186 Hirsch MR, Tiveron MC, Guillemot F, Brunet JF, Goridis C (1998) Control of noradrenergic differentiation and Phox2a expression by MASH1 in the central and peripheral nervous system. *Development* 125: 599–608
187 Morin X, Cremer H, Hirsch MR, Kapur RP, Goridis C, Brunet JF (1997) Deficits in sensory and autonomic ganglia and absence of locus coeruleus in mice deficient for the homeobox gene Phox2a. *Neuron* 18: 411–423
188 Pattyn A, Morin X, Cremer H, Goridis C and Brunet JF (1999) The homeobox gene Phox2b is essential for the development of autonomic neural crest derivatives. *Nature* 399: 366–370
189 Guo S, Brush J, Teraoka H, Goddard A, Wilson SW, Mullins MC, Rosenthal A (1999) Development of noradrenergic neurons in the zebrafish hindbrain requires BMP, FGF8, and the homeodomain protein soulless/Phox2a. *Neuron* 24: 555–566
190 Lints R, Emmons SW (1999) Patterning of dopaminergic neurotransmitter identity among Caenorhabditis elegans ray sensory neurons by a TGFβ family signaling pathway and a Hox gene. *Development* 126: 5819–5831
191 Reiriz J, Espejo M, Ventura F, Ambrosio S, Alberch J (1999) Bone morphogenetic protein-2 promotes dissoicated effects on the number and differentiation of cultured ventral mesencephalic dopaminergic neurons. *J Neurobiol* 38: 161–170
192 Daadi M, Arcellanapanlilio MY, Weiss S (1998) Activin co-operates with fibroblast growth factor 2 to regulate tyrosine hydroxylase expression in the basal forebrain ventricular zone progenitors. *Neuroscience* 86: 867–880
193 Fann MJ, Patterson PH (1994) Depolarization differentially regulates the effects of bone morphogenetic protein (BMP)-2, BMP-6, and activin A on sympathetic neuronal phenotype. *J Neurochem* 63: 2074–2079
194 Coulombe JN, Schwall R, Parent AS, Eckenstein FP, and Nishi R (1993) Induction of somatostatin immunoreactivity in cultured ciliary ganglion neurons by activin in choroid cell-conditioned medium. *Neuron* 10: 899–906
195 Paralkar VM, Weeks BS, Yu YM, Kleinman HK, Reddi AH (1992) Recombinant human bone morphogenetic protein 2B stimulates PC12 cell differentiation: potentiation and binding to type IV collagen. *J Cell Biol* 119: 1721–1728
196 Iwasaki S, Hattori A, Sato M, Tsujimoto M, Kohno M (1996) Characterization of bone morphogenetic protein-2 as a neurotrophic factor. *J Biol Chem* 271: 17360–17365

197 Carri NG, Bengtsson H, Charette MF, Ebendal T (1998) BMPR-II expression and OP-1 effects in developing chicken retinal explants. *Neuroreport* 9: 1097–1101
198 Lein P, Johnson M, Guo X, Rueger D, Higgins D (1995) Osteogenic protein-1 induces dendritic growth in rat sympathetic neurons. *Neuron* 15: 597–605
199 Withers GS, Higgins D, Charette M, Banker G (2000) Bone morphogenetic protein-7 enhances dendritic growth and receptivity to innervation in cultured hippocampal neurons. *Eur J Neurosci* 12: 106–116
200 Le Roux P, Behar S, Higgins D, Charette M (1999) OP-1 enhances dendritic growth from cerebral cortical neurons *in vitro*. *Exp Neurol* 160: 151–163
201 Granholm ACh, Sanders LA, Ickes B, Albeck D, Hoffer BJ, Young DA, Kaplan PL (1999) Effects of osteogenic protein-1 on fetal spinal cord transplants to the anterior chamber of the eye. *Cell Transplant* 8: 75–85
202 Guo X, Rueger R, Higgins D (1998) Osteogenic protein-1 and related bone morphogenetic proteins regulate dendritic growth and the expression of microtubule-associated protein-2 in rat sympathetic neurons. *Neurosci Lett* 245: 131–134
203 Chandrasekaran V, Zhai Y, Wagner M, Kaplan PL, Napoli JL, Higgins D (2000) Retinoic acid regulates the morphological development of sympathetic neurons. *J Neurobiol* 42: 383–393
204 Wyatt S, Andreas R, Rohrer H, Davies AM (1998) Regulation of neurotrophin receptor expression by retinoic acid in mouse sympathetic neuroblasts. *J Neurosci* 19: 1062–1071
205 Guo X, Metzler-Northrup J, Lein P, Rueger D, Higgins D (1997) Leukemia inhibitory factor and ciliary neurotrophic factor regulate dendritic growth in cultures of rat sympathetic neurons. *Dev Brain Res* 104: 101–110
206 Guo X, Chandrasekaran V, Lein P, Kaplan PL, Higgins D (1999) Leukemia inhibitory factor and ciliary neurotrophic factor cause dendritic retraction in cultured rat sympathetic neurons. *J Neurosci* 19: 2113–2121
207 Purves D, Snider WD, Voyvodic JT (1998) Trophic regulation of nerve cell morphology and innervation in the autonomic nervous system. *Nature* 336: 123–128
208 Banner LR, Patterson PH (1994) Major changes in the expression of mRNAs for cholinergic differentiation factor/leukemia inhibitory factor and its receptor after injury to adult peripheral nerves and ganglia. *Proc Natl Acad Sci USA* 91: 7109–7113
209 Sun YS, Landis SC, Zigmond RE (1996) Signals triggering the induction of leukemia inhibitory factor in sympathetic superior cervical ganglia and their nerve trunks after axonal injury. *Mol Cell Neurosci* 7: 152–163
210 Perides G, Jensen FE, Edgecomb P, Rueger DC, Charness ME (1995) Neuroprotective effect of human osteogenic protein-1 in a rat model of cerebral hypoxia/ischemia. *Neurosci Lett* 187: 21–24
211 Lin S, Hoffer BJ, Kaplan P, Wang Y (1999) Osteogenic protein-1 protects against cerebral infarction induced by MCA ligation in adult rats. *Stroke* 30: 126–133
212 Kawamata T, Ren J, Chan TC, Charette M, Finklestein, SP (1998) Intracisternal

osteogenic protein-1 enhances functional recovery following focal stroke. *Neuroreport* 9: 1441–1445

213 Ren J, Kaplan PL, Charette MF, Speller H, Finklestein SP (2000) Time window of intracisternal osteogenic protein-1 in enhancing functional recovery after stroke. *Neuropharm* 39: 860–865

214 Wu DD, Lai M, Hughes PE, Sirimanne E, Gluckman PD, Williams CE (1999) Expression of the activin axis and neuronal rescue effects of recombinant activin A following hypoxic-ischemic brain injury in the infant rat. *Brain Res* 835: 369–378

215 Krieglstein K, Suter-Crazzolara C, Hotten G, Phol J, Unisicker K (1995) Trophic and protective effects of growth/differentiation factor 5, a member of the transforming growth factor-β superfamily, on midbrain dopaminergic neurons. *J Neurosci Res* 42: 724–732

216 Espejo M, Cutillas B, Ventura F, Ambrosio S (1999) Exposure of foetal mesencephalic cells to bone morphogenetic protein-3 enhances the survival of dopaminergic neurons in rat striatal grafts. *Neurosci Lett* 275: 13–16

217 Sullivan AM, Pohl J, Blunt SB (1998) Growth/differentiation factor 5 and glial cell line-derived neurotrophic factor enhance survival and function of dopaminergic grafts in a rat model of Parkinson's disease. *Eur J Neurosci* 10: 3681–3688

218 Hughes PE, Alexi T, Williams CE, Clark RG, Gluckman PD (1999) Administration of recombinant activin A has powerful neurotrophic effects on select striatal phenotypes in the quinolinic acid lesion model of Huntington's disease. *Neuroscience* 92: 197–209

219 Tretter YP, Hertel M, Munz B, Bruggencate GT, Werner S, Alzheimer S (2000) Induction of activin A is essential for the neuroprotective action of basic fibroblast growth factor *in vivo*. *Nature Medicine* 6: 812–815

# Index

## The PIR-Series
## Progress in Inflammation Research

Homepage: http://www.birkhauser.ch

Up-to-date information on the latest developments in the pathology, mechanisms and therapy of inflammatory disease are provided in this monograph series. Areas covered include vascular responses, skin inflammation, pain, neuroinflammation, arthritis cartilage and bone, airways inflammation and asthma, allergy, cytokines and inflammatory mediators, cell signalling, and recent advances in drug therapy. Each volume is edited by acknowledged experts providing succinct overviews on specific topics intended to inform and explain. The series is of interest to academic and industrial biomedical researchers, drug development personnel and rheumatologists, allergists, pathologists, dermatologists and other clinicians requiring regular scientific updates.

**Available volumes:**

*T Cells in Arthritis*, P. Miossec, W. van den Berg, G. Firestein (Editors), 1998
*Chemokines and Skin*, E. Kownatzki, J. Norgauer (Editors), 1998
*Medicinal Fatty Acids*, J. Kremer (Editor), 1998
*Inducible Enzymes in the Inflammatory Response*, D.A. Willoughby, A. Tomlinson (Editors), 1999
*Cytokines in Severe Sepsis and Septic Shock*, H. Redl, G. Schlag (Editors), 1999
*Fatty Acids and Inflammatory Skin Diseases*, J.-M. Schröder (Editor), 1999
*Immunomodulatory Agents from Plants*, H. Wagner (Editor), 1999
*Cytokines and Pain*, L. Watkins, S. Maier (Editors), 1999
In Vivo *Models of Inflammation*, D. Morgan, L. Marshall (Editors), 1999
*Pain and Neurogenic Inflammation*, S.D. Brain, P. Moore (Editors), 1999
*Anti-Inflammatory Drugs in Asthma*, A.P. Sampson, M.K. Church (Editors), 1999
*Novel Inhibitors of Leukotrienes*, G. Folco, B. Samuelsson, R.C. Murphy (Editors), 1999
*Vascular Adhesion Molecules and Inflammation*, J.D. Pearson (Editor), 1999
*Metalloproteinases as Targets for Anti-Inflammatory Drugs*, K.M.K. Bottomley, D. Bradshaw, J.S. Nixon (Editors), 1999
*Free Radicals and Inflammation*, P.G. Winyard, D.R. Blake, C.H. Evans (Editors), 1999
*Gene Therapy in Inflammatory Diseases*, C.H. Evans, P. Robbins (Editors), 2000
*New Cytokines as Potential Drugs*, S. K. Narula, R. Coffmann (Editors), 2000
*High Throughput Screening for Novel Anti-inflammatories*, M. Kahn (Editor), 2000
*Immunology and Drug Therapy of Atopic Skin Diseases*, C.A.F. Bruijnzeel-Komen, E.F. Knol (Editors), 2000
*Novel Cytokine Inhibitors*, G.A. Higgs, B. Henderson (Editors), 2000
*Inflammatory Processes. Molecular Mechanisms and Therapeutic Opportunities*, L.G. Letts, D.W. Morgan (Editors), 2000

*Cellular Mechanisms in Airways Inflammation*, C. Page, K. Banner, D. Spina (Editors), 2000
*Inflammatory and Infectious Basis of Atherosclerosis*, J.L. Mehta (Editor), 2001
*Muscarinic Receptors in Airways Diseases*, J. Zaagsma, H. Meurs, A.F. Roffel (Editors), 2001
*TGF-β and Related Cytokines in Inflammation*, S.N. Breit, S. Wahl (Editors), 2001
*Nitric Oxide and Inflammation*, D. Salvemini, T.R. Billiar, Y. Vodovotz (Editors), 2001
*Neuroinflammatory Mechanisms in Alzheimer's Disease. Basic and Clinical Research,* J. Rogers (Editor), 2001
*Disease-modifying Therapy in Vasculitides,* C.G.M. Kallenberg, J.W. Cohen Tervaert (Editors), 2001
*Inflammation and Stroke*, G.Z. Feuerstein (Editor), 2001
*NMDA Antagonists as Potential Analgesic Drugs*, D.J.S. Sirinathsinghji, R.G. Hill (Editors), 2002
*Migraine: A neuroinflammatory disease?* E.L.H. Spierings, M. Sanchez del Rio (Editors), 2002
*Mechanisms and mediators of neuropathic pain*, A.B. Malmberg, S.R. Chaplan (Editors), 2002